ENGINEERING HYDROLOGY of ARID and SEMI-ARID REGIONS

ENGINEERING HYDROLOGY of ARID and SEMI-ARID REGIONS

MOSTAFA M. SOLIMAN

Professor, Faculty of Engineering
Ain Shams University
Cairo, Egypt

CRC Press
Taylor & Francis Group
Boca Raton   London   New York

CRC Press is an imprint of the
Taylor & Francis Group, an **informa** business

CRC Press
Taylor & Francis Group
6000 Broken Sound Parkway NW, Suite 300
Boca Raton, FL 33487-2742

First issued in paperback 2017

© 2010 by Taylor and Francis Group, LLC
CRC Press is an imprint of Taylor & Francis Group, an Informa business

No claim to original U.S. Government works

ISBN 13: 978-1-138-11444-9 (pbk)
ISBN 13: 978-1-4398-1555-7 (hbk)

Library of Congress Cataloging-in-Publication Data

Soliman, Mostafa M. (Mostafa Mohammed)
  Engineering hydrology of arid and semi-arid regions / Mostafa M. Soliman.
    p. cm.
  Includes bibliographical references and index.
  ISBN 978-1-4398-1555-7 (alk. paper)
  1. Hydrogeology. 2. Hydraulic engineering. I. Title.

GB1005.S64 2010
628.1'14—dc22   2010000994

Visit the Taylor & Francis Web site at
http://www.taylorandfrancis.com

and the CRC Press Web site at
http://www.crcpress.com

Contents

Preface.. xiii
Author xv

Chapter 1 Introduction 1

    1.1   General Remarks 1
    1.2   Engineering Hydrology for Arid and Semi-Arid Regions 2
        1.2.1   Structural Design 3
        1.2.2   Municipal Water Supplies 4
        1.2.3   Irrigation... 5
        1.2.4   Flood Control 6
        1.2.5   Erosion Control 7
        1.2.6   Environmental Impacts 9
    1.3   Hydrologic Cycle 9
    1.4   Hydrologic Systems... 11
    1.5   Wadi Hydrology... 12
    1.6   Modeling.. 13
    References 15
    Bibliography.. 16

Chapter 2 Meteorological Processes and Hydrology 19

    2.1   Introduction 19
    2.2   Solar and Earth Radiation 19
    2.3   Temperature... 20
        2.3.1   Measurement of Temperature.............................. 20
        2.3.2   Terminology 20
    2.4   Humidity.. 21
        2.4.1   Properties of Water Vapor................................... 21
        2.4.2   Terminology 23
            2.4.2.1   Units.. 23
            2.4.2.2   Dew Point 23
            2.4.2.3   Relative Humidity.................. 23
        2.4.3   Measurement of Humidity 25
    2.5   Wind 25
        2.5.1   Measurement of Wind.. 25
        2.5.2   Geographic Variation of Wind............................. 27
    2.6   Climate Change 28
        2.6.1   Climate and Human Activity 28
        2.6.2   Solar Variability and Climate Change 30
    References 32

Chapter 3   Precipitation 35
          3.1   Introduction 35
          3.2   Forms of Precipitation 35
                 3.2.1   Rain 36
                 3.2.2   Snow 36
                 3.2.3   Drizzle 36
                 3.2.4   Glaze 36
                 3.2.5   Sleet 36
                 3.2.6   Hail 36
          3.3   Types of Precipitation 37
                 3.3.1   Front 37
                 3.3.2   Cyclone 37
                 3.3.3   Anticyclones 38
                 3.3.4   Convective Precipitation 38
                 3.3.5   Orographic Precipitation 39
                 3.3.6   Precipitation Enhancement 39
          3.4   Measurement of Precipitation 40
                 3.4.1   Nonrecording Gauges 41
                 3.4.2   Recording Gauges 41
                          3.4.2.1   Tipping-Bucket Gauges 41
                          3.4.2.2   Weighing-Bucket Gauges 42
                          3.4.2.3   Float-Type Gauges 42
                 3.4.3   Advantages and Disadvantages of Recording
                          Gauges 43
                 3.4.4   Radar Measurement of Rainfall 43
                 3.4.5   Weather Satellites 44
                          3.4.5.1   Polar-Orbiting or Active Satellites 44
                          3.4.5.2   Geostationary Satellites 45
          3.5   Precipitation Gauge Network 46
          3.6   Interpretation of Precipitation Data 48
                 3.6.1   Estimating Missing Precipitation Data 48
                 3.6.2   Double-Mass Analysis 49
          3.7   Average Precipitation over an Area 50
                 3.7.1   Arithmetic Mean 50
                 3.7.2   Thiessen Method 50
                 3.7.3   Isohyetal Method 50
          3.8   Design Storms 52
                 3.8.1   Frequency Analysis of Point Rainfall 53
                 3.8.2   Rainfall Duration 57
                 3.8.3   Rainfall Depth 57
                 3.8.4   Intensity–Duration–Frequency Relationship 57
                          3.8.4.1   Partial-Duration Series 58
                 3.8.5   Depth–Duration–Frequency Relationship in
                          Arid Regions 58
                 3.8.6   Probable Maximum Precipitation 60

3.8.7 Temporal Distribution 63
References 64
Bibliography..65

Chapter 4 Precipitation Losses 67

4.1 Introduction 67
4.2 Evaporation...67
    4.2.1 Evaporimeters...68
        4.2.1.1 Class A Evaporation Pan 69
        4.2.1.2 U.S. Geological Survey Floating Pan 69
    4.2.2 Pan Coefficient, C_p 69
    4.2.3 Evaporation Station Network 70
4.3 Empirical Evaporation Equations.............................70
4.4 Estimation of Evaporation by Analytical Methods 72
    4.4.1 Water-Budget Method 73
    4.4.2 Energy-Budget Method 73
    4.4.3 Mass-Transfer Method.....................................75
4.5 Reservoir Evaporation and Methods for Its Reduction 75
4.6 Evaporation and Transpiration 76
    4.6.1 Transpiration 76
    4.6.2 Evapotranspiration 77
    4.6.3 Measurement of Evapotranspiration 77
    4.6.4 Evapotranspiration Equations 78
4.7 Interception...80
4.8 Surface Retention Loss...81
4.9 Recommended Methods for Estimating Rainfall Losses........81
    4.9.1 Holton Infiltration Equation 81
    4.9.2 Green–Ampt Infiltration Equation...........................82
    4.9.3 NRCS Curve-Number Model.............................86
    4.9.4 Initial Loss Plus Uniform Loss Rate 91
4.10 Measurement of Infiltration...................................95
    4.10.1 Flooding-Type Infiltrometers 95
    4.10.2 Rainfall Simulators 95
4.11 Infiltration Indexes 96
References 97
Bibliography..97

Chapter 5 Catchment Characteristics and Runoff 99

5.1 Introduction 99
5.2 Catchment Characteristics.....................................99
    5.2.1 Stream Density.......................................101
    5.2.2 Drainage Density..101
    5.2.3 Shape of a Drainage Basin 102
    5.2.4 Stream Order...103
    5.2.5 Channel Slope 103

	5.2.6	Mean and Median Elevation	104
	5.2.7	Hydraulic Characteristics of Streams	104
	5.2.8	Classification of Streams	106
		5.2.8.1 Influent and Effluent Streams	106
		5.2.8.2 Intermittent and Perennial Streams	107
	5.2.9	Time of Concentration	107
		5.2.9.1 Kinematic-Wave Equation	107
		5.2.9.2 NRCS Method	109
		5.2.9.3 Kirpich Equation	110
		5.2.9.4 Izzard Equation	111
		5.2.9.5 Kerby Equation	111
	5.2.10	Isochrones	114
	5.2.11	Factors Affecting Runoff	114
5.3	Estimation of Runoff		115
	5.3.1	Empirical Formulas, Curves, and Tables	115
	5.3.2	Infiltration Method	116
	5.3.3	Rational Method	116
References			118
Bibliography			119

Chapter 6	Stream Flow Measurement		121
6.1	Introduction		121
6.2	Measurement of Stages		121
	6.2.1	Staff Gauges	121
	6.2.2	Automatic Stage Recorders	122
	6.2.3	Stage Data	123
6.3	Discharge Measurement		124
	6.3.1	Velocity Measurement by Floats	124
	6.3.2	Chemical Gauging for Stream Flow Measurement	125
	6.3.3	Electromagnetic Method	126
	6.3.4	Ultrasonic Method	127
6.4	Flow-Measuring Structures		128
	6.4.1	Weirs	128
		6.4.1.1 Clear over Fall Weir	129
		6.4.1.2 Standing Wave Weir	130
	6.4.2	Cut-Throat Flumes	131
References			132

Chapter 7	Stream-Flow Hydrographs		133
7.1	Introduction		133
7.2	Characteristics of the Hydrograph		133
7.3	Hydrograph Separation		138
7.4	Unit Hydrograph Concept		139

7.4.1   Duration of Rain... 140
7.4.2   Time–Intensity Pattern.. 140
7.4.3   Areal Distribution of Runoff..................................... 140
7.4.4   Amount of Runoff 141
7.5   Derivation of Unit Hydrographs from Simple Hydrogaphs ... 141
7.6   Derivation of Unit Hydrographs from Complex Storms 142
7.7   Unit Hydrographs for Various Durations 144
7.8   Synthetic Unit Hydrographs 147
7.8.1   Snyder Method 147
7.8.2   NRCS Dimensionless Unit Hydrograph................... 150
7.8.3   Transposing Unit Hydrographs 152
7.9   Hydrograph of Overland Flow... 154
7.9.1   Design of the Side Channel of the Gutter 160
References 161
Bibliography... 161

Chapter 8   Flood Routing... 163

8.1   Introduction 163
8.2   Hydraulic Routing Techniques 163
8.2.1   Equations of Motion.. 163
8.3   Hydrologic Routing Techniques 165
8.3.1   Storage Equation 165
8.3.2   Channel Routing... 165
8.3.3   Development of the Muskingum Routing
        Equation 166
8.3.4   Muskingum–Cunge Channel Routing 168
        8.3.4.1   Development of Equations........................ 169
        8.3.4.2   Data Requirements 169
8.3.5   Reservoir Routing... 172
8.3.6   Modified Puls Reservoir Routing............................. 172
8.4   Case Study: Flood Routing for the High Aswan
      Dam Reservoir... 175
8.4.1   Model Development 175
        8.4.1.1   Model Constraints................................... 177
        8.4.1.2   Model Results 178
References 182
Bibliography... 182

Chapter 9   Groundwater Hydrology.. 183

9.1   Introduction 183
9.2   Distribution of Subsurface Water 183
9.3   Groundwater Flow Theories.. 184
9.3.1   Steady-State Groundwater Flow in Aquifers 186
9.3.2   Unsteady-State Groundwater Flow in
        Confined Aquifers 188

          9.3.2.1   Basic Modified Equation 189
          9.3.2.2   Adjustment of the Modified Equations
                    for Free-Aquifer Conditions 191
          9.3.2.3   Recovery Equation................................. 192
          9.3.2.4   Drawdown Equation for Water-Table
                    Conditions....................................... 193
          9.3.2.5   Unsteady-State Flow in Semiconfined
                    Aquifers 197
     9.3.3   Effects of Partial Penetration of a Well.................. 198
9.4 Hydraulics of the Well and Its Design................................201
     9.4.1   Specific Capacity...................................... 201
     9.4.2   Effective Radius 201
     9.4.3   Well Screens...202
     9.4.4   Velocity Distribution 202
9.5 SLUG Tests...203
9.6 Groundwater Recharge...206
9.7 Application 210
     9.7.1   Unsteady-State Well Formulas........................... 210
          9.7.1.1   Confined Aquifer 210
          9.7.1.2   Semiconfined Aquifer.................. 212
          9.7.1.3   Water-Table Condition 213
     9.7.2   Groundwater Recharge Application....................... 214
9.8 Groundwater Pollution.. 217
     9.8.1   Migration of Pollutants in Aquifers 218
References 219
Bibliography..220

Chapter 10   Sediment Yield from Watersheds....................................... 221

10.1 Introduction 221
10.2 Sediment-Yield Theories 221
     10.2.1   Determination of the Soil-Erodibility Factor (K).....223
     10.2.2   Determination of the Slope Length-and-Gradient
              Factor...224
     10.2.3   Determination of the Parameters Influencing
              Erosion-Control Practices 224
10.3 Reservoir Sedimentation 225
References 229
Bibliography..229

Chapter 11   Hydraulic Structures 231

11.1 Introduction 231
11.2 Crossing Works..232
     11.2.1   Hydraulic Design of a Bridge................................233
          11.2.1.1   Calculating the Heading Up 233
     11.2.2   Hydraulic Design of Culverts................................235

11.2.3 Hydraulic Design of Siphons...................................238
11.2.4 Hydraulic Design for Aqueducts.............................239
11.3 Control and Storage Works...240
11.3.1 Introduction...240
11.3.2 Weirs..241
11.3.2.1 Percolation or Seepage...........................241
11.3.2.2 Uplift..243
11.3.2.3 Precautions against Scouring of
Downstream Weir Structures...................245
11.3.3 Storage Works...245
11.3.3.1 Rainfall-Harvesting Storage System........246
11.3.3.2 Check Dams..247
11.3.3.3 Cistern Systems.......................................248
References...252
Bibliography...252

Chapter 12 Case Studies 255

12.1 Case Study 1—Water Resources Management in Wadi
Naghamish at the North Coastal Zone of Egypt....................255
12.1.1 Introduction...255
12.1.2 Model Description..256
12.1.3 Hydrologic Modeling Module HEC-1......................257
12.1.4 Methodology..257
12.1.5 Hydrologic Studies...259
12.1.6 Rainfall Analysis..260
12.1.7 Overview of Rainfall-Runoff Regime.......................262
12.1.8 Application of WMS Program..................................262
12.1.9 Conclusions...266
References...267
12.2 Case Study 2—Urbanization Impacts on the
Hydrological System of Catchments in Arid and
Semi-Arid Regions...267
12.2.1 Introduction...267
12.2.2 Wadi El-Arish Study Case.......................................268
12.2.2.1 Catchment Characteristics......................268
12.2.2.2 Rainfall Analysis.....................................270
12.2.2.3 Runoff Analysis.......................................272
12.2.3 Wadi Adai Study Case...276
12.2.3.1 General Remarks.....................................276
12.2.3.2 Rainfall Analysis.....................................276
12.2.3.3 Runoff Analysis.......................................278
12.2.3.4 Cyclone Gonu Analysis...........................280
12.2.3.5 Urbanized Study Area.............................282
12.2.4 Conclusion...285
References...287

12.3   Case Study 3—Runoff Simulation Using Different
        Precipitation Loss Methods 287
        12.3.1   Introduction 287
        12.3.2   Initial and Constant Loss Method 288
        12.3.3   SCS Loss Method 289
        12.3.4   Green–Ampt Method 290
        12.3.5   Characteristics of the Flood Events Studied 291
        12.3.6   Development of Models 292
        12.3.7   Optimization of Models 293
        12.3.8   Calibration Phase 293
        12.3.9   Validation Phase 296
        12.3.10 Discussion and Conclusion 297
        References 299
12.4   Case Study 4—Design of Salboukh Flood Control
        System, in Riyadh City, Kingdom of Saudi Arabia 299
        12.4.1   Introduction 299
        12.4.2   Data Processing and Site Reconnaissance 300
                   12.4.2.1   LandSat Image 300
                   12.4.2.2   Hydrometeorological Data 300
        12.4.3   Hydrologic Analysis 307
                   12.4.3.1   Construction of the Hydrologic Model 307
                   12.4.3.2   System Hydrologic Parameters 308
                   12.4.3.3   Rainfall Input 308
                   12.4.3.4   Design Flood Hydrographs 311
        12.4.4   Flood Control Works 313
                   12.4.4.1   Design Criteria 313
                   12.4.4.2   Design Concept 314
                   12.4.4.3   Minimum Hydraulic Dimensions 314
                   12.4.4.4   Proposed Alternatives for Main
                               Wadis A and B 317
                   12.4.4.5   Alternatives for Small Wadis 318
                   12.4.4.6   Recommended Flood Protection Scheme ... 320
                   12.4.4.7   Fill Methodology 321
                   12.4.4.8   Comments on the Preliminary
                               Proposed Master Plan 322
                   12.4.4.9   Effect on the Downstream 323
        References 323
        Bibliography 323

Appendix A: Conversion Tables 325

Appendix B: Glossary 329

Appendix C: Statistics and Stochastic Analysis in Hydrology 345

Appendix D: Software Manual for Hydrograph Development Using a
               Simplified Model in Arid and Semi-Arid Regions 373

Index 389

Preface

The demand for water is increasing worldwide due to social and economic development activities. Meeting the development goals of the new millennium puts additional pressure on natural water systems, especially in countries located in arid and semi-arid regions of the world. Countries facing water stress are trying to rely on alternative sources to augment their water availability, provide safe and adequate water supply, and meet the increasing food demand for an increasing population.

Natural scarcity is due to low rainfall, aridity, high evaporation rates, and widely random temporal and spatial variation in the occurrences and distributions of surface and groundwater resources. These factors make it difficult to achieve a reliable water supply.

This already difficult water situation is further complicated by man-made factors, such as increased pollution levels, wasteful utilization, limited databases, and a shortage of institutional and human resources in most countries.

Rainfall is usually less predictable; large floods are common; and recharge events are difficult to quantify. In addition, much water is abstracted from nonrenewable aquifers. Such degradation of water quantity at both national and regional levels is further exacerbated by a lack of regulatory and legal institutions to manage, protect, and monitor water resources, especially in developing countries. The effect of global warming on climate change and its possible effect on rainfall will also be discussed.

An adequate scientific understanding of the hydrologic processes in arid and semi-arid areas is the key component to formulate and implement integrated management approaches in catchment systems. These objectives can be achieved by undertaking basic and applied research and sustained education and training programs. Additional results can be achieved by taking advantage of new techniques in hydraulic structures and instrumentations with the support of adequate financial resources and human resources development.

Research efforts and networking during the last two decades have contributed to the enhanced state of knowledge of the hydrology of arid and semi-arid areas, especially the watershed system. Initial efforts in these directions came from related books enhancing the knowledge of hydrologic principles and their application to satisfy the needs of civil engineering students and hydrologists. These objectives are being achieved through capacity-building processes, education, and training, as well as institutional development.

Watershed systems in arid regions form ephemeral streams that dry up completely in rainless periods. These streams are called wadis in Arab regions. This term is increasingly recognized as an international name used in most hydrologic publications all over the world. Yet, despite their great role as a vital source of water supply, as well as the threat of catastrophic floods in many countries, scientific understanding and knowledge about wadis' hydrologic processes are poor in most of the arid and semi-arid countries all over the world.

The book begins with an introduction to the field of engineering hydrology of arid and semi-arid regions in Chapter 1. Chapter 2 covers the meteorological processes and hydrology in arid and semi-arid zones. Chapter 3 covers precipitation, and Chapter 4 presents precipitation losses. Chapter 5 covers the catchment characteristics and runoff estimation methods. Chapter 6 covers stream flow measurements, and Chapter 7 presents stream flow hydrographs. Chapter 8 covers flood routing, while Chapter 9 covers groundwater hydrology, including the basic equations of groundwater flow, analytic solutions describing flow in aquifers, pumping tests, and salt water intrusion. Chapter 10 covers sediment yield in arid and semi-arid watersheds and its streams. Chapter 11 discusses the design of hydraulic structures that serve to protect and manage the water resources systems in arid and semi-arid regional catchments. Chapter 12 presents some selected case studies in different arid and semi-arid watersheds in order to demonstrate a variety of engineering water resources management projects. All chapters include solved problems to demonstrate and explain the usage of different theories and equations presented in these chapters.

The appendices include Appendix A, conversion tables; Appendix B, a glossary of important hydrologic terms; Appendix C, statistic and stochastic analysis theories; and Appendix D, a software manual for hydrologic models designed by the author, including an example explaining how to use the hydrologic model in order to get the flood hydrograph of a catchment in a simple way. The software can be downloaded from the publisher's site with permission.

This textbook can serve both undergraduate and graduate students who are studying or researching the hydrologic sciences. It can also be used as a guide for both hydrologic professionals and hydraulic engineers.

Prof. Dr. Mostafa M. Soliman

Author

Mostafa M. Soliman received his Bachelor of Science in civil engineering from Cairo University in 1953, his Master of Science in civil engineering from Colorado State University in 1957, and his PhD in civil engineering from Utah State University in 1959.

He has published several books in the field of irrigation engineering and was a senior author of *Environmental Hydrogeology* (1998) and of the second edition of the same book, published by Lewis/CRC of New York. This book is a highly recommended text used by many universities.

Dr. Soliman has acted as a chairman of six International Conferences on Environmental Hydrology (in 1995, 1999, 2002, 2005, 2007, and 2009) in Cairo, Egypt, and he was also the editor of those conference proceedings.

Dr. Soliman has undertaken a variety of assignments in the area of water sciences. His contribution to the field of water resources, hydrologic modeling, and its uses include promoting evolution in the fields of water resources' quality protection, reuse, and capacity building in the Middle East, Europe, and the United States. He has held several academic and administrative positions as a lecturer, assistant professor, and professor at the University of Ain Shams, followed by his appointment as department head of irrigation and hydraulics. Later, he was appointed as the chairman of environmental engineering at the Institute of Environmental Studies and Research at the university. He had also worked as a visiting professor at Washington State University in Pullman, Washington, for more than 2 years, the University of Mississippi, and the University of Alabama. He lectures short courses in hydrological sciences at many international institutes sponsored by UNESCO in Europe and the Middle East.

As a consultant, Dr. Soliman has emerged as one of Egypt's most progressive engineers, combining a sound understanding of problems and technology with an enlightened vision of what the region needs and which developments offer promise. He has experience in research and consultation with both UNESCO and FAO and a substantial share in designing major reclamation projects, pumping stations, and their hydraulic structures. He also conducts hydrological research in Egypt and other Arab countries.

Dr. Soliman has received numerous awards, including the Ideal Engineer Medal from the Egyptian Syndicate of Engineers (1984), the Arid Land Hydraulic Prize from the American Society of Civil Engineers (1998), the Selected Distinguished Member Award from the American Society of Civil Engineers (2000), the Ain Shams University Award in Engineering Sciences (2003), and the Egyptian

Academy of Research Sciences and Technology Award in Advanced Engineering Technology Sciences (2007). Dr. Soliman also holds the following volunteer positions: distinguished member of the American Society of Civil Engineers, president of the American Society of Civil Engineers (Egypt section), president of the Egyptian Society of Irrigation Engineers, member of the Board of Egyptian Society of Engineers, the Board of Directors of the National Water Research Center, and member of many other scientific societies for many years, such as the Academy of Science, the American Scientist Society, the American Institute of Hydrology, the International Association of Hydrological Sciences, the International Association for Hydraulic Research, and the International Association of Hydrogeology.

1 Introduction

1.1 GENERAL REMARKS

Water is considered a scarce resource in arid and semi-arid areas, which occupy approximately one-third of the land, yet the demand for water is growing as populations expand and economies develop. At the same time, threats to existing water resources are also increasing due to pollution, overexploitation of resources, and climate change. The importance of water to the natural environment and the need for conservation of resources to protect valuable ecosystems are increasingly recognized in arid and semi-arid areas. Water is also a major hazard. Floods remain one of the most damaging and dangerous natural hazards globally, and flood risk is increasing with increasing development, as housing, industry, and infrastructure, which are increasingly encroaching on flood-prone areas, expand. Climate change is expected to lead to intensification of extreme events and further increases in risk. This will be discussed in Chapter 2.

Although there is no agreement among hydrologic experts on the distinct classifications of arid and semi-arid regions based on their annual rainfall, the following categories may be generally identified (Soliman 2008):

1. Areas where the annual total rainfall is less than 70 millimeters per year and evaporation exceeds the yearly rainfall may be classified as extreme desert areas. Two-thirds of the Middle East region can be classified as desert.
2. Areas where the annual total rainfall is between 70 and 200 millimeters per year with sparse vegetation are called arid regions.
3. Areas where the annual total rainfall is between 200 and 450 millimeters per year are classified as semi-arid regions. The Mediterranean Sea coast is classified as a Mediterranean zone with rainy and moderately warm winters and dry summers and can be considered to be between arid and semi-arid regions.

Clearly, there is an unprecedented need for effective, appropriate, and sustainable management of water to protect populations and the natural environment and to provide secure water supplies. This requires a sound foundation of scientific understanding of the natural hydrologic systems and appropriate tools to support water management and optimize the sustainable use of water resources.

Hydrology principles generally relate to the waters of the earth, their occurrence, circulation, and distribution, their chemical and physical properties, and their reaction with the environment, including their relation to living things. The domain of hydrology embraces the full life history of water on earth. Engineering hydrology includes segments pertinent to the design and operation of engineering projects for

1

the control and use of water. The boundaries between hydrology and other earth sciences such as meteorology, oceanography, and geology are indistinct, and there is no need to define them rigidly. Likewise, the distinctions between engineering hydrology and other branches of applied hydrology are vague. Indeed, an engineer owes much of his or her present knowledge of hydrology to agriculturists, foresters, meteorologists, geologists, and others in a variety of fields. With this definition, we may think of hydrology as being bounded above by meteorology, below by geology, and at land's end by oceanography, but without any distinct lines of demarcation. The several sciences are not blocks to be fitted into individual compartments, and perhaps none of them represents a distinct body of subject matter as much as it represents a different point of view. The entire cycle of water movement, from clouds to earth to sea and back to clouds, is of interest to meteorologists, hydrologists, geologists, and oceanographers alike, but each treats it from a different aspect. Meteorologists concern themselves primarily with the precipitation phase of the cycle; they analyze the movements of air masses and make short-term predictions of temperature and wind and the occurrence of precipitation. Hydrologists measure the flow of streams and groundwater, analyze the regimen of rivers, and then, on the basis of their own measurements and the predictions of meteorologists, they make both short- and long-term quantitative predictions of flood, drought, and normal flow; hydrologists measure the water intake into the soil and groundwater reservoirs and are concerned with evaporation and transpiration losses; they also study the methods of developing water resources and predict the effects of proposed improvements on the regimen of streams and the groundwater reservoir. Geologists (LaMoreaux et al. 2008) focus on the earth's structure and consider water both a mechanical and chemical agent that produces changes in physiography and internal structure by erosion, transportation, and deposition of sediment, freezing and thawing, and chemical action; hydrologists depend on geologists for information on the probable location, extent, and source of recharge of groundwater reservoirs. Oceanographers are concerned largely with tidal movements, wave actions, ocean currents, and similar phenomena.

Clearly, all these sciences have a large body of subject matter in common; the list of related sciences could be extended still further if we so choose. Thus, meteorologic, hydrologic, and geologic data provide the basis for climatology, which, with the introduction of biological aspects, blends into plant ecology; and from ecology, it is but a step to agronomy. To complete the cycle, hydrologists make use of their knowledge of ecology to deduce the depth to groundwater, seasonal fluctuations in the groundwater table, and probable rates of evapotranspiration and work with agronomists to develop adequate conservation practices that will protect the soil from erosion and the streams from silting and will provide water where it is needed and remove it where it is in excess.

1.2   ENGINEERING HYDROLOGY FOR ARID AND SEMI-ARID REGIONS

A brief review of some of the practical applications of engineering hydrology (Linsley et al. 1975) may provide a helpful background for a more detailed study of the subject. In this section, we consider a few of its uses related to structural design, water supply, irrigation, flood control, erosion control, and environmental impacts on water resources, and finally its applications on water resources management (Mays 2005; Chin 2006) in arid and semi-arid regions.

1.2.1 STRUCTURAL DESIGN

On highways and railroads that cross a stream of a catchment in an arid or semi-arid area, the construction cost of crossing works such as bridges, culverts, or Irish crossings adds up to an appreciable part of the total cost of such projects. Sometimes, culverts or bridges are located so that we need not expect them to be destroyed by a flood that exceeds, even by a considerable amount, their design capacity; in such cases, a temporary interruption of service due to flooding of the road or rails is only an inconvenience. Figure 1.1 illustrates the flash flood water damage at one of the crossing works a stream in Sinai, Egypt, which is an arid region where there is occasional rainfall during winter (Kotb and El Belasy 2007). Here again an economic balance must be struck between an occasional interruption of service, on the one hand, and the costs of larger drainage structures, on the other. Hydrology can provide the maximum design flow in order to get the minimum vent sizes for such

FIGURE 1.1 Flash flood damage to one of the highways crossing a stream at the Wadi Watir Sinai catchment, Egypt. (Adapted from Egyptian Water Resources Research Institute [WRRI]. 2005–2006. Report prepared for the Development of Wadi Watir in Sinai, Egypt to the Egyptian Ministry of Water Resources and Irrigation.)

structures, while hydraulics principles provide the basis for analyzing the effect of the structures on the flow beside the hydraulic design of the structure size and its protection needed.

Closely akin to the problem of highway and railroad culvert design are those of storm sewer design, though the hydrology of urban areas is a study in itself and is much more complex than the apparent similarity of such areas would suggest.

In any type of reservoir, provisions must be made for passing flood flows over or around the dam. The spillway structure is often the most expensive portion of the dam, and hence, it should be as small as possible. On the other hand, overtopping of the nonoverflow section is a serious matter, and in an earth dam it is almost certain to cause failure, with destruction and frequently death in the areas downstream. The best hydrologic design is a prerequisite to safe and economical dam design. The hydrologist must evaluate not only the probability of floods of various magnitudes but also the effect of the reservoir upon the distribution of the flood volume. This latter point is worthy of emphasis, because it is sometimes overlooked and unnecessary expenditures may be made on "overdesigned" structures to protect a dam on it from being overtopped, because of the pond action of the reservoir behind the dam.

The reader should note the distinction between hydrologic designs and hydraulic designs (Chow et al. 1988). The hydrologic design determines the quantity of water that must be handled, whereas the hydraulic design proceeds from there and determines the structure best suited for the job. The engineer is no more warranted in undertaking the hydraulic design of a structure without considering the hydrologic design than he or she would be trying to determine the size of the structural members without first ascertaining the load that may come upon them.

1.2.2 MUNICIPAL WATER SUPPLIES

Even in regions where water is in the most abundant supply, the location and development of sources adequate to the needs of urban and rural areas (Figure 1.2) and industries is a matter of increasing concern. Most laymen do not realize, and some engineers fail to appreciate, that the availability of water is often the limiting factor in the growth of municipalities.

Drought is another problem. First, hydrologists should recognize that the drought is less severe in some areas than in other "meteorologically similar" areas—in other words, a drought of the same overall intensity and probability of occurrence could, "next time," hit many catchments in arid regions harder than they were hit before. Next, hydrologists have at their disposal plenty of geological and botanical evidence that nature has come far from doing her worst in the dry years. Finally, hydrologists know that in many areas the increased pumping rates in certain years may already have lowered groundwater levels to the point that an even less severe drought might set new minima in stream flow and underground water reserves. Application of this knowledge in a specific case and producing a reliable quantitative answer is one of the most complex and important problems of hydrology.

FIGURE 1.2 Dam constructed across one of the streams of Sinai, Egypt. Notice the flood water that covers a farm and a road downstream. (Adapted from Egyptian Water Resources Research Institute [WRRI]. 2005–2006. Report prepared for the Development of Wadi Watir in Sinai, Egypt to the Egyptian Ministry of Water Resources and Irrigation.)

1.2.3 IRRIGATION

The hydrologic problems in irrigation are similar to those in water supply but occur on a bigger scale and particularly in arid and semi-arid regions. A few years ago, the hydrology of irrigation was relatively simple. Today, an equally summary analysis may still be adequate in some regions, however, we increasingly find ourselves confronted with limiting conditions and the complexities of the problem correspondingly increase. In some streams, appropriations of water far exceed the total discharge, and the downstream projects depend on "return flow" from upstream projects. Elsewhere, irrigation projects depend on groundwater reservoirs, artificially recharged by "water spreading" on permeable areas, and an elaborate system of hydrologic bookkeeping is required to keep inventory of the supply and determine safe pumping rates (Soliman 1990). A hydrologist is often called upon to evaluate new projects in areas where the margin of safety is already low, to discover new water sources for projects having difficulty, or to develop more economical methods of water use.

In most arid and semi-arid regions, the economy is founded largely upon irrigation, the raising of camels and sheep, and some limited industry. The continued security and development of this entire region will be increasingly dependent on the vision and skills of hydrologists.

1.2.4 Flood Control

The flood-control methods in arid and semi-arid regions include using reservoirs to hold back flash flood waters (Figure 1.3), channel improvements to speed them on their way, and diversions to transfer them to channels not available to them in nature (WRRI 2005–2006). Flood-control projects range from small improvements, such as localized dredging or channel straightening undertaken by municipalities, to large, basin-wide developments involving tens of millions of dollars.

The design of any flood-control project must be based on reliable hydrologic studies. First, the probable frequency of floods of various magnitudes must be statistically analyzed so that potential future flood losses may be reasonably predicted. Next, a "design flood" must be synthesized, and a variety of preliminary plans must be prepared for works that might protect against it. Then, a number of the more promising alternatives must be studied in detail, analytically, by means of hydrologic models, or by a combination of both. Flood-control studies are complicated by the

FIGURE 1.3 Detention dam constructed across one of the streams at Wadi Watir (Sinai, Egypt) as a flood control project. (Adapted from Egyptian Water Resources Research Institute [WRRI]. 2005–2006. Report prepared for the Development of Wadi Watir in Sinai, Egypt to the Egyptian Ministry of Water Resources and Irrigation.)

fact that they modify the natural regimen of the stream and thus, in the process of protecting one area, may increase flood damage in another. As a final step, the best alternatives must be investigated even more intensively, with an objective of seeing how they can be expected to behave when subjected to floods other than the design flood. Some types of works, for example, will give complete protection against floods smaller than the design flood, while others may not go into effective action on any flood except a major one. Again, some types of works are capable of withstanding floods greater than the design flood and even giving some protection against them, whereas with other types the occurrence of a super design flood can be expected to result in major damage to the works, complete loss of benefits, and possible additional losses.

Continued development of both the basic theory and the technique of flood-routing studies are essential to intelligent, economical planning of flood-control projects. Methods of study that were adequate in the day of the "local" project do not suffice for the integrated, basin-wide improvements toward which engineering and political thought are increasingly turning.

Flood forecasts are similar to flood control. During high water, we rely upon these forecasts for planning evacuations of threatened areas, organizing standby crews for emergency work on railroads and highways, and putting emergency controls into effect at municipal water plants and other public utilities. These forecasts, which have a remarkably high degree of accuracy, are the joint work of hydrologists and meteorologists.

1.2.5 EROSION CONTROL

Soil is an exhaustible resource and provides the impetus for intensive study and the development of soil conservation practices in arid and semi-arid regions. Despite—or perhaps because of—the aggressive political and economic arguments that accompanied such developments, a sound foundation for truly scientific conservation practices rapidly evolved. In this development, the hydrologists, ecologists, and agronomists worked hand in hand.

From the standpoint of hydrology, erosion control problems center around the phenomena of overland flow and infiltration. How effective is a given cover of vegetation in protecting a soil from erosion? With given soil and given cover on a given slope, what is the critical distance from the crest beyond which erosion may be expected to begin? Under given initial conditions of soil and cover, at what rate of rainfall will surface runoff begin? How will the infiltration rate vary as rain continues? How many tons of soil will be lost per acre per year with various crops and cropping procedures? What effects will land conservation practices have on flood flows and low-water flows of streams? What types of structure are best suited for preventing erosion in ditches and arresting the development of gullies?

Loss of topsoil to streams means a deposition of that soil in other areas (Figure 1.4). In a natural state, a rough longtime balance of values may exist between the loss and gain; we need only consider the building up of alluvial plains and delta lands to see this. However, where there are existing farms, reservoirs, or canalized reaches, deposition of sediment is as much an economic loss as erosion, and the

FIGURE 1.4 Silt deposition upstream: a storage dam at one of the streams in Sinai, Egypt, during a low water level. Notice the high water level mark at the upstream face of the dam. (Adapted from Egyptian Water Resources Research Institute [WRRI]. 2005–2006. Report prepared for the Development of Wadi Watir in Sinai, Egypt to the Egyptian Ministry of Water Resources and Irrigation.)

loss from the topsoil of a headwater farm is not offset economically by the deposition in river bottoms or in the delta. Every dam creates a settling basin, in which the stream that has been momentarily brought to a halt drops its suspended load. Thus, the life of a reservoir is fixed, to a great extent, by the rate at which erosion is taking place in the tributary drainage area. Estimation of this "useful life" has become an important phase of the hydrologic investigation of proposed reservoir projects, but much work remains to be done both in establishing a sound analytical basis for such estimates and in developing procedures for the desilting of existing reservoirs whose life is threatened. A fascinating corollary of the reservoir-silting problem is the effect of clear water on the channel downstream. For every velocity and depth of flow, a river has a certain capacity for carrying suspended matter, and when robbed of its load by a reservoir, it tends to pick up a new charge of sediment from its bed and banks after leaving the pool. This may result in a lowering of the bed, radical changes in channel alignment, and steepening of the slopes of tributaries—in short, a complete upset of the physiographic balance that must either work itself out or be arrested by additional control works. Analysis of such problems is still in its infancy and is a stimulating challenge to the combined efforts of hydrologists and geologists.

1.2.6 ENVIRONMENTAL IMPACTS

The growth of urban areas in arid and semi-arid regions has seen many environmental impacts on public health problems, not the least important of which is the pollution of streams. Many streams flowing downstream from cities have become open sewers, dangerous to public health, and destructive to wildlife and natural beauty. In less serious instances, stream pollution creates public nuisance. Table 1.1 provides classes of contaminants that are found in many streams. Pollution control is largely a sanitary engineering problem to be solved by the enforcement of strict laws, and involves vast expenditures of public funds for sewage and industrial waste treatment. The disposal of a certain amount of sewage by dilution is usually considered permissible, particularly after the second treatment, functioning through bacterial action and aeration. Complete prevention of stream pollution, although possible in some streams, is not economically feasible. It is here that the hydrologist can assist the sanitary engineer. A complete stream pollution control study must include an investigation of stream flow, particularly the magnitude and duration of low flows. In some instances, the augmentation of low flows by means of reservoirs has proved to be at least as important to the control of stream pollution as investments in additional sewage treatment plants.

1.3 HYDROLOGIC CYCLE

The hydrologic cycle deals with the movement of water, in its three phases—gas, liquid, and solid—from the ocean, land, or forests into the atmosphere by evaporation and transpiration. It passes through complicated atmospheric phenomena, generalized as the precipitation process, back to the earth's surface, upon and within which it moves in a variety of ways, and is incorporated into nearly all compounds and organisms. The relevant sciences that deal with the hydrologic cycle are astronomy, solar physics, cloud physics, meteorology, climatology, hydrometeorology, environmental sciences, geography, engineering, agriculture–biology, economics, surface

---

TABLE 1.1
Classes of Water Contaminants

Contaminant Class	Example
Oxygen-demanding wastes	Plant and animal materials
Infectious agents affecting plants	Bacteria and viruses
Nutrients	
Organic chemicals	Fertilizers such as nitrates and phosphates
Inorganic chemicals	Pesticides, detergent molecules
	Acids from coal mine drainage, inorganic chemicals
Sediment from land erosion	Clay silt on streambed
Radioactive substances	Waste products from mining and processing of radioactive material, radioactive isotopes after use
Heat from industry	Cooling water used in steam generation of electricity

water hydrology, limnology, oceanography, soil physics, ground water hydrology, and geology.

The hydrologic cycle, as given in Figure 1.5 (Soliman 2008), starts with the evaporation of water from the oceans by solar energy. The evaporated water (i.e., water vapor) rises by convection, condenses in the atmosphere to form clouds, and precipitates onto land and ocean surfaces, predominantly as rain or snow. Precipitation on land surfaces is partially intercepted by surface vegetation, partially stored in surface depressions, partially infiltrated into the ground, and partially flows over land into drainage channels and streams that ultimately lead back to the sea. Precipitation that is intercepted by surface vegetation eventually evaporates into the atmosphere; water held in surface depressions either evaporates or filters into the ground; and water that filters into the ground contributes to the recharge of groundwater, which either is utilized by plants or becomes subsurface flow that ultimately emerges as recharge to streams or directly to the ocean. Groundwater is defined as the water below the land surface; water above the land surface (in liquid form) is called surface water. The ground surface in urban areas is typically much less pervious than in rural areas, and surface runoff is mostly controlled by constructing drainage systems. Surface water and groundwater in urban areas also tend to be significantly influenced by the water supply and wastewater removal systems that are an integral part of urban development. Since human-made systems are part of the hydrologic cycle, hydrologic engineers must ensure that systems constructed for water use and control are in harmony with the needs of the natural environment in arid and semi-arid regions.

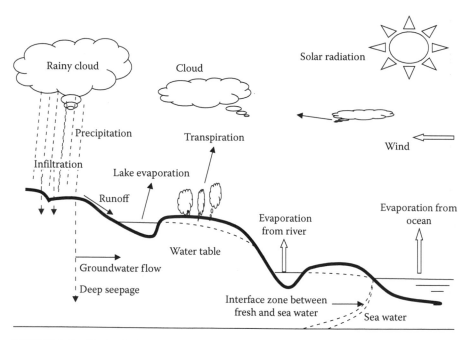

FIGURE 1.5   The hydrologic cycle.

The quality of water varies considerably as it moves through the hydrologic cycle, with contamination resulting from several sources. Classes of common contaminants were stated before in Section 1.2.6. The effects of the quantity and quality of water on the health of terrestrial ecosystems and the value of these ecosystems in the hydrologic cycle are often overlooked.

1.4 HYDROLOGIC SYSTEMS

Chow et al. (1988) defined the hydrologic system as a volume domain in space, surrounded by a boundary that accepts water and other inputs, operates on them internally, and produces them as outputs. The domain (for surface or subsurface flow) or volume in space (for atmospheric moisture flow) is the totality of the flow paths through which the water may pass from the point it enters the system to the point it leaves. The boundary is a continuous surface defined in three dimensions enclosing the volume. A working medium enters the system as input, interacts with the domain and other media, and leaves as output. Physical, chemical, and biological processes operate on the working media within the system; the most common working media involved in the hydrologic analysis are water, air, and heat energy.

The global hydrologic cycle can be represented as a system containing three subsystems: the atmospheric water system, the surface water system, and the subsurface water system (Figure 1.6). Another example is the storm–rainfall–runoff process on a watershed, which can be represented as a hydrologic system. The input is rainfall distributed in time and space over the watershed, and the output is stream flow at the

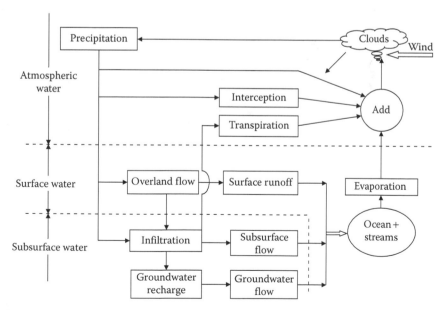

FIGURE 1.6 Block diagram showing the hydrologic system. (Adapted from Chow, V. T. et al. 1988. *Applied Hydrology.* New York: McGraw-Hill.)

watershed outlet. The boundary is defined by the watershed divide, and it extends vertically upward and downward to the horizontal planes.

The science that deals with the atmospheric water system is meteorology, more specifically hydrometeorology. The components of the hydrologic cycle, including the engineering aspects, will be discussed in this book, with enough information to perform water resources management in arid and semi-arid regions.

1.5 WADI HYDROLOGY

Although "wadi" is the name given to seasonal watercourses in the Arab region, it has been increasingly recognized as an international name used in most hydrologic publications all over the world. Yet, despite their great role as vital water supply sources and causes of catastrophic floods in many Arab countries, the scientific understanding and knowledge base of their hydrologic processes are rather poorly understood in most of these countries.

In recognition of this lack of understanding, the fifth cycle of the International Hydrological Program (IHP-V) of UNESCO (UNESCO 1996, 2002) gave great consideration to this topic in many of its eight themes. For example, seasonal watercourses have been explicitly included in various projects related to the theme "Integrated Water Resources Management in Arid and Semi-Arid Zones" (theme 5), while many of its components fit very well, among others, with projects related to the theme "Groundwater Resources at Risk" (theme 3) and "Knowledge, Information, and Technology" (theme 8). It is therefore logical that the above three themes have also been selected as priority areas for the region by the participants of the Sixth Regional Meeting of the IHP committees of Arab Countries, held in Jordan in December 1995. During the same meeting, wadi hydrology was specifically outlined as a target project. Both this project and the regional priority themes coincide with concentration areas of the UNESCO Cairo Office, that is, rainfall water management and groundwater protection. The author of this book, as a member of the Egyptian National Committee of IHP, was the first to suggest that wadi hydrology be used as a general term the hydrology of arid and semi-arid regions. This term may be applied to the catchments that are typically referred to as small drainage basins, but no specific area limits have been established. Wadi hydrology may also include the hydrologic studies of the coastal streams along both the Mediterranean Sea and the Red Sea in addition to the inland streams of arid and semi-arid regions. Figure 1.7 shows the streams' patterns at a coastal area near the city of Mersa Matruh on the north coast of Egypt. The stream catchments at this coastal area look almost the same. Figure 1.8 shows one of the catchments (Wadi Naghamish) at the coastal area of the Mediterranean Sea north of Egypt.

Because of the similarity of the coastal catchments' characteristics—rainfall and water management practices on the Mediterranean coast—the author of this book may suggest another term, "coastal hydrology," be applied to the wadis of the Mediterranean coast. Rainfall harvesting by the old Roman wells, galleries, and check dams across the wadis on the Mediterranean coast are still in use. Chapter 12 includes two cases of studies for rainfall harvesting in different arid and semi-arid coastal regions.

6,400   3,200   0   6,400 Meters

N
W   E
S

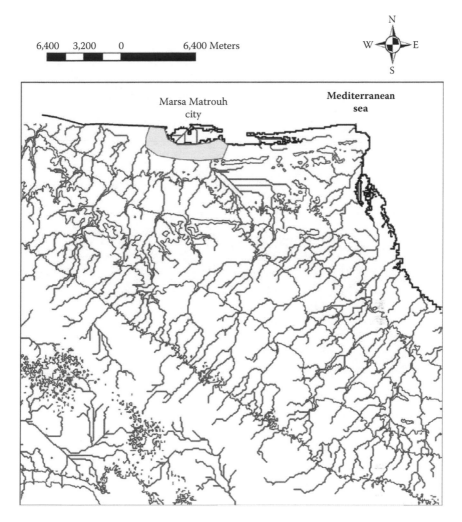

FIGURE 1.7   Stream distribution in the Mersa Matrouh area.

1.6   MODELING

Hydrologic models have been developed by a wide range of researchers and engineers for the management of surface and groundwater quantity and quality. Although these models have proven to be highly effective management tools, they have focused almost entirely on physical processes with little consideration of ecological dynamics. It is generally assume that plant communities have negligible impacts on hydrologic dynamics, but this is not the case in many terrestrial, riparian, and wetland systems. For example, brush invasion in many parts of the world was thought to have resulted in significant reductions in flow in streams and rivers. Expensive control programs have been widely implemented with little insight into long-term cost-effectiveness.

FIGURE 1.8   Wadi Naghamish.

Assessments of the impact of hydrologic alteration on vegetation and of vegetation on water supply can be greatly facilitated by linking existing hydrologic models with general ecosystem models designed to make long-term projections of ecosystem dynamics.

In arid and semi-arid regions, several surface and groundwater models have been applied with a considerable degree of success as long as the data needed for any catchment has been simulated and processed properly. This means that the results of the model should be verified with the observed values in order to obtain dependable results for flood forecasting or groundwater management.

The surface hydrologic modeling system is designed to simulate the rainfall–runoff processes of watershed systems. Its design is applicable to a wide range of geographic areas for solving diverse problems including large river basin water supply and flood hydrology and small urban or natural watershed runoff. A model of the watershed is constructed by separating the hydrologic cycle into manageable pieces and constructing boundaries around the watershed of interest. In most cases, several model choices are available for representing each water pathway in the cycle. Each mathematical model included in the program is suitable in different environments and under different conditions. Making the correct choice requires knowledge of the watershed and the goals of the hydrologic study and engineering judgment. The model features a completely integrated work environment including a database, data entry utilities, computation engine, and results reporting tools. The following popular software models can be downloaded free through the Internet, from different sources:

1. WMS 7 Environmental Modeling System Incorporated (www.ems-i.com)
2. HEC-HMS by the Hydrologic Engineering Center
3. Hydrocad model (www.hydrocad.net)
4. Mod-flow (older version) for groundwater flow modeling
5. Simplified model of surface water hydrology and a groundwater model designed by the author, included at the end of this book

Of course, there are more hydrologic model software designed by researchers and institutes around the world. Users should select the models that suit their problems and their budget. Geographic information systems (GIS), together with Arc Hydro, helps a lot in organizing hydrologic databases by base map preparation, land use, soil type, drainage map, and contour map, needed for any hydrologic model.

REFERENCES

Chin, D. A. 2006. *Water Resources Engineering.* 2nd ed. Upper Saddle River, NJ: Prentice Hall
Chow, V. T. et al. 1988. *Applied Hydrology.* New York: McGraw-Hill.
Egyptian Water Resources Research Institute (WRRI). 2005–2006. Report prepared for the Development of Wadi Watir in Sinai, Egypt to the Egyptian Ministry of Water Resources and Irrigation.
Kotb, G. and A. M. El Belasy. 2007. Protecting the economical area north west Suez gulf against flash flood hazards (south part). Fifth International Symposium on Environmental Hydrology. Cairo, Egypt.
LaMoreaux, P. E. et al. 2008. *Environmental Hydrogeology.* Boca Raton, FL: CRC Press.
Linsley, R. K., M. A. Kohler, and J. L. H. Paulhus. 1975. *Hydrology for Engineers.* 2nd ed. New York: McGraw-Hill.
Mays, L. W. 2005. *Water Resources Engineering.* 2nd ed. New York: John Wiley & Sons, Inc.
Soliman, M. M. 1990. Watershed modeling of the western coast wadis of Egypt ASCE Proceedings of the International Symposium on Hydraulics/Hydrology of Arid Lands (H2AL), San Diego, CA.
Soliman, M. M. 2008. Surface water hydrology in arid and semi-arid regions. Lecture notes presented at the International Course of Environmental Hydrology sponsored by UNESCO and the Egyptian Ministry of Water Resources and Irrigation (MWRI).
Weshah, R. 2002. *Water Resources of Wadi Systems in the Arab World—Arab Wadi Hydrology Network (AWHN).* UNESCO Cairo Office (IHP) No. 12.

BIBLIOGRAPHY

METEOROLOGICAL REFERENCES

Ackerman, S. et al. 2006. *Meteorology: Understanding the Atmosphere*. 2nd ed. Pacific Grove, CA: Brooks Cole.

Ahrens, C. D. 2006. *Meteorology Today: An Introduction to Weather Climate And the Environment*. 8th ed. Pacific Grove, CA: Brooks Cole.

Danielson, E. W. et al. 2003. *Meteorology*. 2nd ed. New York: McGraw-Hill.

Eagleman, J. R. 1980. *Meteorology: The Atmosphere in Action*. D. Van Nostrand ed. Ann Arbor, MI: University of Michigan.

Holton, J. R. 1992. *An Introduction to Dynamic Meteorology*. 3rd ed. New York: Academic Press.

Lutgens, F. K. et al. 2006. *The Atmosphere an Introduction to Meteorology*. 10th ed. Upper Saddle River, NJ: Prentice Hall.

Milham, W. I. 1912. *Meteorology: A Text Book on the Weather, the Causes of its Changes and Weather Forecasting for the Student and General Reader, The Macmillan Company*. Berkeley, CA: University of California.

Stull, R. B. 1999. *Meteorology for Scientists and Engineers*. 2nd ed. Pacific Grove, CA: Brooks Cole.

Weisberg, J. S. 1976. *Meteorology: The Earth and Its Weather*. Boston: Houghton Mifflin.

HYDROLOGICAL REFERENCES

Bedient, P. B., W. C. Huber, and B. E. Vieux. 2007. *Hydrology and Floodplain Analysis*. 4th ed. Englewood Cliffs, NJ: Prentice Hall.

Biswas, A. K. et al. 1979. *Water Management for Arid Lands in Developing Countries (Workshop Sponsored by UNEP) Water Development, Supply and Management*, vol. 13. Pergram Press.

Bras, R. L. 1988. *Hydrology: An Introduction to Hydrologic Science*. Reading, MA: Addison Wesley.

Chow, V. T. 1964. *Handbook of Applied Hydrology*. New York: McGraw-Hill.

Cross, W. P., and D. Johnston. 1949. *Elements of Applied Hydrology*. New York: The Ronald Press Company.

Decaurs, M., and G. F. Gifford. 1986. Applicability of the Green and Ampt infiltration equation to rangelands. *Water Resour Bull* 22(1):19–27.

Fahmi, A. H. 1995. *Development of a Water Shed Model in Arid Regions*. M.Sc. thesis, Ain Shams University, Cairo, Egypt.

Flood Control District of Maricopa County. 1992. *Drainage Design for Maricopa County*. vol. 1. Phoenix, AZ: Maricopa County.

Gupta, R. S. 1989. *Hydrology and Hydraulic Systems*. Englewood Cliffs, NJ: Prentice Hall.

Gupta, R. S. 2007. *Hydrology and Hydraulic Systems*. 3rd ed. Waveland Pr Inc.

Ibrahim, K. T. 2006. *Water Resources Management for Wadi Water South Sinai*. Ph.D. Thesis. Ain Shams University.

Koutsoyiannis, D., and Th. Xanthopoulos. 1999. *Engineering Hydrology*. 3rd ed. Athens: National Technical University of Athens.

Lineke, J. M. et al. 1996. *Workshops Sponsored by UNESCO on "Wadi Hydrology" and "Groundwater Protection."* UNESCO Publications: Cairo, Egypt.

Mays, L. W. 2005. *Water Resources Engineering*. 2nd ed. New York: John Wiley & Sons, Inc.

McCuen, R. H. 1998. *Hydrologic Analysis and Design*. 2nd ed. Englewood Cliffs, NJ: Prentice Hall.

Ponce, V. M. 1999. *Engineering Hydrology*. 2nd ed. New York: Prentice Hall.

Prakash, A. 2004. *Water Resources Engineering: Handbook of Essential Methods and Design Technology and Engineering*. New York: ASCE.

Roberson, J. A., J. J. Cassidy, and M. H. Chaudhry. 1998. *Hydraulic Engineering*. 2nd ed. New York: John Wiley & Sons, Inc.

Sen, Z. 2008. *Wadi Hydrology*. Boca Raton, FL: CRC Press.

Servat, E. 2003. *Hydrology of Mediterranean and Semi-Arid Regions International Association of Hydrological Sciences*. Paris: UNESCO Hydrology.

Simonvic, S. P. 2002. *Non Structural Measures for Water Management Problems*. 56th ed. Paris: UNESCO.

Singh, V. P. 1992. *Elementary Hydrology*. Englewood Cliffs, NJ: Prentice Hall.

Soliman, M. M. et al. 1998. *Environmental Hydrology*. New York: Lewis Publishers.

Todd, D. K., and L. Mays. 2008. *Ground Water Hydrology*. 3rd ed. New York: Wiley.

U.S. Army Corps of Engineers. 1973. *Hydrologic Engineering Centre, HEMC Flood Hydrograph, Package*. Davis, CA: User's Manual.

U.S. Department of Agriculture, Soil Conservation Service. 1986a. *A Method for Estimating Volume and Rate of Runoff in Small Watersheds*. Tech Release No. 55. Washington, DC.

U.S. Department of Agriculture, Soil Conservation Service. 1986b. *Urban Hydrology for Small Watersheds*. Tech Release No. 55. Washington, DC.

U.S. Department of Commerce. 1997. *Probable Maximum Precipitation Estimates for the Colorado River and Great Basins Drainages*. Hydrometeor Logical Report No. 49. Silver Spring, MD: NOAA, National Weather Service.

U.S. National Academy of Sciences. 1983. *Safety of Existing Dams: Evaluation and Improvement*. Washington, DC: National Academy Press.

Ward, A. D., and S. W. Trimble. 1998. *Environmental Hydrology*. 2nd ed. Boca Raton, FL: CRC Press.

Wheater, H. et al. 2007. *Hydrological Modelling in Arid and Semi-Arid Areas*. 1st ed. Cambridge, UK: Cambridge University.

Wilson, E. M. 1990. *Engineering Hydrology*. 4th ed. Ann Arbor, MI: University of Michigan.

CONFERENCES ON WADI HYDROLOGY

1st International Conference on Wadi Hydrology, 2000. National Water Research Centre, Cairo, Egypt.

2nd International Conference on Wadi Hydrology, 2003. UNESCO, Amman, Jordan.

3rd International Conference on Wadi Hydrology, 2006. UNESCO, Sanaa, Yemen.

4th International Conference on Wadi Hydrology, UNESCO, Muscat, Oman.

ASCE & ESIE 5th International Symposium on Environmental Hydrology (theme: Wadi Hydrology), 2007. Cairo, Egypt.

HYDROLOGIC MODELS

Hydrological Model and Forecasting System (WATFLOOD)
Hydrologic Modeling System (HEC-HMS)
Hydrologic Simulation Model (HSIMHYD)
Illinois Hydrodynamic Watershed Model III (IHW-III)
Illinois Urban Catchment Runoff Simulation (ILUCAT)
Illinois Urban Storm Runoff Model (IUSR)
Integrated Hydro Meteorological Model (IHMM)

Interactive River-Aquifer Simulation Program (IRAS)
MIKE 11 RR (Rainfall Runoff)
MIKE SWMM
Rainfall-Runoff Modelling Toolbox (RRMT) and Monte Carlo Analysis Toolbox (MCAT)
Soil Conservation Service Curve Number Model (SCS-CN)
Storm Water Management Model (SWMM)

2 Meteorological Processes and Hydrology

2.1 INTRODUCTION

The hydrologic characteristics of an arid or semi-arid region are determined largely by its climate, geology, and geography. The climatic factors that establish the hydrologic features of these regions include the amount and distribution of precipitation and the effects of wind, temperature, and humidity on evaporation and evapotranspiration. Meteorology plays an important role in hydrologic problems such as determining probable maximum precipitation conditions for spillway design, forecasting precipitation for reservoir operation, and determining probable maximum winds over water surfaces (used for evaluating the resulting waves in connection with the design of dams and levees). Obviously, a hydrologist should have some understanding of the meteorological processes that determine a regional climate. The general features of climatology and climate change are discussed in this chapter.

2.2 SOLAR AND EARTH RADIATION

Solar radiation, the earth's chief energy source, determines the world's weather and climates. Both the earth and the sun radiate essentially as blackbodies; that is, for every wavelength they emit almost the theoretical maximum amount of radiation for their temperatures.

Radiation wavelengths are usually given in micrometers (μm; 10^{-6} m) or in angstroms (Å; 10^{-10} m). The maximum energy of solar radiation is in the visible range of 0.4–0.71 μm, and that of earth radiation is about 10 μm. Radiation from the sun is called shortwave radiation and that from the earth is called long wave radiation.

The rate at which solar radiation reaches the upper limits of the earth's atmosphere on a surface normal to the incident radiation and at the earth's mean distance from the sun is called the solar constant (Anderson and Baker 1967). Measurements of the solar constant range from 1.89–2.05 Langleys per minute (Langley is abbreviated Ly; 1 Ly = 1 calorie per square centimeter); most of the uncertainty in the measurement values results from corrections for atmospheric effects rather than from fluctuations in solar activity, which are considered relatively small. High-altitude observations with airborne instruments, which minimize atmospheric effects, indicate a range of 1.91–1.95 Langleys per minute, and 1.94 Langleys per minute is often used as the solar constant.

A large part of the solar radiation reaching the outer limits of the atmosphere is scattered and absorbed into the atmosphere or reflected from clouds and the earth's surface. The scattering of radiation by air molecules is most effective for the shortest wavelengths. With the sun over head and a clear sky, more than 50% of the radiation in the blue range (short wavelengths about 0.45 µm) is scattered, thus accounting for the blue sky. Very little radiation in the red range (about 0.65 µm) is scattered. Estimates of the amount of radiation scattered to space average about 8% of the incident solar radiation. Clouds reflect much incident solar radiation to space; the amount reflected depends on the amount and type of clouds and their reflectivity.

In general, good emitters of radiation are also good absorbers of radiation at specific wavelength bands; this is especially true of gases. This phenomenon is responsible for the earth's greenhouse effect. Likewise, weak emitters of radiation are also weak absorbers of radiation at specific wavelength bands. This is referred to as Kirchhoff's law. Some objects in nature can almost completely absorb and emit radiation and are called blackbodies. The radiation characteristics of the sun and the earth are very close to that of black bodies.

Wien's law states that the wavelength of maximum emission of any body is inversely proportional to its absolute temperature. Thus, the higher the temperature, the shorter the wavelength of maximum emission. The following equation describes this law:

$$\lambda \max = \frac{C}{T}$$

where C is a constant equal to 2897 µm*K and T is temperature in kelvin (Pidwirny et al. 2008).

According to this equation, the wavelength of maximum emission for the sun (5800 K) is approximately 0.7 µm, whereas the wavelength of maximum emission for the earth (288 K) is approximately 10 µm.

2.3 TEMPERATURE

2.3.1 MEASUREMENT OF TEMPERATURE

To measure the air temperature properly, thermometers must be placed where air circulation is relatively unobstructed, and they must be protected from direct sun rays and from precipitation. The U.S. National Weather Service recommendation, which has been adopted by many countries, is that thermometers be placed in white, louvered, wooden instrument shelters through which the air can move readily. For better performance, the shelters should be set at about 1.4 meters above the ground. Many kinds of thermometers are available; however, the satellite remote sensing is the most up-to-date technique used to get the Vertical Temperature Profile Radiometer data (Shi et al. 2008).

2.3.2 TERMINOLOGY

The terms "average," "mean," and "normal" are the same as arithmetic means. The first two are used interchangeably, but the term "normal," which is generally used as a standard of comparison, is the average value for a particular date, month, season, or year over a specific 30-year period (WMO 1967). Plans call for recomputing the

30-year normal every decade, dropping off the first 10 years and adding the most recent 10 years.

The mean daily temperature may be computed by several methods (WMO 1960). The most accurate practical method is to average hourly temperatures. Accurate results can be obtained by averaging 3- or 6-hour observations, although the random error for an individual day with irregular variations may be of some importance, especially for 6-hour observations. In some countries, climatological observations are made at selected hours (usually three per day: morning, noon, and evening) to permit computation of acceptable daily means by a formula that gives the mean as a linear function of the observed values, with constants depending on observation time, time of year, and location.

In the United States, the mean daily temperature is the average of the daily maximum and minimum temperatures and is usually less than a degree above the true daily average. Once-daily temperature observations are usually made at 7 AM or 5 PM. Temperatures are published according to the date of the reading, even though the maximum or minimum may have occurred on the preceding day. Mean temperatures computed from evening readings tend to be slightly higher than those from midnight readings. Morning readings yield mean temperatures with a negative bias; the difference is less than the evening readings. The maximum effect of arbitrary changes in observation time on the mean temperature varies with place and season and may exceed 1.6°C (Mitchell 1958).

The normal daily temperature is the average daily mean temperature for a given date computed for a specific 30-year period. The daily temperature range is the difference between the highest and lowest temperatures recorded on a particular day. The mean monthly temperature is the average of the mean monthly maximum and minimum temperatures. The mean annual temperature is the average of the monthly means for the year. The degree day is a departure of one degree for one day in the mean daily temperature from a specified base temperature.

The lapse rate, or vertical temperature gradient, is the rate of temperature change with height in the free atmosphere. The mean lapse rate is a decrease of about 1°C per 150 meters in the vertical direction. The greatest variations in lapse rate are found in the layer of air just above the land surface. The earth radiates heat energy to space at a relatively constant rate, which is a function of its absolute temperature in kelvin. At night, incoming radiation is less than the outgoing radiation, and the temperature of the earth's surface and the air immediately above the surface decreases. For more details, refer to the references. Global warming and cooling and their effects on the precipitation of arid and semi-arid regions will be discussed in Section 2.6.

2.4 HUMIDITY

2.4.1 PROPERTIES OF WATER VAPOR

The process by which liquid water is converted into vapor is called evaporation. Water molecules with sufficient kinetic energy to overcome the attractive forces that tend to hold them within the body of liquid water are projected through the water surface. Because kinetic energy increases and surface tension decreases as temperature rises, the evaporation rate increases with temperature. Most atmospheric vapor is the product of evaporation from water surfaces. Molecules may leave a snow or ice surface in the same manner as they leave a liquid. The process by which a solid is transformed directly into a vapor state, and vice versa, is called sublimation. In a mixture of gases, each gas exerts a partial pressure independent of the other gases. The partial pressure exerted by water vapor is called vapor pressure. If all the water vapor in a closed container filled with moist air with an initial total pressure p were removed, the final pressure of the dry air alone would be less than p_d. The vapor pressure e would be the difference between the pressure of the moist air and that of the dry air $(p - p_d)$.

Practically speaking, the maximum amount of water vapor that can exist in any given space is a function of temperature and is independent of the coexistence of other gases. When the maximum amount of water vapor for a given temperature is contained in a given space, the space is said to be saturated. The more common expression "the air is saturated" is not strictly correct. The pressure exerted by the vapor in a saturated space is called the saturation vapor pressure, which, for all practical purposes, is the maximum vapor pressure possible at a given temperature.

Condensation is the process by which the vapor changes to the liquid or solid state. In a space in contact with a water surface, condensation and vaporization always occur simultaneously. If the space is not saturated, the vaporization rate will exceed the condensation rate, resulting in a net evaporation. If the space is saturated, the vaporization and condensation rates balance, provided that the water and air temperatures are the same.

Vaporization removes heat from the liquid being vaporized, whereas condensation adds heat. The latent heat of vaporization is the amount of heat absorbed by a unit mass of a substance, without a change in temperature, while passing from the liquid to the vapor state. The change from the vapor to the liquid state releases an equivalent amount of heat.

The latent heat of fusion for water is the amount of heat required to convert one gram of ice to liquid water at the same temperature. When one gram of liquid water at 0°C freezes into ice at the same temperature, the latent heat of fusion (79.7 calories per gram) is liberated.

The latent heat of sublimation for water is the amount of heat required to convert one gram of ice into vapor at the same temperature without passing through the intermediate liquid state. It is equal to the sum of the latent heat of vaporization and the latent heat of fusion, and at 0°C, it is about 677 calories per gram. Direct condensation of vapor into ice at the same temperature liberates an equivalent amount of heat.

The specific gravity of water vapor is 0.622 times that of dry air at the same temperature and pressure. The density of water vapor ρ_v in grams per cubic centimeter is

$$\rho_v = 0.662\left(\frac{e}{RT}\right) \tag{2.1}$$

where T is the absolute temperature in kelvin and R is the gas constant (2.87 × 10^3) cm/Kelvin when the vapor pressure e is in millibars. The density of dry air ρ_d in grams per cubic centimeter is

$$\rho_d = \frac{p_d}{RT} \tag{2.2}$$

where p_d is the pressure in millibars. Millibar is the standard unit of pressure in meteorology. It is equivalent to a force of 1000 dynes per square centimeter, 1 bar = 1000 millibars. The mean sea-level air pressure is 1013 millibars (see Appendix A).

The density of moist air is equal to the mass of water vapor plus the mass of dry air in a unit volume of the mixture. If p_a is the total pressure of the moist air, $p_a - e$ will be the partial pressure of the dry air alone. Adding Equations 2.1 and 2.2 and substituting $p_a - e$ for p_d gives

$$\rho_a = \left(\frac{p_a}{RT}\right)\left(1 - 0.378\left(\frac{e}{p_a}\right)\right) \tag{2.3}$$

This equation shows that moist air is lighter than dry air.

2.4.2 TERMINOLOGY

2.4.2.1 Units

Many expressions are used to indicate the moisture content of the atmosphere. Each serves special purposes, but here we discuss only those expressions common to hydrologic uses. Vapor pressure (e), usually expressed in millibars but sometimes in millimeters of mercury, is the pressure exerted by the vapor molecules. In meteorology and hydrology, vapor pressure denotes the partial pressure of the water vapor in the atmosphere.

The saturation vapor pressure is the maximum vapor pressure in saturated space and is a function of temperature alone. At any given temperature below the freezing point, the saturation vapor pressure over liquid water is slightly greater than that over ice. Vapor pressure over water is generally used for most meteorological purposes, regardless of temperature. Table 2.1 gives the water properties and saturated vapor pressure values related to temperatures in Celsius.

2.4.2.2 Dew Point

The dew point (T_d) is the temperature at which space becomes saturated when air is cooled under constant pressure and with constant water vapor content. It is the temperature at which saturation vapor pressure (e_s) is equal to the existing vapor pressure (e).

2.4.2.3 Relative Humidity

The relative humidity (f) is the percentage ratio of the actual to the saturation vapor pressure and is therefore a ratio of the amount of moisture in a given space to the amount the space could contain if saturated.

$$f = 100\left(\frac{e}{e_s}\right) \tag{2.4}$$

TABLE 2.1
Properties of Water and Vapor Pressure in Metric Units

Temperature (°C)	Specific Gravity	Density (g/cm³)	Heat of Vaporization (cal/g)	Viscosity Absolute (Centipoises)[b]	Viscosity Kinematic (Centistokes)[c]	Vapor Pressure mm Hg	Vapor Pressure Millibars	Vapor Pressure g/cm²
0	0.99987	0.99984	597.3	1.79	1.79	4.58	6.11	6.23
5	0.99999	0.99996	594.5	1.52	1.52	6.54	8.72	8.89
10	0.99973	0.99970	591.7	1.31	1.31	9.20	12.27	12.51
15	0.99913	0.99911	588.9	1.14	1.14	12.78	17.04	17.38
20	0.99824	0.99821	586.0	1.00	1.00	17.53	23.37	23.83
25	0.99708	0.99705	583.2	0.890	0.893	23.76	31.67	32.30
30	0.99568	0.99565	580.4	0.798	0.801	31.83	42.43	43.27
35	0.99407	0.99404	577.6	0.719	0.723	42.18	56.24	57.34
40	0.99225	0.99222	574.7	0.653	0.658	55.34	73.78	75.23
50	0.98807	0.98804	569.0	0.547	0.554	92.56	123.40	125.83
60	0.98323	0.98320	563.2	0.466	0.474	149.46	199.26	203.19
70	0.97780	0.97777	557.4	0.404	0.413	233.79	311.69	317.84
80	0.97182	0.97179	551.4	0.355	0.365	355.28	473.67	483.01
90	0.96534	0.96531	545.3	0.315	0.326	525.89	701.13	714.95
100	0.95839	0.95836	539.1	0.282	0.294	760.00	1013.25	1033.23

[a] Maximum density is 0.999973 g/cm³ at 3.98°C.
[b] Centipoise: g/cm s 10².
[c] Centistokes: cm²/s × 10².

Relative humidity can also be computed directly from air temperature T and dew point T_d by an approximate formula that is convenient for computer use:

$$f = \left[\frac{(112 - 0.1T + T_d)}{(112 + 0.9T)} \right]^8 \tag{2.5}$$

where the temperature is in Celsius. This formula approximates relative humidity to be within 1.2% in the meteorological range of temperatures and humidity to be within 0.6% in the range of −25–45°C (−13–113°F).

2.4.3 MEASUREMENT OF HUMIDITY

In general, humidity (Stine 1965) in the surface layers of the atmosphere is measured with a psychrometer, which consists of two thermometers, one with its bulb covered by a jacket of clean muslin saturated with water. The thermometers are ventilated by whirling or a fan. Because of the cooling effect of evaporation, the moistened, or wet-bulb, thermometer reads lower than the dry thermometer. The difference is called the wet-bulb depression. The air and wet-bulb temperatures are used to obtain various expressions of humidity by reference to the psychrometric table provided in Table 2.2 (USA Weather Bureau 1972). Instruments such as hair hygrometers and dew point hygrometers can be used to measure humidity, but they give an appreciable error if they are not used in a proper manner. Fortunately, several electronic instruments that can accurately measure the relative humidity are available and can be easily ordered through the Internet.

2.5 WIND

Wind is a very influential factor in several hydrometeorological processes. Moisture and heat are readily transferred to and from air, which tends to adopt the thermal and moisture conditions of the surfaces with which it is in contact. Stagnant air in contact with a water surface eventually assumes the vapor pressure of the surface so that no evaporation takes place. Similarly, stagnant air over a snow or ice surface eventually assumes the temperature and vapor pressure of the surface so that melting by convection and condensation ceases. Consequently, wind exerts a considerable influence on evaporation and is also important in the production of precipitation, since precipitation can be maintained only through sustained inflow of moist air into a storm.

2.5.1 MEASUREMENT OF WIND

Wind has both speed and direction. The wind direction is the direction from which it is blowing. For surface winds, direction is usually expressed in terms of 16 compass points (N, NNE, NE, ENE, etc.), and for winds aloft in degrees from the north, the direction is measured clockwise. Wind speed is usually given in miles per hour, meters per second, or knots (kn; 1 meter per second = 2.237 miles per hour = 1.944 knots; and 1 knot = 1.151 miles per hour = 0.514 meters per second).

TABLE 2.2

Variation of Relative Humidity (%) with Temperature and Wet-Bulb Depression (°C)

Air Temperature (°C)	Wet-Bulb Depression (°C)															
	0	1	2	3	4	5	6	7	8	9	10	11	12	13	14	15
−10	91	60	31	2												
−8	93	65	39	13												
−6	94	70	46	23	0											
−4	96	74	53	32	11											
−2	98	78	58	39	21	3										
0	100	81	63	46	29	13										
2	100	84	68	52	37	22	7									
4	100	85	71	57	43	29	16									
6	100	86	73	60	48	35	24	11								
8	100	87	75	63	51	40	29	19	8							
10	100	88	77	66	55	44	34	24	15	6						
12	100	89	78	68	58	48	39	29	21	12	4					
14	100	90	79	70	60	51	42	34	26	18	10	3				
16	100	90	81	71	63	54	46	38	30	23	15	8				
18	100	91	82	73	65	57	49	41	34	27	20	14	7			
20	100	91	83	74	66	59	51	44	37	31	24	18	12	6		
22	100	92	83	76	68	61	54	47	40	34	28	22	17	11	6	
24	100	92	84	77	69	62	56	49	43	37	31	26	20	15	10	5
26	100	92	85	78	71	64	58	51	46	40	34	29	24	19	14	10
28	100	93	85	78	72	65	59	53	48	42	37	32	27	22	18	13
30	100	93	86	79	73	67	61	55	50	44	39	35	30	25	21	17
32	100	93	86	80	74	68	62	57	51	46	41	37	32	28	24	20
34	100	93	87	81	75	69	63	58	53	48	43	39	35	30	26	23
36	100	94	87	81	75	70	64	59	54	50	45	41	37	33	29	25
38	100	94	88	82	76	71	66	61	56	51	47	43	39	35	31	27
40	100	94	88	82	77	72	67	62	57	53	48	44	40	36	33	29

Source: Adapted from U.S. Weather Bureau. 1972. Relative Humidity and Dew Point Tables. Washington, DC: U.S. Weather Bureau.

Wind speed is measured by anemometers, of which there are many types. The three- or four-cup anemometer with a vertical axis of rotation is commonly used for official observations. It tends to register a high mean speed in a variable wind because the cups accelerate faster than they lose speed. Vertical currents (turbulence) tend to rotate the cups and cause overregistration of horizontal speeds. Most cup anemometers will not record speeds below 1 or 3 kilometers per hour because of starting friction. The propeller anemometer has a horizontal axis of rotation. Pressure-tube anemometers operate on the pilot-tube principle.

Although wind speed varies greatly with the height above the ground, no standard anemometer level has been adopted. Differences in wind speed resulting from differences in anemometer height, which may range anywhere from 10 to several hundred meters above the ground, usually exceed the errors from instrumental deficiencies. However, approximate adjustments can be made for differences in height.

2.5.2 GEOGRAPHIC VARIATION OF WIND

In winter, surface winds tend to blow from the colder interior of land masses toward the warmer oceans. Conversely, in summer the winds tend to blow from the cooler water bodies toward the warmer land. Similarly, diurnal land and sea breezes may result from temperature contrasts between land and water.

On mountain ridges and summits, wind speeds at 10 meters (30 feet) or more above the ground are stronger than in the free air at corresponding elevations because of the convergence of the air that is forced by the orographic barriers. On lee slopes and in sheltered valleys, wind speeds are light. Wind direction is greatly influenced by the orientation of orographic barriers. With a weak pressure system, diurnal variation of wind direction may occur in mountain regions; the winds blow upslope in the daytime and downslope at night.

Wind speeds are reduced and directions are deflected in the lower layers of the atmosphere because of the friction produced by trees, buildings, and other obstacles. These effects become negligible above 600 meters (2000 feet), and the lower layer is referred to as the friction layer. Over land, the surface wind speed averages about 40% of that just above the friction layer, and at sea averages is about 70%.

The variation of wind speed with height in the friction layer, called the wind profile, is usually expressed by one of two general relationships: the logarithmic velocity profile (Ruggles 1970) or the power-law profile (Johnson 1959). In hydrology, these relationships are most often used to estimate the wind speed in the surface boundary layer, that is, the thin layer of air between the ground surface and the anemometer level, which is usually about 10 meters (30 feet), but is often lower at special test sites or experimental stations. The common requirement for computations of snowmelt and evaporation is the wind speed above a snow or a water surface.

For most meteorological purposes, the power-law profile is usually expressed as

$$\frac{v}{v_1} = \left(\frac{z}{z_1}\right)^k \tag{2.6}$$

where exponent k varies with surface roughness and atmospheric stability and usually ranges from 0.1 to 0.6. v and v_1 are the average velocities at height z and z_1, respectively (Geiger 1965).

Some investigators (Frost 1947, 1948; Huss and Portman 1949) consider the power-law profile to be more representative of the wind profile in the layer from several meters to about 100 meters (300 feet) above the ground. The results of some investigations that were based on 1-hour averaging periods indicate that, for this range of elevation, exponent k increases with surface roughness and atmospheric stability except for large superadiabatic lapse rates, in which case it increases with the instability of atmosphere. Exponent k also varies with the height of the layer considered. Higher layers have lesser k values when the lapse rate is superadiabatic, and they have greater k values when the lapse rate is adiabatic or less. For adiabatic and superadiabatic lapse rates, k ranges from 0.1 to 0.3, and the variation is in proportion to surface roughness. A value of 1/7 has been found to be applicable for a wide range of conditions in the 0–10 meters (0–30 feet) layer. Over a snow surface under conditions favoring melting, a value of 1/6 may be more appropriate.

2.6 CLIMATE CHANGE

2.6.1 CLIMATE AND HUMAN ACTIVITY

The global industrial revolution changed our species' world forever. Fossil fuels are a legacy bequeathed to us by this biosphere of the distant past. On an ancient, warmer earth with a high concentration of carbon dioxide (CO_2) in the atmosphere, photosynthetic organisms (algae and higher plants) absorbed CO_2 and used it to produce abundant organic material. When these organisms died, they were buried deep within the earth and slowly turned into coal and oil.

Since the 1800s, we have been burning vast quantities of fossil fuels to power our developing technological and global civilization. As a result, we have been releasing the CO_2 trapped in the fuels in the form of energy-rich organic molecules back into the atmosphere, increasing the atmospheric concentration of CO_2. This is not a concern by itself. Carbon dioxide comprises a very small proportion of the atmosphere, and no projected increase would affect our breathing. However, CO_2 has another significant property. As we discussed in Section 2.2 on the greenhouse effect, CO_2 absorbs heat, but other major component gases of the earth's atmosphere such as oxygen (O_2) and nitrogen (N_2), do not.

Since the 1800s, CO_2 concentrations worldwide have increased from approximately 280 ppm (0.028%) to around 365 ppm (0.0365%). Though the increase seems trivial, approximately 3 gigatons (3 billion metric tons) of CO_2 are being added to the atmosphere every year. Because CO_2 is a powerful greenhouse gas, the earth's temperature will increase as CO_2 concentrations increase. Based on weather data collected all around the world, climatologists have detected a steady but small increase in global average temperatures over the last few decades. Six of the last ten decades were the hottest years on record.

Regardless of the causes of the warming, we understand enough about the global climate to predict that as the temperature increases, the entire global climate system, which is powered by heat energy, will also change, although the magnitude and direction of the changes are uncertain. Will the climate change for the worse because of our actions? No one knows for sure. Most atmospheric scientists believe that the global climate is warming at least partially because of a buildup of CO_2 from fossil fuel use, but what that means to humans and natural ecosystems is largely unknown. The climate is vastly complex and strongly influenced by many factors other than greenhouse gas concentrations, such as solar variability and other effects. This makes it extremely difficult to link any climatic events or characteristics to a single cause. As a result, controversy exists as to the magnitude and danger of global warming induced by greenhouse gases. Many scientists take the issue very seriously and support efforts to slow or reverse the buildup of atmospheric CO_2, with the expectation that this will slow global warming. Others, however, contend that CO_2 may not affect the climate and that the changes are part of natural, long-term climatic cycles. They suggest that efforts to reduce CO_2 emissions are unnecessary and are dangerous to economic growth and development. While the controversy rages, researchers around the world continue to gather atmospheric data, develop and refine predictive computer models, and try to reduce the uncertainty in our understanding of the earth's climate.

In this respect, they explore the critical issues in climate change, such as sources and sinks (or reservoirs) of CO_2, the nature of climate change and predictions of future changes, and the elements of the scientific and political debates that will ultimately determine how we respond to climate change.

We know that the earth's climate has changed over time. Throughout the earth's history, there have been periods of glaciations followed by warming trends during which the glaciers retreated toward higher altitudes and latitudes. Today's concerns focus on the current and projected rate of climate change based, in large part, on human activities.

Brown (2002) studied the global temperature rise and reported that 2001 was the second warmest year since recordkeeping began in 1867. Following the all-time high of 1998, the recent high temperatures extend a strong trend of rising temperatures that began around 1980. The 15 warmest years since 1867 have all happened since 1980. Monthly global temperature data compiled by NASA's Goddard Institute for Space Studies in a series based on meteorological station estimates going back to 1867 show that September 2001 was the warmest September on record (Figure 2.1; Eissa 2007). November 2001 also set an all-time high. Six recent months—August and December 2001 and January, March, April, and May 2002—were each the second warmest respective months on record.

The global average temperature for 2001 was 14.52°C (58.1°F). The all-time high in 1998 was 14.69°C. Over the last century, the average global temperature increased from 13.88°C in 1899–1901 to 14.44°C in 1999–2001, an increase of 0.56°, but four-fifths of this increase occurred in the century's last two decades. This increase during the last century is quite small compared with projections of a rise in temperature rise of 1.4–5.8°C (2.5–10.4°F) this century by the Intergovernmental Panel on Climate Change (IPCC).

The contrast between the sea-level rise for the last century and that projected for this century is similarly worrying. During the last century, the sea level rose

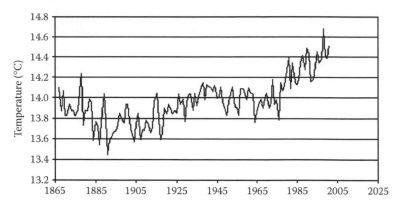

FIGURE 2.1   Average global temperature (1867–2001). (Adapted from Eissa, M. M. 2007. *Meteorol Res Inst Bull* 22.)

an estimated 10–20 centimeters (4–8 inches). The IPCC projects that during this century the sea level will rise 9–88 centimeters (4–36 inches). Rising temperature is not an irrelevant abstraction. It brings many physical changes—from more intense heat waves, more severe droughts, and faster-melting ice to more powerful storms, more destructive floods, and a rising sea level. These changes in turn affect not only the food security and habitability of low-lying regions but also the species composition of local ecosystems.

Climate change affects food security in many ways. In 2000, the World Bank published a map of Bangladesh showing that a 1-meter rise in sea level would inundate 50% of that country's rice land. Bangladesh would lose not only 50% of its rice supply but also the livelihoods of a large share of its population. The combination of a population of 134 million that is expanding by 2.7 million a year and a shrinking cropland base is not a reassuring prospect for Bangladesh.

Climate change also triggers widespread changes in ecosystems. In recent years governments and environmental organizations have invested heavily to protect particular ecosystems by converting them into parks or reserves. However, if the rise in temperature cannot be stopped, there is no ecosystem on the earth that can be saved. Everything will change.

An additional year of temperature data reinforces the concerns expressed by the team of eminent scientists who produced the latest IPCC report, "Climate Change 2001." This data makes clear what it is now becoming obvious even to nonscientists: burning fossil fuels is changing the earth's climate.

2.6.2   SOLAR VARIABILITY AND CLIMATE CHANGE

Variability in the amount of energy from the sun has caused climate changes in the past. It is now accepted that the global cooling during the ice ages was the result of changes in the distribution and amount of sunlight that reaches the earth. During the last ice age, the globally averaged temperature of the earth was about 6°C colder than that temperature today. While this difference may not sound like much, the effect of

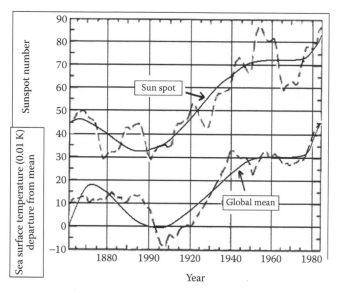

FIGURE 2.2   The effect of the sunspots on global sea surface temperature.

this temperature drop was to cover large parts of Canada, Alaska, and Siberia with huge sheets of ice up to a mile thick.

Even the climate changes of the twentieth century may have a significant solar component. Figure 2.2 shows comparisons of globally averaged temperatures and solar activity. Many scientists find that these correlations are convincing evidence that the sun has contributed to the global warming of the twentieth century, and some say that one-third of global warming may be the result of an increase in solar energy. So, while it is becoming clear that human activity changes the climate, solar activity may also be contributing to climate change and probably changed the climate in the past.

In order to accurately predict how future human activities will change the climate, it is critical to understand the variability of the natural system. Therefore, even though solar activity may not be the dominant factor in global warming, understanding how the climate responds to small changes in solar irradiance will help scientists predict the climate changes caused by human activity.

The National Oceanic and Atmospheric Administration (NOAA) Space Environment Center (SEC) combines scientific research and an operational Space Weather Center to maintain a vigilant watch on solar activity. SEC's primary mission is studying the effects of a variable sun on the upper atmosphere and the near-earth space environment. Monitoring and understanding solar effects on the middle and lower atmosphere is a new component of SEC's mission. Present NOAA–SEC activities include monitoring the sun in x-ray and ultraviolet wavelengths as well as sunspots. NOAA recognizes the need for new efforts in this area and will include solar extreme ultraviolet measurements on the next generation of GOES spacecraft and total solar irradiance and solar spectral irradiance measurements as part of its upcoming NPOESS spacecraft mission.

FIGURE 2.3   Total station including one recording rain gauge, a wind speed anemometer, and an instrument shelter with thermometers in Sinai, Egypt.

From the statistical model results of both Panofsky and Brier (1968) and Essenwanger (1976), Eissa (2007) concluded that the increase in the global temperature in the twentieth century was a natural increase and the twenty-first century will see a global cooling temperature.

A typical total station located in Sinai, Egypt, representing an arid region that gathers meteorological data on temperature measurements, wind speed, and rainfall is shown in Figure 2.3. In conclusion, any global temperature changes will undoubtedly affect the climate and physical aspects of both arid and semi-arid regions.

REFERENCES

Anderson, E. A., and D. R. Baker. 1967. Estimating incident terrestrial radiation under all atmospheric conditions. *Water Resour Res* 3(4):975–98.
Brown, L. R. 2002. *Global Temperature Rising*. Washington, DC: Earth Policy Institute Magazine.
Eissa, M. M. 2007. New statistical study for global temperature. *Meteorol Res Inst Bull* 22: 1687–1014.
Essenwanger, D. 1976. *Applied Statistics in Atmospheric Science*. Huntsville, AL: University of Alabama.
Frost, R. 1947. The velocity profile in the lowest 400 feet. *Meteorol Mag* 74:14–18.
Frost, R. 1948. Atmospheric turbulence. *Q J R Meteorol Soc* 74:316.
Geiger, R. 1965. *The Climate Near the Ground*. Cambridge, MA: Harvard University Press.

Huss, P. O., and D. J. Portman. 1949. "Study of natural wind and computation (Austausch turbulence constant)." Daniel Guggenheim Airship Institute Report.

Johnson, O. 1959. An examination of the vertical wind profile in the lowest layers of the atmosphere. *J Meteorol* 16:144–8.

Mitchell, J. M., Jr. 1958. Effect of changing observation time on mean temperature. *Bull Am Meteorol Soc* 39:83–9.

NASA Goddard Institute for Space Studies. 2005. Indicator: Global and Canadian temperature variations from Global Source. New York: NASA Goddard Institute for Space Studies.

Panofsky, H. A., and G. W. Brier. 1968. *Some Applications of Statistics to Meteorology.* Pennsylvania, USA: University Park.

Pidwirny, M. et al. 2008. Electromagnetic radiation. In *Encyclopedia of Earth,* ed. C. J. Cleveland. Washington, DC: Environmental Information Coalition, National Council for Science and the Environment.

Ruggles, K. W. 1970. The vertical mean wind profile over the ocean for light to moderate winds. *Appl Meteorol* 9:389–95.

Shi, L. et al. 2008. Extending the satellite sounding archive back in time: the vertical temperature profile radiometer data. *J Appl Remote Sens* vol. 2

Stine, S. L. 1965. Carbon humidity elements: Manufacture, performance, and theory. In *Humidity and Moisture: Measurement and Control of Science and Industry,* ed. A. Wexler, vol. 1, 316–30. New York: Reinhold.

U.S. Weather Bureau. 1972. Relative Humidity and Dew Point Tables. Washington, DC: U.S. Weather Bureau.

World Meteorological Organization. 1960. Guide to Climatological Practices, WMO no. 100, Technical Paper 44, Geneva, V.9-V.11 pp.

World Meteorological Organization. 1967. A Note on Climatological Normal, Technical Note 84, Geneva.

3 Precipitation

3.1 INTRODUCTION

Hydrologists have long known that only about one-fourth the total precipitation that falls on continental areas returns to the seas by direct runoff and underground flow. Evaporation from ocean surfaces is the chief source of moisture for precipitation, and probably no more than about 10% of continental precipitation can be attributed to continental evaporation.

The term "precipitation" denotes all forms of water that reach the earth from the atmosphere. The usual forms of precipitation are rainfall, snowfall, hail, frost, and dew. Of these, only rainfall and snowfall contribute significant amounts of water. Rainfall is the predominant form of precipitation causing stream flow, especially the flood flow that happens in a majority of rivers all over the world. The magnitude of precipitation varies with time and space. The magnitude of rainfall in different parts of a country at a given time and the rainfall in the same place at various times of the year obviously vary. This variation is responsible for many hydrological problems, such as floods and droughts. The study of precipitation forms a major portion of hydrometeorology. This chapter will familiarize the engineer with important aspects of rainfall and, in particular, the collection and analysis of rainfall data in arid and semi-arid regions.

For precipitation to occur: (1) the atmosphere must have moisture; (2) sufficient nuclei must be present to aid condensation; (3) weather conditions must be optimal for the condensation of water vapor to take place; and (4) the products of condensation must reach the earth. Under proper weather conditions, the water vapor condenses over the nuclei to form tiny water droplets of less than 0.1 millimeter in diameter. The nuclei are usually salt particles or products of combustion that are normally available in the atmosphere. Wind speed facilitates the movement of clouds, while its turbulence retains the water droplets in suspension. Water droplets in a cloud are somewhat similar to the particles in a colloidal suspension. Precipitation results when water droplets come together and coalesce to form larger drops that then fall. A considerable part of this precipitation evaporates back to the atmosphere. The net precipitation and its form at a given place depend upon a number of meteorological factors, such as the weather elements like wind, temperature, humidity, and pressure in the region enclosing the clouds.

3.2 FORMS OF PRECIPITATION

Some of the common forms of precipitation are rain, snow, drizzle, glaze, sleet, and hail.

3.2.1 RAIN

Rain is the principal form of precipitation in arid and semi-arid regions. The term "rainfall" is used to describe precipitation in the form of water drops larger than 0.5 millimeters. The maximum size of a raindrop is about 6 millimeters; any drop larger than this tends to break up into drops of smaller sizes during its fall from the clouds. Based on its intensity, rainfall is classified internationally as shown in Table 3.1.

3.2.2 SNOW

Snow is another important form of precipitation consisting of ice crystals that usually combine to form flakes. When it is new, snow has an initial density varying from 0.06–0.15 grams per cubic centimeter, but it usually assumes an average density of 0.1 grams per cubic centimeter. A few semi-arid areas such as Saint Katherine Mountain in Sinai, Egypt, get snow.

3.2.3 DRIZZLE

Drizzle is a fine sprinkle of numerous water droplets less than 0.5 millimeters in diameter and with an intensity of less than 1 millimeter per hour. The drops are so small that they appear to float in the air.

3.2.4 GLAZE

When rain or drizzle comes in contact with cold ground at around 0°C, the water drops freeze to form an ice coating called glaze or freezing rain.

3.2.5 SLEET

Sleet is frozen raindrops of transparent grains that forms when rain falls through the air at subfreezing temperatures. It occurs mostly in cold regions.

3.2.6 HAIL

Hail is a showery precipitation in the form of irregular pellets or lumps of ice more than 8 millimeters in diameter. Hails occur in violent thunderstorms in which vertical currents are very strong.

TABLE 3.1
Types of Rainfall

Type	Intensity
Light rain	Trace to 2.5 mm/h
Moderate rain	2.5–7.5 mm/h
Heavy rain	>7.5 mm/h

3.3 TYPES OF PRECIPITATION

For clouds to form and create subsequent precipitation, the moist air masses must cool to form condensation. This is normally accomplished by adiabatic cooling of moist air while it is lifted to higher altitudes. Some of the terms and processes connected with the weather systems associated with precipitation are given in Sections 3.3.1 to 3.3.6.

3.3.1 FRONT

A front is an interface between two distinct air masses. When a warm air mass and cold air mass meet under ce rtain favorable conditions, the warmer air mass is lifted over the colder one with the formation of a front. The ascending warmer air cools adiabatically and forms clouds and precipitation.

3.3.2 CYCLONE

A cyclone (Figure 3.1) is a large low-pressure region with circular wind motion. Two types of cyclones are recognized: tropical cyclones and extratropical cyclones.

A *tropical cyclone* is a wind system with an intensely strong pressure depression, sometimes below 915 millibars. The normal areal extent of a cyclone is about 100–200 kilometers in diameter. The isobars are closely spaced, and the winds blow anticlockwise in the northern hemisphere. The center of the storm, called the *eye*, may extend to about 10–50 kilometers in diameter and will be relatively quiet. However outside the eye there are very strong winds of up to 200 kilometers per hour. The wind speed gradually decreases toward the outer edge, and the pressure also increases outward. Rainfall is normally heavy in the entire area occupied by the cyclone.

During summer months, tropical cyclones originate in the open ocean at around 5–10° latitude and move at a speed of about 10–30 kilometers per hour to higher latitudes in an irregular path. They get their energy from the latent heat of the condensation of ocean water vapor and increase in size as they move on the ocean. When they

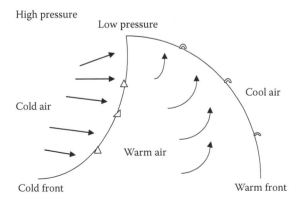

FIGURE 3.1 Cyclonic storm initiation.

move on land, the source of energy is cut off and the cyclone's energy dissipates very fast, rapidly decreasing the intensity of the storm. Tropical cyclones cause intense rainfall and heavy floods in streams and lead to heavy damage to life and property in their land path. Tropical cyclones cause moderate to excessive precipitation for several days over very large areas of about 1000 square kilometers.

Extratropical cyclones are cyclones that form in locations outside the tropical zone. Since they are associated with a frontal system, they possess a strong counterclockwise wind circulation in the northern hemisphere. The magnitude of precipitation and wind velocities are lower than those of a tropical cyclone. However, the duration of precipitation and the areal extent are usually higher.

3.3.3 ANTICYCLONES

Anticyclones are regions of high pressure that usually cover a large area. The weather is usually calm at the center. Anticyclones cause clockwise wind circulations in the northern hemisphere. The winds are of moderate speed and there are clouds and precipitation at the outer edges.

3.3.4 CONVECTIVE PRECIPITATION

In convective precipitation (Figure 3.2), a packet of air that is warmer than the surrounding air due to localized heating rises because of its lesser density. Air from cooler surroundings flows to take up its place, thus setting up a convective cell. The warm air continues to rise, then undergoes cooling and results in precipitation. Depending upon the moisture, thermal, and other conditions, light showers to thunderstorms

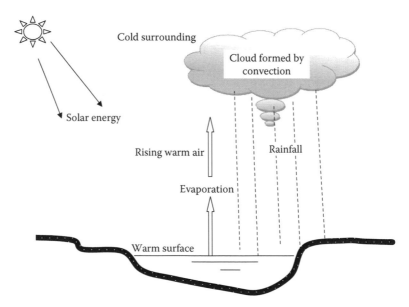

FIGURE 3.2 Convection storm.

can be expected in convective precipitation. Usually, the areal extent of such rains is small, limited to a diameter of about 10 kilometers.

3.3.5 OROGRAPHIC PRECIPITATION

The moist air masses may get lifted up to higher altitudes due to the presence of mountain barriers and consequently undergo cooling, condensation, and precipitation known as orographic precipitation (Figure 3.3). Thus, in mountain ranges, the windward slopes experience heavy precipitation, while the leeward slopes experience light rainfall.

3.3.6 PRECIPITATION ENHANCEMENT

Precipitation enhancement, commonly called *cloud seeding*, artificially stimulates clouds to produce more rainfall or snowfall than they would naturally. In cloud seeding, special substances that enable the formation of snowflakes and raindrops are injected into the clouds. Because of cloud seeding's importance in arid and semi-arid regions, this chapter discusses successful cloud seeding trials in different countries. Precipitation enhancement is the one form of weather modification that is used in California and is also used in other places in the world.

Cloud seeding, a form of weather modification, is the attempt to change the amount or type of precipitation that falls from clouds by dispersing substances into the air that serve as cloud condensation or ice nuclei, altering the microphysical processes within the clouds. The intent is usually to increase rain or snow precipitation, but hail and fog suppression are also widely practiced in airports.

The largest cloud seeding system in the world is that of the People's Republic of China, which believes that it increases the amount of rain over several increasingly arid regions, including its capital city, Beijing, by firing silver iodide rockets into the sky where rain is desired. Political strife is caused by neighboring regions,

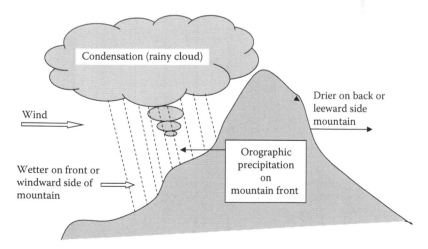

FIGURE 3.3 Orographic storm.

which accuse each other of "stealing rain" using cloud seeding. About 24 countries currently practice weather modification operationally. China used cloud seeding in Beijing just before the 2008 Olympic Games in order to clear the air of pollution, but there are disputes regarding the Chinese claims. In February 2009, China also blasted iodide sticks over Beijing and areas of northern China to artificially induce snowfall after 4 months of drought. The snowfall in Beijing, which rarely experiences snow, lasted for approximately 3 days and led to the closure of 12 main roads around Beijing (China 2009).

In the United States, cloud seeding is used to increase precipitation in areas experiencing drought, to reduce the size of hailstones that form in thunderstorms, and to reduce the amount of fog in and around airports. Cloud seeding is also occasionally used by major ski resorts to induce snowfall. Eleven western states and one Canadian province (Alberta) have ongoing weather modification operational programs (Steinhoff and Ives 1976). In January 2006, a cloud seeding project began in Wyoming to examine the effects of cloud seeding on snowfall over Wyoming's Medicine Bow, Sierra Madre, and Wind River mountain ranges.

We can conclude that scientific proof of the efficacy of weather modification was lacking until 2009 and that a large sustained research program should be developed to reduce the uncertainties of this technology. We have seen progress in seeding agent formulation and targeting, although there is a need for more research on these aspects.

3.4 MEASUREMENT OF PRECIPITATION

Precipitation is expressed in terms of the depth to which rainfall water would stand on an area if all the rain were collected on it. Thus, 1 centimeter of rainfall over a catchment area of 1 square kilometer represents a volume of water equal to 10^4 square meters. In the case of snowfall, an equivalent depth of water is used as the depth of precipitation. The precipitation is collected and measured in a *rain gauge*, also called hyetometer.

A rain gauge consists of a cylindrical vessel assembly that is kept in the open to collect rain. The rainfall catch of the rain gauge is affected by its exposure conditions. Adopting the following standards enables the catch of a rain gauge to accurately represent the rainfall in the surrounding area:

1. The ground must be level and open and the instrument must have a horizontal catch surface.
2. The gauge must be set as near to the ground as possible to reduce the effect of wind, but must be sufficiently high to prevent splashing, flooding, and so on.
3. The instrument must be surrounded by an open fenced area of at least 5.5×5.5 meters. No object should be nearer to the instrument than 30 meters or twice the height of the obstruction.

Rain gauges can be broadly classified into two categories: (1) nonrecording rain gauges and (2) recording rain gauges.

3.4.1 Nonrecording Gauges

The nonrecording gauges extensively used in many arid regions are the standard U.S. National Weather Service (NWS) rain gauges (Figure 3.4). They essentially consist of a circular collecting area that is 20.3 centimeters (8 inches) in diameter and is connected to a funnel or collector. The rim of the collector is set in a horizontal plane at a height of 30.5 centimeters above the ground level. The rain passes from the collector into a cylindrical measuring tube inside the overflow can. The measuring tube has a cross-sectional area one-tenth that of the collector so that a 2.54-millimeter (0.1-inch) rainfall will fill the tube to a depth of 25.4 millimeters (1 inch). Rainfall can be measured to the nearest 0.01 inches (0.25 millimeters) with a measuring stick marked in tenths of an inch or millimeters.

3.4.2 Recording Gauges

Recording gauges produce a continuous plot of rainfall against time and provide valuable data for hydrological analysis of storms on the intensity and duration of rainfall. Sections 3.4.2.1 through 3.4.2.3 outline some of the commonly used recording rain gauges.

3.4.2.1 Tipping-Bucket Gauges

A tipping-bucket type gauge is a 30.5-centimeter rain gauge adopted for use by the U.S. Weather Bureau. The catch from the funnel falls onto one of a pair of small buckets. These buckets are balanced so that when 0.25 millimeters of rainfall collects in one bucket, it tips and brings the other one into position. The water from the tipped bucket is collected in a storage can. The tipping actuates an electrically driven pen to trace a record on a clockwork-driven chart. The water collected in the storage can is measured at regular intervals to provide the total rainfall, and also serves as a check. The record from the tipping bucket gives data on the intensity of rainfall.

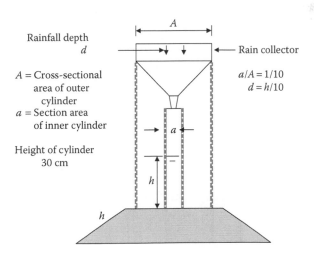

FIGURE 3.4   Standard rain gauge.

FIGURE 3.5   Tipping-bucket recording rain gauge used at Wadi Watir Sinai, Egypt.

Further, the instrument is ideally suited for digitalizing the output signal. Figure 3.5 shows one of the tipping-bucket rain recorders used by the Egyptian Water Resources Institute (WRRI).

3.4.2.2   Weighing-Bucket Gauges

In a weighing-bucket gauge, the catch from the funnel empties into a bucket mounted on a weighing scale. The weight of the bucket and its contents are recorded on a clockwork-driven chart. The clockwork mechanism has the capacity to run for as long as 1 week. This instrument gives a plot of the accumulated rainfall against the elapsed time, that is, the mass curve of the rainfall. In some weighing-bucket gauges, the recording unit is constructed in such a way that the pen reverses its direction at every preset value, for example, 7.5 centimeters (3 inches), so that a continuous plot of storm is obtained.

3.4.2.3   Float-Type Gauges

In a float-type gauge, the rainfall collected by a funnel-shaped collector is led into a float chamber, causing a float to rise. As the float rises, a pen attached to the float through a lever system records the elevation of the float on a rotating drum driven by a clockwork mechanism. A siphon arrangement empties the float chamber when the float has reached a preset maximum level. This type of rain gauge gives a plot of the mass curve of rainfall. It is the standard recording-type rain gauge in many countries.

3.4.3 ADVANTAGES AND DISADVANTAGES OF RECORDING GAUGES

Advantages of recording gauges include the following:

- The rainfall is recorded automatically, thus, there is no need for an attendant.
- The automatic rain gauges also give the rainfall intensity at any instant, while nonrecording gauges give only the quantity of rainfall for the past 24 hours.
- Human error is eliminated.
- They can be set up in the regions where living conditions are difficult.

Disadvantages of recording gauges include the following:

- They are costlier than nonrecording rain gauges.
- Mechanical defects are likely to develop.

3.4.4 RADAR MEASUREMENT OF RAINFALL

The meteorological radar is a powerful instrument for measuring the areal extent, location, and movement of rain storms. The radar emits a regular succession of pulses of electromagnetic radiation in a narrow beam. The amount of rainfall over large areas can be determined with a good degree of accuracy using radar.

The principles of radar and the observation of weather phenomena were established in the 1940s. Since then, great strides have been made in improving equipment, signal and data processing, and its interpretation.

Most meteorological radars are pulsed radars; that is, they transmit electromagnetic waves at fixed preferred frequencies from a directional antenna system into the atmosphere in a rapid succession of short pulses. Any meteorological targets encountered absorb and scatter the short bursts of electromagnetic energy. Between successive pulses, the receiver listens to any return of the wave. The strength of the returned power is directly related to the rainfall rate in the sampled volume (i.e., the rainfall target). The range (i.e., the distance from the radar) of the sampled volume is determined from the time interval the wave requires to arrive back at the receiver. The rainfall target is therefore uniquely defined in space by measurements of the range, azimuth, and elevation angles. A number of conical scans are performed according to the automatic scanning strategy of the radar antenna; then the commercial radar rainfall product is produced. The radar rainfall product is a grid of a geographical extent of approximately 480 × 480 kilometers and a spatial resolution of 1–4 km with the radar at the center of the grid. The time between two products (i.e., the temporal resolution of the radar products) is usually 10 minutes.

Radar measurements may suffer from measuring errors such as radar beam filling, nonuniformity of vertical distribution of precipitation, attenuation by intervening precipitation, beam blocking, signal attenuation, electromagnetic interference, ground and other clutter, anomalous propagation, beam overshooting of precipitation, and lack of antenna accuracy, electronics stability, and processing accuracy. More details of radar observation of weather phenomena can be found in different references such as Skolnik (1970) for engineering and equipment aspects; Sauvageot (1982), Battan (1981) for meteorological phenomena and applications; Atlas (1990) for general review; Rinehart (1991) for modern techniques; and Doviak and Zrnic (1993) for Doppler radar principles and applications.

Present-day developments in the field include online processing of radar data on a computer and the use of Doppler radars for measuring the velocity and distribution of raindrops. Weather radars are not available in most developed countries and in many arid and semi-arid regions.

3.4.5 WEATHER SATELLITES

Weather satellite data is a space measurement because the sensor is located in space (i.e., it uses remote-sensing techniques). There are two types of weather satellites: polar-orbiting satellites and geostationary satellites. Figure 3.6 shows the locations of the two types of satellites with respect to the earth. These two types of satellites are described next.

3.4.5.1 Polar-Orbiting or Active Satellites

Polar-orbiting or active satellites are located approximately 250 kilometers above the earth and rotate in a polar orbit that passes through the north and south poles and crosses the equator every 12 hours. Only a strip around the ground track of the satellite

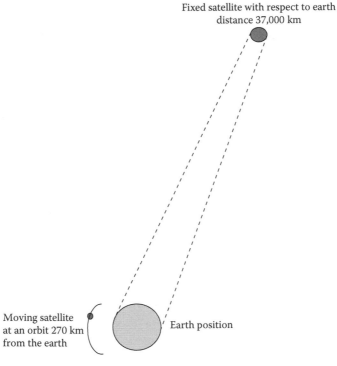

Fixed satellite with respect to earth
distance 37,000 km

Moving satellite
at an orbit 270 km
from the earth

Earth position

FIGURE 3.6   Locations of the two types of satellites.

can be viewed. The satellite's sensor is similar to typical radar (active measurements, i.e., higher measuring accuracy); however, the problem with this type of satellite is that is provides poor spatial and temporal resolutions (30 kilometers and 12 hours). The poor spatial resolution arises because the microwave beam width increases with the distance from the sensor. So, this type of satellite is not useful for short-term applications and those applications requiring a high spatial resolution rainfall input.

3.4.5.2 Geostationary Satellites

Geostationary satellites are located approximately 36,000 kilometers above the equator. This type of satellite rotates in an orbit parallel to the equator at a given speed, so that the satellite is always fixed (i.e., stationary) with respect to a point on the surface of the earth. This enables continuous monitoring of the same face of the earth, which provides a high temporal resolution (30 minutes in case of METEOSAT; Gad 2004a, 2004b). The spatial resolution depends on the precision of the imager (1 kilometer in case of the visible image and 4–7 kilometers in case of the infrared [IR] image).

The images include three main channels: (1) visible, (2) IR, and (3) water vapor. The visible image is the equivalent of taking a black-and-white photo of the earth from space. The bright areas show where the sun is reflected back into space due to clouds or snow cover, which show up white. The thicker the cloud, the brighter the white color. Land surfaces show up as gray and ocean surfaces are nearly black. The major limitation to visible imagery is that it is only valid during daylight.

The IR image shows heat-based radiation from the IR spectrum. In other words, the warmer the surface, the more IR radiation it emits. In a satellite IR image, cooler surfaces are bright and warmer surfaces are dark. Since the atmosphere cools as altitude increases, clouds show up as bright areas and land surfaces as dark areas. In addition, lower clouds will be grayer and higher clouds will be more white. Tall thunderstorm clouds will show up as bright white. An advantage of IR is that it is available 24 hours a day. Currently, there are eight geostationary satellites orbiting above the equator worldwide.

Geostationary satellite data are relatively less accurate than radar data in terms of rainfall estimation accuracy (FMH 1991) but cover a larger geographical area than radar data. Both satellite data and radar are less accurate than the standard rain-gauge data in terms of estimation accuracy but much more accurate in terms of the spatial resolution and the geographical area covered. Recent studies on satellite retrieval methods have shown promising results with regard to the use of satellite IR data for estimating rainfall by itself (Grimes et al. 1993; Vicente et al. 1998). Satellite data are still limited operationally to quality control purposes. In Next Radar Generation (NEXRAD) systems in the United States, for example, satellite data are used to check no-rain areas in order to correct radars errors. If radar reports rainfall in areas where satellite images show no clouds, the rainfall cells in this area are corrected. However, with future research, satellite retrieval algorithms may be able to provide rainfall estimation from only satellite data. This will enable rainfall estimation in areas where radar data are not accurate or not available, in addition to enabling quality control on radar data in areas where radars are available. On the other hand, satellite data can be of great help in terms of tracking the motion of clouds, which

FIGURE 3.7   Remote sensing station receiving data from the European satellite for flash flood warnings in Sinai, Egypt (used by WRRI).

can assist in forecasting rainfall motion, especially on a large scale. The Egyptian Meteorological Authority uses METEOSAT satellite data for recording rainfall data all over Egypt (Gad et al. 2004). Figure 3.7 shows a receiver that uses European remote-sensing techniques for flash flood warnings in Sinai, Egypt.

3.5   PRECIPITATION GAUGE NETWORK

The uses for which precipitation data are intended should determine the network density. A relatively sparse network of stations will suffice for determining annual averages over large areas of level terrain in studies of large general storms. A very dense network is required to determine the rainfall pattern in thunderstorms. The probability that a storm center will be recorded by a gauge varies with the network density. A network should give a representative picture of the areal distribution of precipitation. There should be no concentration of gauges in heavy rainfall areas at the expense of dry areas or vice versa. Unfortunately, the cost of installing and maintaining a network and accessibility of the gauge site to an observer are always important considerations.

Errors in rainfall averages computed from networks of various densities have been investigated. In general, sampling errors in terms of depth tend to increase with increasing areal mean precipitation and decrease with increasing network density, duration of precipitation, and size of the area.

The following four minimum densities of precipitation networks have been recommended by World Meteorological Organization (WMO 1969) for general hydrometeorological purposes:

1. For flat regions of temperate, Mediterranean, and tropical zones: 600–900 square kilometers per station
2. For mountainous regions of temperate, Mediterranean, and tropical zones: 100–250 square kilometers per station
3. For small mountainous islands with irregular precipitation: 25 square kilometers per station
4. For arid and polar zones: 1500–10,000 square kilometers per station

Information on the differences between calculated and true watershed precipitation is of climatological interest but does not answer this hydrologic question: What error results in estimation of runoff due to imperfect precipitation gauging? The answer depends on the climatic characteristics of precipitation in the area (including the effects of watershed topography), the hydrologic characteristics of the watershed, the stream flow characteristics being estimated, and possibly the method used to estimate stream flow greater or much less than the basin average. Stream flow estimated from this index will display a greater variance than observed stream flow. WMO recommends that 10% of the rain gauge stations be equipped with recording gauges to show rainfall intensities.

The accuracy of estimated mean areal precipitation over a catchment is directly related to the density of rain gauges within the area and to the spatial characteristics of storms occurring within the catchment. For arid and semi-arid regions, the U.S. NWS recommends that the minimum number of rain gauges, N, for a local flood-warning network within a catchment of area, A, is given by

$$N = 0.73A^{0.33} \tag{3.1}$$

where A is in square kilometers. However, even if more than the minimum number of gauges is used, not all storms will be adequately gauged. It is interesting to note that precipitation gauges are typically 20–30 centimeters in diameter, and these measurements are routinely used to estimate the average rainfall over areas exceeding 1 square kilometer. Clearly, isolated storms may not be measured well if storm cells are located over areas that are not gauged, or if the distribution of rainfall within storm cells is nonuniform over the catchment area.

A similar formula was adopted by the author to Wadi El Meleha in Sinai Egypt, which has the following value

$$N = 1.2 \, A^{0.43} \tag{3.2}$$

As an example, the area of Wadi El Meleha catchment is 26 square kilometers, and therefore, the least number of recording stations N is

$$N = 1.2 \times (26)^{0.43} = 4.87 \approx 5$$

The number of recording stations used in Wadi El Meleha is five.

3.6 INTERPRETATION OF PRECIPITATION DATA

To avoid erroneous conclusion of the rainfall records of a station, the data must first be checked for continuity and consistency. The continuity of a record may be broken with missing data because of reasons such as damage or fault in the rain gauge during a given period. The missing data can be estimated using the data of the neighboring stations. These calculations use normal rainfall as a standard of comparison. The normal rainfall is the average value of rainfall at a particular date, month, or year over a specified 30-year period. The 30-year normals are recomputed every decade. Thus, the term "normal annual precipitation" at station A means the average annual precipitation at A based on a specified 30 years of records.

3.6.1 ESTIMATING MISSING PRECIPITATION DATA

Many precipitation stations have short breaks in their records for many reasons. These missing records should be estimated. Given the annual precipitation values, P_1, P_2, ...P_n, at neighboring n stations 1, 2, 3, ...n., respectively, the missing annual precipitation P_x at a station X not included in the above n stations must be found. Further, the normal annual precipitations N_1, N_2, ...N_n, at each of the above $(n + 1)$ stations, including station X, are known. If the normal annual precipitations at various stations are within about 10% of the normal annual precipitation at station X, then a simple arithmetic average procedure is followed to estimate P_x. Thus,

$$P_x = (1/n)\left[P_1 + P_2 + \cdots + P_n\right] \tag{3.3}$$

If the normal precipitations vary considerably, then P_x is estimated by weighing the precipitation at the various stations by the ratios of normal annual precipitations. This method, known as the normal ratio method, gives P_x as

$$P_x = (N_x/n)\left[(P_1/N_1) + (P_2/N_2) + \cdots + (P_n/N_n)\right] \tag{3.4}$$

Or in general form as

$$P_x = (N_x/n) \sum_1^n (P_i/N_i) \tag{3.5}$$

Another method, used by the U.S. NWS (1972) for river forecasting, estimates precipitation at a point as the weighted average of that at four stations, one in each of the quadrants delineated by north–south and east–west lines through that point. Each station is the nearest in its quadrant to the point for which precipitation is being estimated. The weight applicable to each station is equal to the reciprocal of the square of the distance between the point and the station. Multiplying the precipitation for the storm (or other period) at each station by its weighting factor, adding the four weighted amounts, and dividing by the sum of the weights yields the estimated precipitation for the point as given in Equation 3.6. If one or more quadrants contain no precipitation stations, as might be the case for a point in a coastal area, then the estimation involves only the remaining quadrants:

$$P_x = \sum_{i=1}^{i=n}\left(P_i/(d_i)^2\right) \Bigg/ \sum_{i=1}^{i=n} 1/(d_i)^2 \tag{3.6}$$

where P_x is the precipitation at station x, P_i is the precipitation at station i, d_i is the distance between station x and station i, and n is the number of stations around station x ($n \geq 4$ as recommended).

This method can be used to estimate precipitation for a regular network station that failed to report or for points in a grid over a basin or other area so as to permit depth–area studies. In mountainous regions, precipitation values should therefore be expressed in percent of normal, as in Equation 3.4.

3.6.2 DOUBLE-MASS ANALYSIS

Changes in gauge location, exposure, instrumentation, or observational procedure may cause a relative change in the precipitation catch. These changes are often not disclosed in published records. U.S. Environmental Data Service practices can be used for new station identification whenever the gauge location is changed by as much as 8 kilometers (5 miles) or 30 meters in elevation.

Double-mass analysis (Kohler 1949) tests the consistency of the record at a station by comparing its accumulated annual or seasonal precipitation with the concurrent accumulated values of mean precipitation for a group of surrounding stations. In Figure 3.8, for example, a change in slope at a point indicates a change in the precipitation regime. A change due to meteorological causes would not cause a change

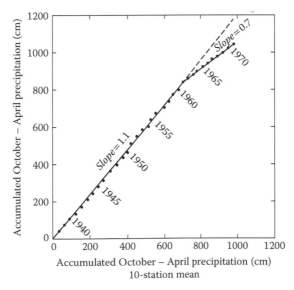

FIGURE 3.8 Adjustment of precipitation data by a double mass curve.

in slope, as all base stations would be similarly affected. The station history for this station discloses a change in gauge location. To make the record prior to this point comparable with that of the more recent location, the records should be adjusted by the ratio of the slopes of the two segments of the double-mass curve (*a/b*). The consistency of the record for each of the base stations should be tested, and those showing inconsistent records should be dropped before other stations are tested or adjusted.

3.7 AVERAGE PRECIPITATION OVER AN AREA

3.7.1 ARITHMETIC MEAN

The average depth of precipitation over a specific area, on a storm, seasonal, or annual basis, is required in many types of hydrologic problems. The simplest method for obtaining the average depth is to arithmetically average the gauged amounts in the area. This method yields accurate estimates in flat country if the gauges are uniformly distributed and the individual gauge catches do not vary widely from the mean as given in Equation 3.7. These limitations can be partially overcome if topographic influences and areal representation are considered in the selection of gauge sites (Wilm et al. 1939)

$$P_{av} = (1/n)\sum_{1}^{n} P_i \tag{3.7}$$

where P_{av} is the average precipitation, P_i is the precipitation at station i, and n is the number of stations.

3.7.2 THIESSEN METHOD

The Thiessen method (Thiessen 1911) attempts to allow for nonuniform distribution of gauges by providing a weighting factor for each gauge. The stations are plotted on a map and connecting lines are drawn between them (Figure 3.9). Perpendicular bisectors of these connecting lines form polygons around each station. The sides of each polygon are the boundaries of the effective area assumed for the station. The area of each polygon is determined by planimetry and is expressed as a percentage of the total area. The weighted average rainfall for the total area is computed by multiplying the precipitation at each station by its assigned percentage of area. The result is usually more accurate than that obtained by simple arithmetical averaging. The greatest limitation of the Thiessen method is its inflexibility; a new Thiessen diagram is required every time there is a change in the gauge network, and this method does not allow for orographic influences. It simply assumes linear variation of precipitation between stations and assigns each area segment to the nearest station.

3.7.3 ISOHYETAL METHOD

The most accurate method for averaging precipitation over an area is the isohyetal method. Station locations and amounts are plotted on a suitable map and contours of

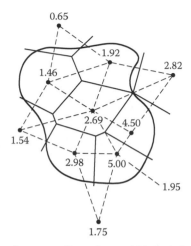

*Area of corresponding polygon within basin boundary

Observed precipitation (cm)	Area (sq km)	Percent total area	Weighted precipitation (col. 1 × col. 3)
0.65	7	1	0.01
1.46	120	19	0.28
1.92	109	18	0.35
2.69	120	19	0.51
1.54	20	3	0.05
2.98	92	15	0.45
5.00	82	13	0.65
4.50	<u>76</u>	12	0.54
	626	100	2.84

Average precipitation = 2.84 cm

FIGURE 3.9   Thiessen method and application.

equal precipitation called isohyets are drawn (Figure 3.10). The average precipitation for an area is then computed by weighting the average precipitation between successive isohyets (usually taken as the average of the two isohyetal values) by the area between isohyets, totaling these products, and dividing by the total area.

The isohyetal method permits the use and interpretation of all available data and is well adapted to display and discussion. In constructing an isohyetal map, the analyst can make full use of his knowledge of orographic effects and storm morphology, and in this case, the final map should give a more realistic precipitation pattern than could be obtained from the gauged amount alone. The accuracy of the isohyetal method is highly dependent upon the skill of the analyst. Moreover, an improper analysis can lead to serious error.

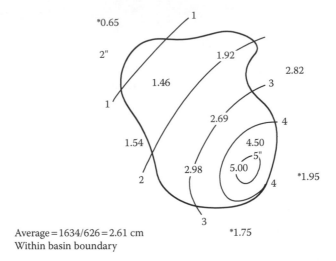

Average = 1634/626 = 2.61 cm   *1.75
Within basin boundary

Isohyet (cm)	Area enclosed (sq km)	Net area (sq km)	Avg. precip. (cm)	Precipitation volume (col. 3×col. 4)
5	13	13	5.3	69
4	90	77	4.6	354
3	206	116	3.5	406
2	402	196	2.5	490
1	595	193	1.5	290
<1	626	31	0.8	25
				1634

FIGURE 3.10   Isohyetal method and application.

3.8   DESIGN STORMS

A hypothetical rainfall event corresponding to a specified return period is usually the basis for the design and analysis of storm water management systems. However, the return period of a rainfall event is not equal to the return period of the resulting runoff, and therefore the reliability of surface-water management systems will depend on factors such as the antecedent moisture conditions in the catchment and its geographical location, for example in an arid or semi-arid region. In contrast to the single-event design storm approach, a continuous simulation approach is sometimes used, in which a historical rainfall record is input to a rainfall–runoff model; the resulting runoff is analyzed to determine the hydrograph corresponding to a given return period. The design storm approach is more widely used in engineering practice than the continuous simulation approach. Design storms can be either synthetic or actual (historic), with synthetic storms determined from historical rainfall statistics.

Synthetic design storms are characterized by their frequency or return period, duration, depth, temporal distribution, and spatial distribution. The selection of these quantities for design purposes is described in Sections 3.8.1 through 3.8.6.

3.8.1 FREQUENCY ANALYSIS OF POINT RAINFALL

In many hydraulic engineering applications, such as those concerned with floods, the probability of a particular extreme rainfall of a certain duration will be important. In practice, dividing the rainfall duration into three groups—namely, durations lasting from 1 minute to 1 hour as "short," from 1 to 24 hours as "intermediate," and more than 24 hours as "long"—has led to some meaningful interpretations and works on the regionalization of intensity duration frequency relationships in different geographical areas for several countries (Froehlich 1995a, 1995b, 1995c; Garcia-Bartual and Schneider 2001). For arid and semi-arid regions that commonly have short-duration rainfall, a maximum duration of 2 hours is recommended. Such information is used for frequency analysis of the point rainfall data. The rainfall at a given place is a random hydrologic process, and the rainfall data at a given place when arranged in chronological order constitutes a time series. One of the commonly used data series is the annual series, which is composed of annual values such as annual rainfall. If the extreme values for a specified event occurring in each year are listed, the series also constitutes an annual series. Thus, for example, the maximum 2-hour rainfall occurring in a year at a station can be listed to prepare an annual series of 2-hour maximum rainfall values in arid and semi-arid regions. The probability of occurrence of an event in this series is studied by frequency analysis of the annual data series. A brief description of the terminology and a simple method of predicting the frequency of an event are described next, and for details, see Appendix C. The analysis of annual series, though described with rainfall as a reference, is equally applicable to any other random hydrological process as flood flow.

First, the terminology used in frequency analysis must be understood. The probability of occurrence of an event (e.g. rainfall) whose magnitude is equal to or in excess of a specified magnitude X is denoted by P. The return period (also known as recurrence interval) is defined as

$$T = 1/P \tag{3.8}$$

This represents the average interval between the occurrence of a rainfall of a magnitude equal to or more than X. Thus, the return period of rainfall of 20 millimeters in 2 hours is 10 years at a certain station A implies that on an average, rainfall magnitudes equal to or greater than 20 millimeters in 2 hours occur once in 10 years. However, this does not mean that every 10 years one such event is likely, that is, periodicity is not implied. The probability of a rainfall of 20 millimeters in 2 hours occurring in any one year at station A is

$$P = 1/T = 1/10 = 0.1 \tag{3.9}$$

If the probability of an event occurring is P, the probability of the event not occurring in a given year is $q = (1 - P)$.

The purpose of frequency analysis of an annual series is to obtain a relationship between the magnitude of the event and its probability of exceedance. The probability analysis may be made either by the empirical method, which is called plotting position formulas, or by analytical and statistical methods, which will be given in Appendix C.

A simple empirical technique is to arrange the given annual extreme series in descending order of magnitude and to assign an order number m. Thus, for the first entry, $m = 1$, for the second entry, $m = 2$, and so on until the last event for which $m = N =$ number of years on record. The probability P of an event equaled or exceeded is given by the Weibull formula:

$$P = m/(N+1) \qquad (3.10)$$

and the return period $T = (N+1)/m$.

Equation 3.10 is an empirical formula, and there are several other such empirical formulae available to calculate P as shown in Table 3.2. Equation 3.10 is the most popular plotting position formula.

Having calculated P (and hence T) for all the events in the series, the variation of the rainfall magnitude is plotted against the corresponding T on a semilog or log-log paper (Figure 3.11). By suitably extrapolating this plot, the rainfall magnitude of specific duration for any recurrence interval can be estimated. This simple empirical procedure can yield accurate results for small extrapolations, but the errors increase with the amount of extrapolation.

The return period of a design rainfall should be selected on the basis of economic efficiency (ASCE 1992). In practice, however, the return period is usually selected on the basis of level of protection. Typical return periods are given in Table 3.3, although longer return periods are sometimes used. For example, return periods of 10–30 years in arid and semi-arid regions are commonly used for designing storm sewers in commercial and high-value districts. Local drainage regulations should be reviewed and followed when selecting the return period for a particular project. An implicit assumption in designing drainage systems for a given return period of rainfall is that the return period of the resulting runoff is equal to the return period of the design rainfall. Based on this assumption, the risk of failure of the drainage system is taken to be the same as the exceedance probability of the design rainfall.

TABLE 3.2
Plotting Position Formulas

Method	P
California	m/N
Hazen	$(m - 0.5)/N$
Weibull	$(m)/N + 1$
Chegodayev	$(m - 0.3)/(N + 0.4)$
Cunnane	$(m - 0.4)/(N + 0.2)$

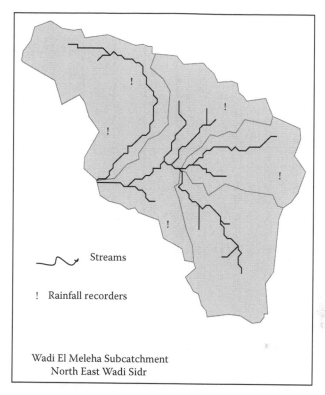

FIGURE 3.11   Wadi El Meleha catchment.

TABLE 3.3
Typical Return Periods

Land Use	Design Storm Return Period (Years)
Minor drainage systems	
Residential	2–5
High-value general commercial area	2–10
Airports (terminals, roads, aprons)	2–10
High-value downtown business areas	5–10
Major drainage-system elements	Up to 100 years

Source: Adapted from ASCE (1992).

EXAMPLE 3.1

Table 3.4 contains selected rainfall data for storms in Wadi El Meleha, Sinai (Figure 3.11). Calculate the probabilities and the return periods using three selected empirical formulas. Plot the results on semilog paper for Weibull formulas to get the maximum rainfall for the return periods of 25 and 50 years, respectively.

Solution

The data are arranged in descending order. The probability and the return periods of various events are calculated as indicated in Table 3.5 for three methods (i.e., Weibull, Chegodayev, and Cunnane). It is clear that the Weibull results are quite close to Chegodayev results; however, the Weibull formula is considered the most popular by many hydrologists for its simplicity when compared with the other empirical formulas.

A graph is then plotted between the rainfall magnitudes and the return period T on semilog paper (Figure 3.12). A smooth curve or mostly a straight line can be drawn through the plotted points, and the straight line can be extended to give the needed values.

From Figure 3.12, the values of the maximum rainfall for the return periods of 25 and 50 years were found to be 9 and 10 millimeters, respectively.

TABLE 3.4
Annual Maximum (60-Minute) Rainfall Storms at Wadi El Meleha (26 km²), Sinai

Month	23/3/91	24/12/92	27/12/93	11/3/94	6/2/95
Rainfall, d (mm)	6.3	5.4	5.8	6.4	5.0

TABLE 3.5
Plotting Position Formulas for 1-Hour Rainfall in Wadi El Meleha, Sinai

Rank	Max Rainfall Depth (mm)	Weibull		Chegodayev		Cunnane	
		P	T (years)	P	T (years)	P	T (years)
1	6.4	0.167	6	0.14	7.14	0.115	8.67
2	6.3	0.333	3	0.314	3.18	0.308	3.25
3	5.8	0.5	2	0.5	2	0.5	2
4	5.4	0.667	1.5	0.685	1.46	0.694	1.44
5	5	0.833	1.2	0.85	1.15	0.885	1.13

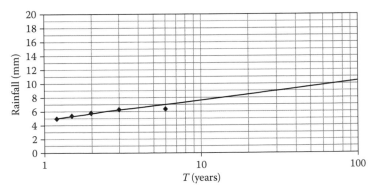

FIGURE 3.12   Rainfall versus return period of 1-hour rainfall in Wadi El Meleha in Sinai.

3.8.2 Rainfall Duration

The design duration of a storm is usually selected based on the time response charac-teristics of the catchment. The time response of a catchment is measured by the travel time of surface runoff from the most remote point of the catchment to the catchment outlet and is called the time of concentration. The duration of design storms used to design water control systems must generally equal or exceed the time of concentra-tion of the area covered by the control system. On small urban catchments (<1 square kilometer), the current practice is to select the duration of the design rainfall to be equal to the time of concentration. This approach usually leads to the maximum peak runoff for a given return period. For the design of detention basins, however, the duration causing the largest detention volume is most critical, and several different storm durations may be needed to identify the most critical design storm duration. For catchments with high infiltration losses, the duration of the critical rainfall asso-ciated with the maximum peak runoff may be shorter than the time of concentration (Chen and Wong 1993). Many drainage districts require that the performance of drainage systems for arid regions be analyzed.

3.8.3 Rainfall Depth

The design rainfall depth for a selected return period and duration is obtained directly from the intensity–duration–frequency (IDF) curve of the catchment. The IDF curve can be estimated from rainfall measurements derived from NWS publications, or obtained as given by Fahmi et al. (2005).

3.8.4 Intensity–Duration–Frequency Relationship

The intensity of storms decreases with an increase in storm duration. Further, a storm of any given duration will be more intense if its return period is large. In many design problems related to watershed management, such as runoff disposal and erosion con-trol, rainfall intensities of different durations and different return periods must be known. The interdependency among the intensity (i, centimeters per hour), duration (t, hours), and return period (T, years) is commonly expressed in a general form as

$$i = \frac{KT^x}{(t+a)^n} \tag{3.11}$$

where K, x, a, and n are constants for a given catchment. The values of these con-stants based on rainfall data obtained from small catchments in Sinai are

$$K = 1.8, \; x = 0.1, \; a = 1.2, \; n = 1.2$$

EXAMPLE 3.2

For the catchment of Example 3.1, $T = 25$ years, $t = 1$ hour, then

$$i = \frac{1.8 \times 25^1}{(1+1.2)^{1.2}} = 0.902 \text{ cm/h}$$

From Figure 3.12, i was found to be 9 millimeters per hour.

3.8.4.1 Partial-Duration Series

In the annual hydrologic data series of floods or rainfall, only one maximum value of flood or rainfall per year is selected as the data point. In some catchments, there may be more than one independent event in a year and many of these may be of appreciably high magnitude. A flood magnitude larger than an arbitrary selected base value is included in the analysis so all large flood peaks may be considered. Such a data series is called a partial-duration series.

When using the partial-duration series, it is necessary to establish that all events considered are independent. Hence, the partial-duration series is adopted mostly for rainfall analysis in which the conditions of independency of events are easy to establish, and its use in flood studies is rather rare. The recurrence interval of an event obtained by the annual series (T_a) and by the partial-duration series (T_p) are related by

$$T_p = \frac{1}{\ln T_a - \ln(T_a - 1)} \tag{3.12}$$

From this, we can see that the difference between T_a and T_p is significant for $T_a < 10$ years and that for $T_a > 20$, the difference is negligibly small.

3.8.5 Depth–Duration–Frequency Relationship in Arid Regions

Depth–duration–frequency (DDF) relationships have a significant impact on peak discharge estimation. Unfortunately, in many regions, only 24-hour rainfall data are available. Therefore, a designer of flood protection works is constrained either to assume a uniform rainfall distribution over the total storm duration, which leads to underestimation of peak discharges, or to adopt predetermined ratios between short-duration rainfall values derived for other countries and other climates.

The Egyptian Water Resources Institute (WRRI) studied DDF relationships in the Sudr Region, Sinai Peninsula, Egypt. Available short duration rainfall records were analyzed for six stations, all located in the Wadi El Meleha subcatchment (Figure 3.1; Fahmi et al. 2005). Short-duration ratios were derived using two approaches: the first, using all independent storms in the records to maximize the use of scarce information and the second, by performing frequency analysis on peak-over-threshold (POT) series for each storm duration and then deriving ratios for DDF rainfalls at different return periods.

The results of the two approaches are compared with regionalized U.S. and other international short-duration ratios. Finally, generalized DDF equations are developed and compared with equations from previously published studies for both humid and arid climates.

Considerable attention has been paid to the derivation of DDF relationships throughout the world, and many empirical rainfall intensity formulas have been derived. The relatively large amount of U.S. high-intensity rainfall data (used in U.S. Weather Bureau 1955; Hershfield 1961, usually referred to as T_p 40; Miller et al. 1973; updated recently by Frederick et al. 1977, which is usually referred to as Hydro 35) probably contains information of potential value to other countries.

There are various possible generalized relationships for the rainfall frequency values. These could be utilized to reduce some of the limitations due to sampling deficiencies in any individual record. The relationships described by Hershfield (1961), Reich (1963), and Bell (1969) provide useful methods for estimating short-duration rainfall ratios or relationships from long-duration rainfall and other readily available information. To develop these generalized relationships for the Sudr region, the following methodology was implemented:

1. Maximum rainfall depths at different rainfall durations for each of the six Sudr stations were extracted. For each gauging station, every storm was analyzed. For each storm, 6-, 12-, 18-, 30-, 60-, and 120-minute rainfall depths were extracted from the rainfall recorder tracer.
2. Determinations of independent storms and establishment of threshold values for each of the six Sudr stations were analyzed. To carry out POT frequency analysis (namely the partial-duration series) on each of these durations, storms must be independent. Corrections to T_p (POT) frequency analysis were applied based on Equation 3.12, upon which a particular value always has a longer return period with the annual series than with the partial series. However, the difference is only about half a year, which is relatively unimportant when return periods of 25 years or more are the main concern (Chow et al. 1988).
3. Frequency analysis on POT series and the selection of the best distribution fit for each of the 6 El Meleha stations. For consistency, the Gumbel distribution was adopted for all frequency analyses (see Appendix C).
4. The rainfall depth to which all other rainfall depths will be referred is selected. The rainfall of 1-hour duration and 10-year return period is p_{10}^1 and p_T^D is the rainfall depth for return period T and duration D. The ratio of T (year) and D (duration) rainfall depths p_{10}^1/p_T^D is calculated for each station for previously mentioned durations (6-, 12-, 18-, 30-, and 120-minute) and return periods of 2, 5, 10, 25, 50, and 100 years. p_{10}^1 rainfall depth was found to be a good reference base in previous studies, and it can be accurately estimated using a short-length record.
5. Estimation of mean ratios between 6-, 12-, 18-, 30-, and 120-minute rainfall depths and the reference rainfall depth (here the 1-hour depth) were adopted. This is probably the most accurate method, as it also uses the well-established principles of regional frequency analysis (Hosking and Wallis 1993).
6. Estimation of mean ratios considering several stations between 6, 12, 18, 30, and 120 minutes, rainfall depths at different return periods, and the reference rainfall depth (here the 1-hour 10-year rainfall depth). These ratios provide generalized curves for the study area, allowing the development of one generalized equation for the region.
7. One generalized relationship is derived for the set of average DDF ratios following the form chosen by Bell (1969):

$$p_T^D/p_{10}^1 = (a \times \ln T + b)(l \times D^m + n),$$
$$\text{for } 5 \leq D \leq 120 \text{ minutes and } 2 \leq T \leq 100 \tag{3.14}$$

This proposed mathematical form has two components: a *depth–frequency* component with two parameters a and b to fit and a *depth–duration* component with three parameters l, m, and n to fit. The depth–frequency component is chosen in a logarithmic form to suit the behavior of the Gumbel distribution, while the depth–duration form is chosen in an nth square root form to simulate the trend of the short-duration ratios previously established in step 5.

Finally, Fahmi et al. (2005) were able to obtain a generalized equation for an arid region using the same form as that of Bell:

$$p_T^D / p_{10}^1 = (0.255 \times \ln T + 0.397)(0.54 \times D^{0.25} - 0.5),$$
$$\text{for } 5 \leq D \leq 120 \text{ minutes and } 2 \leq T \leq 100 \tag{3.15}$$

The equation derived by Bell (1969) for Australia (Equation 3.16) was found to compare well with Equation 3.15. The ratios are given in Table 3.6.

$$p_T^D / p_{10}^1 = (0.21 \times \ln T + 0.52)(0.54 \times D^{0.25} - 0.5),$$
$$\text{for } 5 \leq D \leq 120 \text{ minutes and } 2 \leq T \leq 100 \tag{3.16}$$

This comparison is shown in Figure 3.13.

3.8.6 Probable Maximum Precipitation

In the design of any hydraulic structures such as spillways in dams or highway crossings, the hydrologist and hydraulic engineer should try to keep the failure probability as low as possible, that is, virtually zero. This is because such failure will cause heavy damages to life, property, economy, and national morale. In the design and analysis of such structures, the maximum possible precipitation that

TABLE 3.6
Generalized Average Ratios $\left(p_T^D\right)/\left(p_{10\text{-Year}}^{1\text{-Hour}}\right)$ of Rainfall Depth for Sudr Region

Return Period (year)	Storm Duration, D (minutes)					
	6	12	18	30	60	120
2	0.19	0.27	0.32	0.42	0.55	0.65
5	0.29	0.40	0.48	0.64	0.82	1.02
10	0.35	0.49	0.59	0.78	1.00	1.26
25	0.43	0.60	0.73	0.96	1.23	1.56
50	0.49	0.68	0.83	1.09	1.39	1.79
100	0.55	0.76	0.93	1.22	1.56	2.00

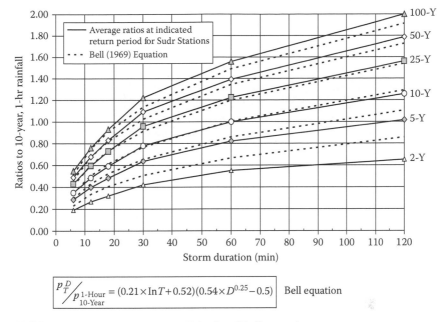

$$\left. \frac{p_{T}^{D}}{p_{1\text{-Hour}}^{10\text{-Year}}} = (0.21 \times \ln T + 0.52)(0.54 \times D^{0.25} - 0.5) \right| \quad \text{Bell equation}$$

FIGURE 3.13 Comparison between Fahmi and Bells equations.

can reasonably be expected at a given location is used because there is a physical upper limit to the amount of precipitation that can fall over a specified area in a given time.

The probable maximum precipitation (PMP) is defined as the greatest or most extreme rainfall possible for a given duration over a station or basin. From an operational point of view, PMP can be defined as rainfall over a basin that would produce a flood flow with virtually no risk of being exceeded. Obtaining the PMP for a given region is an involved procedure and requires the knowledge of an experienced hydrometeorologist. Two approaches are used: (1) meteorological methods and (2) the statistical study of rainfall data. Details of meteorological methods that use storm models are available in published literature (Chow et al. 1988). Statistical studies indicate that PMP can be estimated as

$$p_{m} = \bar{p} + k\sigma_{p} \tag{3.17}$$

where p_{m} is the mean of annual maximum rainfall in a given duration, $\bar{p}$ and σ_{p} are the mean and standard deviation of the annual maximum rainfall for a given duration, and k is the frequency factor, which depends upon the statistical distribution of the series, number of years of record, and the return period (Hershfield 1961). The value of K ranges between 15 and 20 for 24-hour rainfall. For any other duration, K_{T} should be calculated using extreme value Type I distribution.

TABLE 3.7
Gumbel Extreme Value Frequency Factors

Sample Size	Return Period								
	2.33	5	10	20	25	50	75	100	1000
15	0.065	0.967	1.703	2.410	2.632	3.321	3.721	4.005	6.265
20	0.052	0.919	1.625	2.302	2.517	3.179	3.563	3.386	6.006
25	0.044	0.888	1.575	2.235	2.444	3.088	3.463	3.729	5.842
30	0.038	0.866	1.541	2.188	2.393	3.026	3.393	3.653	5.727
40	0.031	0.838	1.495	2.126	2.326	2.943	3.301	3354	5.476
50	0.026	0.820	1.466	2.086	2.283	2.889	3.241	3.491	5.478
60	0.023	0.807	1.446	2.059	2.253	2.852	3.200	3.446	5.410
70	0.020	0.797	1.430	2.038	2.230	2.824	3.169	3.413	5.359
75	0.019	0.794	1.423	2.029	2.220	2.812	3.155	3.400	5.338
100	0.015	0.779	1.401	1.998	2.187	2.770	3.109	3.340	5.261
00	−0.067	0.720	1.305	1.866	2.044	2.592	2.911	3.137	4.900

Source: Adapted from Viessman, W., Jr., and G.L. Lewis. 2003. *Introduction to Hydrology*. 5th ed. Upper Saddle River, NJ: Prentice Hall.

In the case of an extreme value Type I distribution, the frequency factor, K_T, can be written as

$$K_T = -\frac{\sqrt{6}}{\pi}\left\langle 0.5772 + \ln\left[\ln\left\{\frac{T}{T-1}\right\}\right]\right\rangle \tag{3.18}$$

This expression for K_T is valid only in the limit as the number of samples approaches infinity. For a finite sample size, K_T varies with the sample size, as shown in Table 3.7.

The following example will compare the results obtained by Equation 3.15 and those obtained by the values from Table 3.7.

EXAMPLE 3.3

Table 3.8 includes maximum 1-hour rainfall depths from 1986 to 1994 in Wadi El Meleha, Sinai. Calculate the maximum rainfall depth for a return period of 100 years. Compare this result with the PMP of the same records using Equation 3.17 and Table 3.7.

SOLUTION

Figure 3.14 was plotted using the Weibull equation. By extending the curve, the value of rainfall depth $p_{10}^1 = 12.5$ mm.

From Table 3.6, $\left(p_T^D\right)/\left(p_{10\text{-Year}}^{1\text{-Hour}}\right) = 1.56$ and $P(100 \text{ year}) = 12.5 \times 1.56 = 19.5$ m.

From Table 3.7, K_T for 100 years, the lowest sample number was found to be 4.

Using Equation 3.17 and Table 3.9, $\bar{P}$ was found to be 5.79875 mm and $\sigma_p = 3.447237$.

$p_m = 5.79875 + 3.447237 \times 4.00 = 19.8$ mm, a result that is a little higher than that obtained by Equation 3.15.

TABLE 3.8
1-Hour Rainfall Depths from 1986 to 1994 at Wadi El Meleha

Years	87	88	89	90	91	92	93	94
Rainfall (mm)	6.5	4.72	9.4	2.36	5.74	2.54	3.1	12.03

TABLE 3.9
PMP (mm) for Wadi El Meleha

M	Rainfall (mm)	N + 1/M
1	12.03	9
2	9.4	4.5
3	6.5	3
4	5.74	2.25
5	4.72	1.8
6	3.1	1.5
7	2.54	1.285714
8	2.36	1.125

$N = 8$, Av = 5.79875 mm, Std = 3.447237, and PMP = 19.6188 mm.

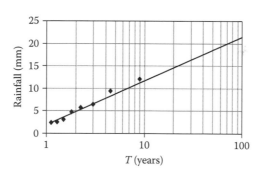

FIGURE 3.14 Relation between rainfall and return period.

3.8.7 TEMPORAL DISTRIBUTION

Realistic temporal distributions of rainfall within design storms are best determined from historical rainfall measurements. In many cases, however, either the data are not available or such a detailed analysis cannot be justified. Under these conditions, the designer must resort to empirical distributions. For arid and semi-arid regions, the designer simply can apply the triangular method with the ratio of $t_d/t_p = 3$, where t_d is the storm duration and t_p is the time to peak.

REFERENCES

American Society of Civil Engineers. 1992. *Design and Construction of Urban Stormwater Management Systems*. New York: ASCE.

Atlas, D., ed. 1990. *Radar in Meteorology: Battan Memorial and 40th Anniversary Radar Meteorology Conference*. Boston, MA: American Meteorological Society.

Battan, L. J. 1981. *Radar Observation of the Atmosphere*. Chicago, IL: University of Chicago Press.

Bell, F. C. 1969. Generalized rainfall depth-duration-frequency relationships. *J Hydraul Div ASCE* 95(HY1):311–27.

Chen, C. N. and T. S. W. Wong. 1993. Critical rainfall duration for maximum discharge from overland plane. *Journal of Hydraulic Engineering* 119(9):1040–1045.

Chow, V. T., D. R. Maidment, and L. W. Mays. 1988. *Applied Hydrology*. New York: McGraw-Hill.

Doviak, R. J., and D. S. Zrnic. 1993. *Doppler Radar and Weather Observations*. 2nd ed. San Diego, CA: Academic Press.

Fahmi, A. H. et al. 2005. Generalized depth duration frequency relationship in arid regions. Fourth International Symposium on Environmental Hydrology in Arid and Semi-Arid Region, Cairo, Egypt.

FMH 1991. *Federal Meteorological Handbook No. 11: Doppler Radar Meteorological Observations-Part C-WSR88D Products and Algorithms*. Washington D.C: U.S Department of Commerce/National Oceanic and Atmospheric Administration.

Frederick, R. H., V. A. Meyers, and E. P. Auciello. 1977. *Five- to 60-Minute Precipitation Frequency for the Eastern and Central United States*. Washington, DC: NOAA Technical Memorandum, NWS HYDRO-35.

Froehlich, D. C. 1995a. Intermediate-duration rainfall equations. *J Hydraul Eng ASCE* 121(10):751–6.

Froehlich, D. C. 1995b. Long-duration rainfall intensity equations. *J Irrig Drain Eng* 121(3):248–52.

Froehlich, D. C. 1995c. Short-duration rainfall intensity equations for drainage design. *J Irrig Drain Eng* 121(4):310–1.

Gad, M. A. 2004a. *Developing GIS for the Estimation and Analysis of Rainfall Fields over Egypt using Meteosat Image*. Cairo, Egypt: ITI Fifth Arab Conference on GIS.

Gad, M. A. 2004b. The ESPP: The ellipsoid-based satellite projection method for georeferencing data from geostationary weather satellites. *J Hydroinf*.

Garcia-Bartual, R., and M. Schneider. 2001. Estimating maximum expected short-duration rainfall intensities from extreme convective storms. *Phys Chem Earth Part B* 26(9):675–81.

Grimes, D. I. F., J. R. Milford, and G. Dugdale. 1993. *The use of satellite rainfall estimates in hydrological modelling*. Presented at the First International Conference of the African Meteorlogical Society, Nairobi, Kenya.

Hershfield, D. M. 1961. Rainfall frequency atlas of the United States. Weather Bureau Technical Paper No. 40.

Hosking, J. R. M., and J. R. Wallis. 1993. Some statistics useful in regional frequency analysis. *Water Resour Res* 29:271–81.

Kohler, M. A. 1949. Double-mass analysis for testing the consistency of records and for making required adjustments. *Bull Am Meteorol Soc* 30:188–9.

Miller, J. F., R. H. Frederick, and R. J. Tracey. 1973. *Precipitation Frequency Analysis of the Western United States*. Washington DC: National Weather Service, NOAA.

Reich, B. M. 1963. Short duration rainfall intensity estimates and other design aids for regions of sparse data. *J Hydrol* 1(1).

Rinehart, R. E. 1991. *Radar for Meteorologists*. Grand Forks, ND: University of North Dakota, Department of Atmospheric Sciences.

Sauvageot, H. 1982. *Radar Meteorology*. Paris: Eyrolles.

Skolnik, M., ed. 1970. *Radar Handbook*. New York: McGraw-Hill.

Steinhoff, H. W., and J. D. Ives, eds. 1976. *Ecological Impacts of Snowpack Augmentation in the San Juan Mountains*. Colorado: Final Report to the Bureau of Reclamation.

Thiessen, A. H. 1911. Precipitation for large areas. *Mon Weather Rev* 39:1082–4.

U.S. National Weather Service. 1972. National Weather Service. River Forecast System Forecast Proceedings, NOAA Tech. Memo. NWS HYDRO 14:3.1–3.14.

U.S. Weather Bureau. 1955. Rainfall intensity frequency duration curves for selected stations. Technical Paper No. 25.

Vicente, G. A., R. A. Scofield, and W. P. Menzel. 1998. The operational GOES infrared estimation technique. *Bull Am Meteorol Soc* 79:1883–98.

Viessman, W., Jr., and G.L. Lewis. 2003. *Introduction to Hydrology*. 5th ed. Upper Saddle River, NJ: Prentice Hall.

Wilm, H. C., A. Z. Nelson, and H. C. Storey. 1939. An analysis of precipitation measurements on mountain watersheds. *Mon Weather Rev* 67:163–72.

World Meteorological Organization. 1969. *Manual for Depth-Area-Duration Analysis of Storm Precipitation*. Geneva: WMO No. 237, Tech. Paper 129, 1–31 pp.

BIBLIOGRAPHY

ASCE Manual No. 81. 1995. *Guidelines for Cloud Seeding to Augment Precipitation*.

ASCE Policy Statement No. 275. 2003. *Atmospheric Water Resources Management*.

ASCE/EWRI 42-04. 2004. *Standard Practice for the Design and Operation of Precipitation Enhancement Projects*.

Bellon, D. E., and G. L. Austin. 1976. SHARP (Short-term automatec radar prediction) on a real time test. In *Seventeenth Conference on Radar Meteorology*, 522-5. Boston, MA: American Meteorological Society.

Bobée, B., and P. F. Rasmussen. 1995. Recent advances in flood frequency analysis. *Rev Geophys* 33:1111–6.

Cunnane, C. 1986. Review of statistical methods for flood frequency estimation. In *International Symposium on Flood Frequency and Risk Analyses: Hydrologic Frequency Modeling*, ed. V. P. Singh, 49–85. Baton Rouge, LA: D. Reidel Publishing Company.

FAO. 1981. *Irrigation and Drainage Paper, No. 37: Arid Zone Hydrology*. Rome, Italy: FAO.

Givati, A., and D. Rosenfeld. 2004. Quantifying precipitation suppression due to air pollution. *J Appl Meter* 43:1038–1056.

Hassan, F. A. 2000. Practical approach for rainfall data processing in wadi Feiran, South Sinai, Egypt. Presented at the International Conference on Wadi Hydrology, Sharm El-Sheikh, Egypt.

Klein, D. A. 1978. *Environmental Impacts of Artificial Ice Nucleating Agents*. Stroudsburg, PA: Dowden, Hutchinson & Ross.

Linsley, R. K. et al. 1975. *Hydrology for Engineers*. 2nd ed. New York: McGraw-Hill.

NRC report. 2003. *Critical Issues in Weather Modification Research*. Washington, DC: The National Academies Press.

Paulhus, J. L. H., and M. A. Kohler. 1952. Interpolation of missing precipitation records. *Mon Weather Rev* 80:129–33.

Ramanathan, P. J. et al. 2001. *Aerosols, Climate, and the Hydrologic Cycle*. Science magazine: 2001, http://www.ramanathan.ucsd.edu/abc.

Ramanathan, V., and M. V. Ramana. 2003. Atmospheric brown clouds, long range transport and climate impacts. *EM* 28–33, http://www-c4.ucsd.edu.

Singh, R. 1968. Double-mass analysis on the computer. *Hydraul Div ASCE* 94:139–42.

Stedinger, J. R., R. M. Vogel, and E. Foufoula-Georgiou. 1993. Frequency analysis of extreme events. In *Handbook of Hydrology*, Chapter 18, ed. D. R. Maidment, 66. New York: McGraw-Hill.

The Weather Modification Association's Response to the NRC Report. 2004. Critical issues in weather modification research, report of a review panel. *J Weather Modif* 53–82.

UNEP and Center for Clouds, Chemistry and Climate. 2002. *The Asian Brown Cloud: Climate and Other Environmental Impacts*. Nairobi: UNEP.

Wan, P. 1976. Point rainfall characteristics of Saudi Arabia. *Proc Inst Civil Eng* 61(2):179–87.

Wheater, H. S., P. Larentis, and G. S. Hamilton. 1989. Design rainfall characteristics for southwest Saudi Arabia. *Proc Inst Civil Eng* 87(2):517–38.

Williams, B. 2004. *Assessment of the Environmental Toxicity of Silver Iodide and Indium Iodide*. Adelaide University. Cooma, NSW, Australia: Snowy Hydro Limited.

World Meteorological Organization. 1970. *Guide to Hydro Meteorological Practices*. 2nd ed. Geneva: WMO No. 168, Tech. Pap. 82.

4 Precipitation Losses

4.1 INTRODUCTION

Before rainfall reaches the outlet of a basin as runoff, certain demands of the catchment must be met, such as interception, depression storage, and infiltration. These diversions are called losses. Precipitation equals, by a mass balance, rainfall excess plus rainfall losses.

Rainfall excess is the portion of the total rainfall that drains directly from the land surface via overland flow. When performing a flood analysis using a rainfall-runoff model, the determination of rainfall excess is of utmost importance. Rainfall excess integrated over the entire watershed yields the runoff volume, and the temporal distribution of the rainfall excess along with the hydraulics of runoff will determine the peak discharge. Therefore, estimation of the magnitude and time distribution of rainfall losses should be performed with the best available practical technology, considering the objective of the analysis, economics of the project, and consequences of inaccurate estimates.

Rainfall losses are generally considered to result from the evaporation of water from the land surface, interception of rainfall by the vegetal cover, depression storage on the land surface (paved or unpaved), or infiltration of water into the soil matrix. A schematic representation of rainfall losses for a uniform intensity rain is shown in Figure 4.1. Such losses are treated in detail in this chapter.

4.2 EVAPORATION

Evaporation is the process by which a liquid changes to a gaseous state at the free surface at a temperature below the boiling point through the transfer of heat energy. Consider a body of water such as a pond. The molecules of water are in constant motion with a wide range of instantaneous velocities. Adding heat causes this range of velocities and the average speed to increase. When some of the molecules possess sufficient kinetic energy, they may cross over the water surface into the atmosphere. Similarly, the atmosphere in the immediate vicinity of the water surface contains water molecules within the water vapor that are in motion, and some of them may penetrate the water surface. The net escape of water molecules from the liquid state to the gaseous state is evaporation. Evaporation is a cooling process in that the latent heat of vaporization must be provided by the water body. The rate of evaporation is dependent on (1) the vapor pressures at both the water surface and air above it, (2) the temperatures of the air and water, (3) wind speed, (4) atmospheric pressure, (5) quality of the water, and (6) size of the water body. The rate of evaporation is proportional to the difference between the saturation

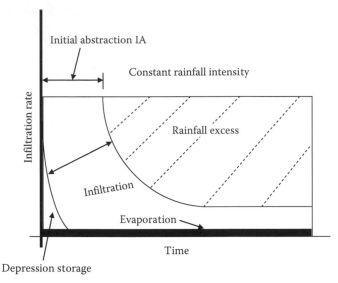

FIGURE 4.1   Components of rainfall losses.

vapor pressure of water at a specific temperature, e_s, and the actual vapor pressure in the air, e_a. Thus

$$E_v = C(e_s - e_a) = Ce_s(1 - \text{RH}/100) \tag{4.1}$$

where E_v is the rate of evaporation (millimeters per day), C is a constant, and RH is the relative humidity percentage, e_s and e_a are in millimeters of mercury. Equation 4.1 is known as Dalton's law of evaporation. Evaporation continues until $e_s = e_a$. If $e_s > e_a$, condensation takes place.

Dissolving a solute in water causes a reduction in the rate of evaporation. The percent reduction in evaporation approximately corresponds to the percentage increase in the specific gravity of the solution. For example, under identical conditions, evaporation from seawater is about 2–3% less than that from fresh water.

4.2.1 EVAPORIMETERS

Estimation of evaporation is of utmost importance in many hydrologic problems associated with the planning and operation of reservoirs and irrigation systems. In arid zones, this estimation is particularly important to conserve scarce water resources. However, the exact measurement of evaporation from a large body of water is one of the most difficult tasks in hydrology.

The amount of water evaporating from a water surface can be estimated by the following methods: (1) using data from an evaporimeter, (2) empirical evaporation equations, or (3) analytical methods. Evaporimeters are water-containing pans that are exposed to the atmosphere, with the loss of water by evaporation measured at regular intervals. Meteorological data, such as humidity, wind movement, the temperatures of the air and water, and precipitation, are noted along with evaporation measurements. Many types of evaporimeters are in use, and a few commonly used pans are described in Sections 4.2.1.1 and 4.2.1.2.

4.2.1.1 Class A Evaporation Pan

Class A evaporation pans are standard pans 1210 millimeters in diameter and 255 millimeters in depth that are used by the U.S. Weather Bureau. The depth of water is maintained between 18 and 20 centimeters (Figure 4.2). The pan is normally made of unpainted galvanized iron sheet and is placed on a wooden platform that is 15 centimeters above the ground to allow free circulation of air below the pan. Evaporation measurements are made by measuring the depth of water with a hook gauge in a stilling well.

Another type of land pan is the ISI standard pan, which is roughly the same shape as that of the Class A pan.

4.2.1.2 U.S. Geological Survey Floating Pan

This square pan (900 millimeters long on each side and 450 millimeters deep) is supported by drum floats in the middle of a raft (4.25 × 4.87 meters) and is set afloat in a lake to simulate the characteristics of a large body of water. The water level in the pan is maintained at the same level as the lake, leaving a rim of 75 millimeters. Diagonal baffles reduce surging in the pan due to wave action. Its high cost of installation and maintenance, together with the difficulty involved in performing measurements, are its main disadvantages.

4.2.2 PAN COEFFICIENT, C_P

Evaporation pans are not exact models of large reservoirs and have the following principal drawbacks:

1. They differ in heat-storing capacity and heat transfer from the sides and bottom. The sunken pan and floating pan aim to reduce this deficiency. Because of this factor the evaporation from a pan depends, to a certain extent, on its size. Although a pan 3 meters in diameter gives a value that is about the same as that from a neighboring large lake, a pan only 1 meter in diameter indicates about 20% more evaporation than that of a pan 3 meters in diameter.

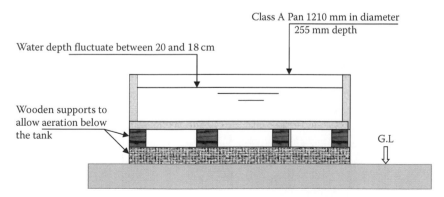

FIGURE 4.2 Class A evaporation pan used by the U.S. Weather Bureau.

TABLE 4.1
Values of the Pan Coefficient, C_p

S. No.	Type of Pan	Average Value	Range
1	Class A Land pan	0.70	0.60–0.80
2	ISI pan (modified Class A)	0.80	0.65–1.10
3	Colorado sunken pan	0.78	0.75–0.86
4	U.S. Geological Survey floating pan	0.80	0.70–0.82

2. The height of the rim in an evaporation pan affects the wind action over the surface. It also casts a shadow of variable magnitude over the water surface.
3. The heat-transfer characteristics of the pan material are different than those of the reservoir.

Because of these drawbacks, the evaporation observed from a pan must be corrected to obtain the evaporation from a lake under similar climatic and exposure conditions. Thus, a coefficient is introduced as follows:

$$\text{Lake evaporation} = C_p \times \text{pan evaporation}$$

where C_p is the pan coefficient. The values of C_p in use for different pans are given in Table 4.1.

4.2.3   EVAPORATION STATION NETWORK

Evaporation pans are usually installed in locations where other meteorological data are also simultaneously collected. The World Meteorological Organization recommends a minimum network of evaporimeter stations as follows:

1. Arid zones: One station for every 30,000 square kilometers
2. Humid, temperate climates: One station for every 50,000 square kilometers
3. Cold regions: One station for every 100,000 square kilometers

4.3   EMPIRICAL EVAPORATION EQUATIONS

A large number of empirical equations are available to estimate lake evaporation using commonly available meteorological data. Most formulas are based on the Dalton-type equation and can be expressed in the following general form:

$$E_L = kf(u)(e_s - e_a) \tag{4.2}$$

where E_L is the lake evaporation in millimeters per day, e_s is the saturated vapor pressure at a specific water-surface temperature, in millimeters of mercury (Table 4.2),

TABLE 4.2

Saturation Vapor Pressure of Water

Temperature (°C)	Saturation Vapor Pressure, e_s(mm of Hg)
0	4.58
5.0	6.54
7.5	7.78
10.0	9.21
12.5	10.87
15.0	12.79
17.5	15.00
20.0	17.54
22.5	20.44
25.0	23.76
27.5	27.54
30.0	31.82
32.5	36.68
35.0	42.81
37.5	48.36
40.0	55.32
45.0	71.20

e_a is the actual vapor pressure of the overlying air at a specified height in millimeters of mercury, $f(u)$ is the wind speed–correction function, and K is a coefficient. The term e_a is measured at the same height at which wind speed is measured. Two commonly used empirical evaporation formulas are as follows:

1. Meyer's formula

$$E_L = K_M(e_s - e_a)\left(1 + \frac{u_g}{16}\right) \tag{4.3}$$

where E_L, e_s, and e_a are as defined in Equation 4.2, u_g is the monthly mean wind velocity in kilometers per hour at about 9 meters above the ground, and K_M is the coefficient accounting for various other factors, with values of 0.36 and 0.50 for large, deep waters and small, shallow waters, respectively.

2. Rohwer's formula, which considers a correction for the effect of pressure, in addition to the wind-speed effect, and is given by

$$E_L = 0.771(1.465 - 0.000732 p_a)(0.44 + 0.0733 u_g)(e_s - e_a) \tag{4.4}$$

where E_L, e_s, and e_a are as defined earlier in Equation 4.2, p_a is the mean barometric reading in millimeters of mercury, u_g is the mean wind velocity in kilometers per hour at ground level, which can be assumed to be the velocity at a height of 0.6 meters above the ground.

These empirical formulas are simple to use and permit the use of standard meteorological data. However, due to the various limitations of the formulas, they can at best be expected to give an approximate magnitude of the evaporation process.

In using the empirical equations, the saturated vapor pressure at a given temperature in Celsius, (e_s) is found from Table 4.2. Often, the wind-velocity data will be available at an elevation other than that needed in the particular equation. However, in the lower part of the atmosphere, up to a height of about 600 meters above ground level, the wind velocity can be assumed to follow the 1/7 power law, stated as

$$u_a = Ch_a^{1/7} \tag{4.5}$$

where u_a is the wind velocity at a height h_a above the ground and C is a constant. This equation can be used to determine the velocity at any desired height if u_a is known.

EXAMPLE 4.1

A reservoir with a surface area of 200 hectares had the following average values for the various parameters during a week: water temperature = 25°C, relative humidity = 50%, and wind velocity at 1 meter above ground = 14 kilometers per hour. Estimate the average daily evaporation from the lake and the volume of water evaporated from the lake in a span of 10 days.

SOLUTION

From Table 4.2, $e_s = 23.76$ millimeters of Hg, $e_a = 0.50 \times 23.76 = 11.88$ millimeters of Hg, and $u_g =$ wind velocity at a height of 9 meters above ground.
From Equation 4.5, $u_g/u_1 = (h_g/h_1)^{1/7}$, $u_g = u_1 (9)^{1/7} = 14.0 (9)^{1/7} = 19.13$ kilometers per hour. By Meyer's formula (Equation 4.3),

$$E_L = 0.36(23.76 - 11.88)\left(1 + \frac{19.13}{16}\right) = 9.4 \text{ mm/day}$$

$$\text{Evaporated volume in 10 days} = 10 \times \frac{9.4}{1000} \times 200 \times 10^4 = 188,000 \text{ m}^3$$

4.4   ESTIMATION OF EVAPORATION BY ANALYTICAL METHODS

The analytical methods for the determination of lake evaporation can be broadly classified into three categories as follows:

1. Water-budget method
2. Energy-balance method
3. Mass-transfer method

4.4.1 WATER-BUDGET METHOD

The water-budget method is the simplest of the three analytical methods but is also the least reliable. It involves writing the hydrological continuity equation for the lake and determining the evaporation from a knowledge or estimation of other variables. Thus, considering the daily average values for a lake, the continuity equation is written as

$$P + V_{is} + V_{ig} = V_{os} + V_{og} + E_L + \Delta S + ET_L \tag{4.6}$$

where P is the daily precipitation, V_{is} is the daily surface inflow into the lake, V_{ig} is the daily groundwater inflow, V_{os} is the daily surface outflow from the lake, V_{og} is the daily seepage outflow, E_L is the daily lake evaporation, ΔS is the increase in lake storage in a day, and ET_L is the daily transpiration loss.

All quantities are in units of volume (cubic meters) or depth (millimeters) over a reference area. Equation 4.6 can be written as

$$E_L = P + (V_{is} - V_{os}) + (V_{ig} - V_{og}) - ET_L - \Delta S \tag{4.7}$$

The terms P, V_{is}, V_{os}, and ΔS can be measured. However, V_{ig}, V_{og}, and ET_L cannot be measured, and therefore can only be estimated. Transpiration losses can be considered insignificant in some reservoirs. If the unit of time is large, for example weeks or months, the estimate of E_L can be more accurate. In view of the various uncertainties in the estimated values and the possibility of errors in the measured variables, the water-budget method cannot be expected to give very accurate results. However, controlled studies, such as that at Lake Hefner in the United States (1952), have given reasonably accurate results using this method.

4.4.2 ENERGY-BUDGET METHOD

The energy-budget method is an application of the law of conservation of energy. The energy available for evaporation is determined by calculating the incoming energy, outgoing energy, and energy stored in the water body over a known time interval.

For a body of water like the one in Figure 4.3, the energy balance at the evaporating surface in a period of one day is given by

$$Q_n = Q_a + Q_e + Q_g + Q_s + Q_i \tag{4.8}$$

where Q_n is the net heat energy received by the water surface; this in turn is equal to $Q_c(1 - r) - Q_b$, in which $Q_c(1 - r)$ is the incoming solar radiation into a surface of reflection coefficient (albedo) r (Table 4.3), Q_b is the back radiation (long wave) from the water body, Q_a is the sensible heat transfer from water surface to air, Q_e is the heat energy used up in evaporation, which is equal to $\rho L E_L$, where ρ is the density of water, L is the latent heat of evaporation, and E_L is the evaporation in millimeters, Q_g is the heat flux into the ground, Q_s is the heat stored in the water body, and Q_t is the net heat conducted out of the system by water flow (advected energy).

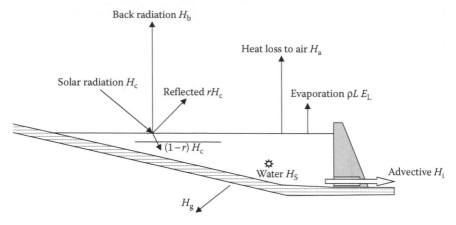

FIGURE 4.3   Energy balance in a water body.

TABLE 4.3
Reflection Coefficient, r

Surface	Range of r Values
Close-grown crops	0.15–0.25
Bare lands	0.05–0.45
Water surface	0.05
Snow	0.45–0.90

All the energy terms are expressed in calories per square millimeter per day. If the time periods are short, the terms Q_s and Q_t can be neglected because they are small. All the terms except Q_a, which cannot be readily measured, are estimated using Bowen's ratio β, given by the expression

$$\beta = \frac{Q_a}{\rho L E_L} = 6.1 \times 10^{-4} \times p_a \frac{T_s - T_a}{e_s - e_a} \tag{4.9}$$

where p_a is the atmospheric pressure, in millimeters of mercury, e_s is the saturated vapor pressure in millimeters of mercury, e_a is the actual vapor pressure of air in millimeters of mercury, T_s is the temperature of the water surface in Celsius, and T_a is the temperature of air, in Celsius. From Equations 4.8 and 4.9, E_L can be evaluated as

$$E_L = \frac{Q_n - Q_g - Q_s - Q_i}{\rho L (1 + \beta)} \tag{4.10}$$

The energy-balance method for estimating evaporation in a lake has been found to give satisfactory results with errors around 5% when applied to periods of less than a week.

4.4.3 Mass-Transfer Method

The mass-transfer method applies the theory of turbulent mass transfer in a boundary layer to calculate the mass water vapor transfer from the surface to the surrounding atmosphere. However, the details of the method are beyond the scope of this book and can be found in published literature.

4.5 RESERVOIR EVAPORATION AND METHODS FOR ITS REDUCTION

Any of the methods mentioned in Sections 4.4.1 through 4.4.3 may be used for the estimation of reservoir evaporation. Although analytical methods provide better results, they also involve parameters that are difficult to assess or expensive to obtain. Empirical equations can at best give approximate values of the correct order of magnitude. Therefore, pan measurements are generally accepted for practical applications. The water volume lost due to evaporation from a reservoir in a month is calculated as

$$V_E = AE_{pm}C_p \qquad (4.11)$$

where V_E is the volume of water lost in evaporation in a month (cubic meter), A is the average reservoir area during the month, E_{pm} is the loss due to pan evaporation in a month in meters, which is equal to E_L (in millimeter per day) $\times$ number of days in the month $\times$ 10^{-3}, and C_p is the relevant pan coefficient.

Evaporation from a water surface is a continuous process. The quantity of stored water lost by evaporation in a year is considerable, as the surface areas of many natural and artificial lakes in arid and semi-arid regions are very large. A small reservoir may have a surface area of about 2000 hectares, whereas in large reservoirs, such as the High Aswan Dam Reservoir, the water loss by evaporation derived using different methods was found to be between 10 and 12 billion cubic meters per year.

Because the construction of various reservoirs as a part of water resource development efforts involves considerable inputs of money, evaporation from the reservoirs is an economic loss. In arid zones where water is scarce, the importance of water conservation through reduction of evaporation is obvious. The various methods available for reduction of evaporation losses can be listed under three categories:

1. *Reduction of surface area*: Because the volume of water lost by evaporation is directly proportional to the surface area of the water body, the reduction of surface area wherever feasible reduces evaporation losses. Measures such as having deep reservoirs instead of wider ones and eliminating shallow areas are also in this category.
2. *Mechanical covers*: Permanent roofs, temporary roofs, and floating roofs such as rafts and lightweight floating particles can be adopted wherever feasible. Obviously, these measures are limited to very small water bodies such as ponds.

3. *Chemical films*: This method consists of applying a thin chemical film to the water surface to reduce evaporation. Currently this is the only feasible method available for reduction of evaporation of moderately-sized reservoirs. Certain chemicals, such as cetyl alcohol (hexadecanol) and stearyl alcohol (octadecanol), form monomolecular layers on a water surface. These layers act as evaporation inhibitors by preventing the water molecules from escaping past them. The thin film formed by these chemical compounds has the following desirable features:

   - The film is strong and flexible and does not break easily due to wave action.
   - If punctured due to the impact of raindrops or birds, insects, and so on, the film will close back up soon after.
   - It is pervious to oxygen and carbon dioxide; therefore, the water quality is not affected by its presence.
   - It is colorless, odorless, and nontoxic.

Cetyl alcohol is the most suitable chemical for use as an evaporation inhibitor. It is a white, waxy, crystalline solid and is available as lumps, flakes, or powder. It can be applied to the water surface in the form of powder, emulsion, or solution in mineral turpentine. About 0.35 kilograms per hectare per day of cetyl alcohol is needed for effective action. The chemical is periodically replenished to make up for the losses due to oxidation, wind sweep of the layer to the shore, and removal by birds and insects. Maximum evaporation reduction can be achieved if a film pressure of 4×10^{-2} N/m^2 (Newton's per square meter) is maintained.

Controlled experiments with evaporation pans have indicated an evaporation reduction of about 60% when using cetyl alcohol. Under field conditions, the reported values of evaporation reduction range from 20–50%. It appears that a reduction of 20–30% can be achieved easily in small lakes ($\leq$1000 hectares) using these monomolecular layers. The adverse effect of heavy wind appears to be the only major impediment affecting the efficiency of chemical films.

4.6   EVAPORATION AND TRANSPIRATION

4.6.1   TRANSPIRATION

Transpiration is the process by which water leaves the body of a living plant and reaches the atmosphere as water vapor. Water is taken up by the plant's root system and escapes through the leaves. The important factors affecting transpiration are atmospheric vapor pressure, temperature, wind, light intensity, and characteristics of the plant, such as the root and leaf systems. For any given plant factors that affect the free-water evaporation also affect transpiration. However, a major difference exists between transpiration and evaporation. Transpiration is essentially confined to daylight hours and the rate of transpiration depends upon the growth periods of the plant. Evaporation, on the other hand, continues throughout the day and night, although the rates are different.

4.6.2 Evapotranspiration

During transpiration the land area on which plants stand also loses moisture via the evaporation of water from the soil and water bodies. In hydrology and irrigation practices, evaporation and transpiration processes can be advantageously grouped under one head as *evapotranspiration*. The term "consumptive use" is also used to denote this loss via evapotranspiration. For a given set of atmospheric conditions, evapotranspiration obviously depends on the availability of water. If sufficient moisture is always available to completely meet the needs of the vegetation covering the area, the resulting evapotranspiration is called potential evapotranspiration (ET_p). Potential evapotranspiration no longer critically depends on soil and plant factors but depends essentially on climatic factors. The real evapotranspiration occurring in a specific situation or location is called actual evapotranspiration (ET_a).

Two terms are important at this stage: *field capacity* and *permanent wilting point*. Field capacity is the maximum quantity of water that the soil can retain against the force of gravity. Any additional input of moisture to soil at field capacity simply drains away. Permanent wilting point is the moisture content of soil at which point moisture is no longer available in sufficient quantities to sustain the plants. At this stage, even though the soil contains some moisture, it will be so firmly held by the soil grains that the roots of the plants are not able to extract moisture in sufficient quantities to sustain the plants and, consequently, the plants wilt. The field capacity and permanent wilting point depend upon the soil characteristics. The difference between these two moisture contents is called *available water*, which is the moisture available for plant growth.

4.6.3 Measurement of Evapotranspiration

The measurement of evapotranspiration for a given type of vegetation can be carried out in two ways: by using lysimeters or by using field plots.

1. *Lysimeters*: A lysimeter is a special watertight tank containing a block of soil and set in a field of growing plants. The plants grown in the lysimeter are the same as those in the surrounding field. Evapotranspiration is estimated in terms of the amount of water required to maintain constant moisture conditions within the tank, measured either volumetrically or gravimetrically through an arrangement in the lysimeter. Lysimeters should be designed to accurately reproduce the soil conditions, moisture content, and the type and size of the vegetation of the surrounding area. They should be buried such that the soil is at the same level both inside and outside the container. Lysimeter studies are important for projects involving water demand.
2. *Field plots*: All the elements of the water budget in a known interval of time are measured in special plots, and the evapotranspiration is determined as

Evapotranspiration = (precipitation + irrigation input − runoff

− increase in soil storage − groundwater loss)

Measurements are usually confined to precipitation, irrigation input, surface runoff, and soil moisture. Groundwater loss due to deep percolation is difficult to measure and can be minimized by maintaining the moisture condition of the plot at the field capacity. This method provides fairly reliable results.

4.6.4 Evapotranspiration Equations

The lack of reliable field data and the difficulties obtaining reliable evapotranspiration data have given rise to numerous methods to predict ET_p using climatological data. Large numbers of formulas, which range from the purely empirical to those backed by theoretical concepts, are available. Numerous empirical formulas are available for the estimation of ET_p based on climatological data. These are not universally applicable to all climatic areas and should be used with caution in areas different from those for which they were derived. One of the simplest empirical equations is the Blaney–Criddle equation. However, for large irrigation projects, other formulas should be adopted.

The Blaney–Criddle Formula is a purely empirical formula based on data obtained from the arid regions of western United States. This formula assumes that the term ET_p is related to both total hours of sunshine and the temperature, which are taken as measures of solar radiation in an area. The potential evapotranspiration in a crop-growing season is given by

$$E_T = KF \text{ inches per month} \tag{4.12}$$

and

$$F = \sum P_k \bar{T}_f / 100 \tag{4.13}$$

where E_T is the ET_p of a crop (inches per month), P_k is the monthly percent of annual daytime hours, which depends on the latitude of the place (Table 4.4), K is an

---

TABLE 4.4
Percentages of Monthly Daytime Hours, P_k, for Use in the Blaney–Criddle Formula

North Latitude	Jan	Feb	Mar	Apr	May	Jun	Jul	Aug	Sep	Oct	Nov	Dec
0°	8.50	7.66	8.49	8.21	8.50	8.22	8.50	8.49	8.21	8.50	8.22	8.50
10°	8.13	7.47	8.45	8.37	8.81	8.60	8.86	8.71	8.25	8.34	7.91	8.10
15°	7.94	7.36	8.43	8.44	8.98	8.80	8.05	8.83	8.28	8.26	7.75	7.88
20°	7.74	7.25	8.41	8.52	9.15	9.00	9.25	8.96	8.30	8.18	7.58	7.66
25°	7.53	7.14	8.39	8.61	9.33	9.23	9.45	9.09	8.32	8.09	7.40	7.42
30°	7.30	7.03	8.38	8.72	9.53	9.49	9.67	9.22	8.33	7.99	7.19	7.15
35°	7.05	6.88	8.35	8.83	9.76	9.77	9.93	9.37	8.36	7.87	6.97	6.86
40°	6.76	6.72	8.33	8.95	10.02	10.08	10.22	9.54	8.39	7.75	6.72	6.52

TABLE 4.5

Values of *K* for Selected Crops

Crop	Value of *K*	Range of Monthly Values
Rice	1.10	0.85–1.30
Wheat	0.65	0.50–0.75
Maize	0.65	0.50–0.80
Sugarcane	0.90	0.75–1.00
Cotton	0.65	0.50–0.90
Potatoes	0.70	0.65–0.75
Natural Vegetation		
Very dense	1.30	
Dense	1.20	
Medium	1.00	
Light	0.80	

TABLE 4.6

Values of *f*, *F*, and Rainfall for Use in Example 4.3

T(h)	*F*(cm/h)	*F*(cm)	Rainfall (cm)
0		0	0
0.1	1.78	0.29	
0.2	1.2	0.43	
0.5	0.72	0.71	0.5
1	0.54	1.02	
1.5	0.44	1.26	
2	0.39	1.47	2

Rainfall intensity = 1 cm/h for 2 hours

Source: Adapted from Rawls, W. J., D. L. Brakensiek, and N. Miller. 1983. *ASCEJ Hydraul Eng* 109(1):62–71.

empirical coefficient that depends on the type of the crop (Table 4.5), *F* is the sum of monthly consumptive-use factors for the period (Table 4.6), and $\bar{T}_f$ is the mean monthly temperature, in Fahrenheit.

The values of *K* depend on the month and locality. The average values for a season for selected crops are given in Table 4.5. The Blaney–Criddle formula is largely used by irrigation engineers to calculate the water requirements of crops and is suitable for both arid and semi-arid regions.

Equation 4.12 can be converted to the metric system to get

$$E_T = 4.57K\ P_k(t+17.8)\ \text{centimeter per month} \tag{4.14}$$

where E_T is in centimeter per month, *t* is in Celsius, and the rest of terms have the same values.

EXAMPLE 4.2

Estimate the E_T of an area for the November season during which maize is grown. The area is located in the Nile Delta at a latitude of 30°N, with mean monthly temperatures of 16.5°C.
Use the Blaney–Criddle formula (metric system).

SOLUTION

From Table 4.5, for maize, $K = 0.65$.
The value of P_k for 30°N is read from Table 4.4 as 7.19%

$$E_T = 4.57 \times 0.65 \times (7.19/100) \times (16.5 + 17.8) = 7.33\,\text{cm (for November)}$$

4.7   INTERCEPTION

When rainfall occurs over a catchment, not all of the precipitation falls directly onto the ground. Before it reaches the ground, some of it may be caught by the vegetation and subsequently evaporated. The volume of water caught in this way is called *intercepted precipitation*, and the process is called *interception*. The intercepted precipitation may follow one of three possible routes.

It may be retained by vegetation as surface storage and returned to the atmosphere by evaporation—a process termed *interception loss*. The amount of water intercepted in a given area is extremely difficult to measure. It depends on the species composition of the vegetation, its density, and the characteristics of the storm. However, in hydrological studies conducted in arid and semi-arid areas that deal with floods, interception loss is rarely significant and is not separately considered. The common practice is to deduct a lump sum value from the initial period of the storm as the initial loss.

To further elucidate the rainfall losses, three periods of rainfall losses are illustrated in Figure 4.1; these must be understood and their implications appreciated before using the procedures in this section. First, there is a period of initial loss when no rainfall excess (runoff) is produced. During this initial period, the losses are a function of the depression storage, interception, and evaporation rates, in addition to the initially high infiltration capacity of the soil. The accumulated rainfall loss during this period of no runoff is called the initial abstraction. This initial period ends with the onset of ponded water on the surface; the time from the start of rainfall to this time is the "time of ponding." Note that losses during this first period are a summation of the losses due to all mechanisms, including infiltration. The second period is marked by a declining infiltration rate and generally very little loss due to other factors. The third and final period occurs for rainfalls of sufficient duration for the infiltration rate to reach the steady state or equilibrium rate of the soil (f_c). The only appreciable loss during the final period is due to infiltration.

The actual loss process is quite complex and the loss mechanisms are dependent both on each other and on the rainfall itself. Therefore, simplifying assumptions are usually made in the modeling of rainfall losses. Figure 4.1 represents a simplified set of assumptions that can be made. Figure 4.1 assumes that surface retention loss is the summation of all losses other than those due to infiltration and that this loss occurs from the start of rainfall until the time the accumulated rainfall equals the magnitude of the surface retention loss capacity. It assumes that infiltration does not occur during this time. After the surface retention is satisfied, infiltration begins. If the infiltration capacity exceeds the rainfall intensity, then no rainfall excess is produced. As the infiltration capacity decreases, it may eventually equal the rainfall intensity. This would occur at the time to ponding, which signals the beginning of surface runoff. As illustrated in Figure 4.1, after the time to ponding, the infiltration rate decreases exponentially and may reach a steady-state equilibrium rate (f_c). These simplified assumptions and processes, as illustrated in Figure 4.1, are modeled by the procedures described in this section.

4.8 SURFACE RETENTION LOSS

Surface retention loss as used herein is the summation of all rainfall losses other than infiltration. The major component of surface retention loss is depression storage; the relatively minor components of surface retention loss are interception and evaporation. Depression storage occurs in two forms. *In-place depression storage* occurs at and in the vicinity of the raindrop impact. The mechanism for this depression storage is the microtopography of the soil and soil cover. The second form of depression storage is the *retention of surface runoff*, which occurs away from the point of raindrop impact in surface depressions, such as puddles, roadway gutters and swales, roofs, and irrigation-bordered fields and lawns. A relatively minor contribution by interception is also considered a part of the total surface retention loss. Estimates of surface retention loss are difficult to obtain and are a function of the physiography and land use of the area. The surface retention loss on impervious surfaces has been estimated to be in the range of 1.6–3.175 millimeters (0.0625–0.125 inches) by Tholin and Reefer (1960).

4.9 RECOMMENDED METHODS FOR ESTIMATING RAINFALL LOSSES

Numerous methods have been developed for estimating rainfall losses. Four methods are considered in this section because of their wide usage in arid and semi-arid regions.

1. Holton infiltration equation
2. Green–Ampt infiltration equation
3. The loss rate estimated by the curve-number (CN) method
4. Initial loss plus uniform loss rate

4.9.1 HORTON INFILTRATION EQUATION

The Horton infiltration equation is an exponential-decay type of equation, in which the rainfall loss rate asymptotically diminishes to the minimum infiltration rate, f_c. Horton (1939) expressed the decay of the infiltration capacity with time as

$$f_{ct} = f_c + (f_{c0} - f_c)e^{-tK_\lambda}, \quad \text{for} \quad 0 \le t \le t_a \tag{4.15}$$

where f_{ct} is the infiltration capacity at any time t from the start of the rainfall, f_{c0} is the initial infiltration capacity at $t = 0$, f_c is the final steady-state value, t_a is the duration of the rainfall, and K_λ is the constant depending on the characteristics of both soil and vegetation cover.

The difficulty in finding a variation in the three parameters f_{c0}, f_c, and K_λ concurrent with the soil characteristics and antecedent moisture conditions precludes the general use of Equation 4.15.

The infiltration-capacity values of soils are subjected to wide variations depending on a large number of factors. Typically, a bare, sandy area will have an infiltration capacity of $f_c \approx 1.2$ centimeters per hour, and a bare, clay area will have a capacity of $f_c \approx 0.15$ centimeters per hour. A good grass or vegetation cover increases these values by as much as 10 times.

The Holton equation is not extensively used; however, it can be used whenever the relevant data are available.

4.9.2 Green–Ampt Infiltration Equation

Consider a vertical column of soil of a unit horizontal cross-sectional area (Figure 4.4), with the control volume defined around the wet soil between the surface and depth L. For a soil initially of moisture content θ_i throughout its entire depth, the moisture content will increase from θ_i to η (the porosity) as the wetting front passes. The moisture content θ_i is the ratio of the volume of water to the total volume within the control surface. $L(\eta - \theta_i)$ is then the increase in the amount of water stored within the control volume, caused by infiltration through a unit cross section. By definition this quantity is equal to F, the cumulative depth of water infiltrated into the soil, so that

$$F(t) = L(\eta - \theta_i) = L\Delta\theta \qquad (4.16)$$

and

$$L \text{ is the length of the wetting front} = F(t)/\Delta\theta \qquad (4.16a)$$

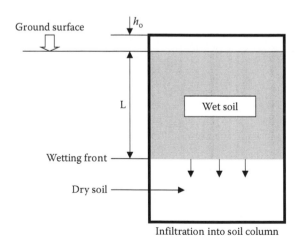

Infiltration into soil column

FIGURE 4.4   Infiltration into a column of soil.

Darcy's law may be expressed as

$$q = K\frac{\partial h}{\partial z} \approx -K\frac{\Delta h}{\Delta z} \qquad (4.17)$$

The Darcy flux q is constant throughout the depth and is equal to $-f$, because q is positive upward, whereas f is positive downward. If points 1 and 2 are located at the ground surface and on the dry side of the wetting front, respectively, then Equation 4.17 can be approximated as

$$f = K\left[\frac{h_1 - h_2}{z_1 - z_2}\right] \qquad (4.18)$$

The head h_1 at the surface is equal to the ponded depth h_0. The head h_2, in the dry soil below the wetting front, equals $-\psi - L$. Darcy's law for this system is written as

$$f = K\left[\frac{h_0 - (-\psi - L)}{L}\right] \qquad (4.19)$$

If the ponded depth h_0 is negligible compared to ψ and L,

$$f = K\left[\frac{\psi + L}{L}\right] \qquad (4.20)$$

The assumption $h_0 = 0$ is usually appropriate for problems involving surface-water hydrology because it is assumed that ponded water becomes surface runoff. From Equations 4.16a and 4.20, we get

$$\frac{dF}{dt} = f = K\left[\frac{\psi\Delta\theta + F}{F}\right] \qquad (4.21)$$

The solution to Equation 4.21 was found by Green–Ampt to be

$$F(t) = Kt + \psi\Delta\theta\ln\left[1 + \frac{F(t)}{\psi\Delta\theta}\right] \qquad (4.22)$$

Given the values of K, t, ψ, and $\Delta\theta$, a trial value F is substituted on the right-hand side until adjustment is reached. As a first trial, consider $F(t) = Kt$. Application of the Green–Ampt method requires estimating the hydraulic conductivity K, the wetting front soil suction head ψ (Table 4.7), and $\Delta\theta$.

The *residual moisture* content of the soil, denoted by θ_r, is the moisture content after the soil has been thoroughly drained. The *effective saturation* is the ratio of the available moisture $(\theta - \theta_r)$ to the maximum possible available moisture content $(\eta - \theta_r)$, which is given as

$$s_e = \frac{\theta - \theta_r}{\eta - \theta_r} \qquad (4.23)$$

where the term $(\eta - \theta_r)$ is called the effective porosity, θ_e.

TABLE 4.7
Green–Ampt Infiltration Parameters for Various Soil Classes

Soil Class	Porosity (η)	Effective Porosity (θ_e)	Wetting-Front Soil-Suction Head (ψ, cm)	Hydraulic Conductivity (K, cm/h)
Sand	0.437	0.417	4.95	11.78
Loamy sand	0.437	0.401	6.13	2.99
Sandy loam	0.453	0.412	11.01	1.09
Loam	0.463	0.434	8.89	0.34
Silt loam	0.501	0.486	16.68	0.65
Sandy clay loam	0.398	0.330	21.85	0.15
Clay loam	0.464	0.309	20.88	0.10
Silty clay	0.471	0.432	27.30	0.10
Sandy clay	0.430	0.321	23.90	0.06
Silty clay loam	0.479	0.423	29.22	0.05
Clay	0.475	0.385	31.63	0.03

Source: Adapted from Rawls, W. J., D. L. Brakensiek, and N. Miller. 1983. *ASCEJ Hydraul Eng* 109(1):62–71.

The effective saturation has the range $0 \leq s_e \leq 1$, provided $\theta_r \leq \theta \leq \eta$. For the initial condition, when $\theta = \theta_i$, cross-multiplying Equation 4.23 gives $\theta_i - \theta_r = s_e\,\theta_e$, and the change in the moisture content when the wetting front passes is written as

$$\Delta\theta = (1 - s_e)\theta_e \qquad (4.24)$$

As the soil becomes finer, moving from sand to clay, the wetting front soil suction head increases while the hydraulic conductivity decreases. Table 4.7 also lists typical ranges for η, θ_e, and ψ. The ranges are not large for η and θ_e, but ψ can vary over a wide range for a given soil. The term K varies along with ψ; therefore, the values given in Table 4.7 for both ψ and K should be considered typical values and may show a considerable degree of variability in application.

EXAMPLE 4.3

Use the Green–Ampt method to evaluate the infiltration rate and cumulative infiltration depth for a silty clay soil at 0.2-hour increments up to 2 hours from the start of infiltration. Assume an initial effective saturation of 20% and continuous ponding.

Solution

From Table 4.7, for a silty clay soil $\theta_e = 0.423$, $\psi = 29.22$ centimeters, and $K = 0.05$ centimeters per hour. The initial effective saturation is $s_e = 0.2$; thus, $\Delta\theta = (1 - s_e)$ $\theta_e = (1 - 0.20) \times 0.423 = 0.338$; and $\psi\Delta\theta = 29.22 \times 0.338 = 9.89$ centimeters.

Assuming continuous ponding, the cumulative infiltration F is found by successive substitutions in Equation 4.22:

$$F(t) = Kt + \psi\Delta\theta\ln\left[1 + \frac{F(t)}{\psi\Delta\theta}\right] = 0.05t + 9.89\ln[1 + F/9.89]$$

Accordingly for $t = 0.2$ hours, the cumulative infiltration converges to a final value $F = 0.42$ centimeters. Infiltration rate f is then computed using Equation 4.21 as follows:

$$f = K\left[1 + \frac{\psi\Delta\theta}{F}\right] = 0.05(1 + 9.89/F) \tag{4.25}$$

As an example, at time $t = 0.2$ hours, $f = 0.05(1 + 9.89/0.42) = 1.2$ centimeters per hour. The infiltration rate and cumulative infiltration are computed in the same manner between 0 and 2 hours at 0.2-hour intervals; the results obtained are listed below:

t (hour)	0.2	0.4	0.5	1	1.5	2.0
F (cm)	0.43	0.63	0.71	1.02	1.26	1.47
f (cm/h)	1.2	0.84	0.72	0.54	0.44	0.39

For constant rainfall, the infiltration rate f and the cumulative infiltration F are related by Equation 4.25. The cumulative infiltration at ponding time t_p is $F_p = it_p$, where i is the constant rainfall intensity (Figure 4.5). Substituting $F_p = it_p$ and the infiltration rate $f = i$ into Equation 4.25 yields

$$i = K\left[1 + \frac{\psi\Delta\theta}{it_p}\right] \tag{4.26}$$

Solving Equation 4.26, we get

$$t_p = \frac{k\psi\Delta\theta}{i(i - k)} \tag{4.27}$$

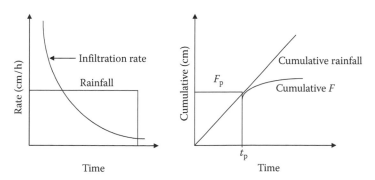

FIGURE 4.5 Ponding time.

EXAMPLE 4.4

For Example 4.3, draw the relationship between rainfall intensity of 1 centimeter per hour for a 2-hour storm. The soil is the same as that of the Example 4.3. Calculate t_p in hours using Equation 4.27. Check this value with that obtained from the graph.

SOLUTION

$$t_p = \frac{k\psi\Delta\theta}{i(i-k)}$$

$$t_p = \frac{0.05 \times 9.89}{1(1-0.05)} = 0.52 \text{ hour}$$

From Figure 4.6, the point of intersection for the F and f curves is located at $t_p = 0.5$ hours. However, the same curve, by the definition for $f = i = 1$ centimeter per hour, gave $t_p = 0.3$ hours, which is lower than the calculated value and hence may be more appropriate to rely on than Equation 4.27.

4.9.3 NRCS CURVE-NUMBER MODEL

The curve-number model was originally developed by the NRCS of the U.S. Department of Agriculture. It was published in 1954 in the first edition of the National Engineering Handbook, which has subsequently been revised several times (SCS 1993). This empirical method is the most widely used method for estimating rainfall excess (which is equal to rainfall minus abstractions) in the United States, and the popularity of this method can be attributed to its ease of application, lack of serious competition, and extensive database of parameters. The curve-number model was originally developed for calculating how daily runoff was affected by land-use

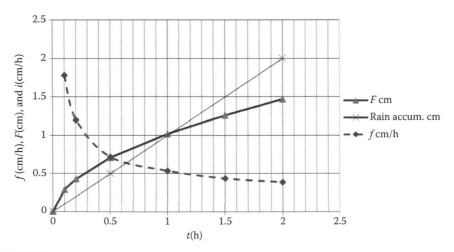

FIGURE 4.6   Components of curve numbers.

practices in small agricultural watersheds; it was later adapted to urban catchments because of its overwhelming success.

The NRCS curve-number model separates rainfall into three components as follows: rainfall excess Q; initial abstraction I_a; and retention F. Figure 4.6 illustrates these components. The initial abstraction includes the rainfall that is stored in the catchment areas before runoff begins. This includes interception, infiltration, and depression storage. If the amount of rainfall is less than the initial abstraction, then runoff does not occur. The retention, F, consists primarily of infiltrated water and is the portion of the rainfall reaching the ground that is retained by the catchment. The basic assumption of the NRCS model is that for any rainfall event, the precipitation P, runoff Q, retention F, and initial abstraction I_a, are related by the equation

$$\frac{F}{S} = \frac{Q}{P - I_a}$$
(4.28)

where S is the potential maximum retention and measures the retention capacity of the soil. The maximum retention S does not include I_a. The rationale for Equation 4.26 is that for any rainfall event, the portion of available storage (S), that is, the infiltration F, is equal to the portion of available rainfall ($P - I_a$) appearing as the runoff, Q. Equation 4.28 is, of course, applicable only when $P > I_a$. Conservation of mass requires that

$$F = P - Q - I_a$$
(4.29)

Eliminating F from Equations 4.28 and 4.29 yields

$$Q = \frac{(P - I_a)^2}{(P - I_a) + S} \quad P > I_a$$
(4.30)

Empirical data indicate that the initial abstraction, I_a, is directly related to the maximum retention, S. The following relation is commonly assumed:

$$I_a = 0.2S$$
(4.31)

Recent research has indicated that a factor of 0.2 is probably adequate for large storms in rural areas, but is likely an overestimate for small to medium storms and is probably too high for urban areas (Schneider and McCuen 2005; Singh 1992). However, because the storage capacity of many catchments is determined based on Equation 5.68, the factor 0.2 should be retained when using storage estimates calibrated from field measurements of rainfall and runoff.

Combining Equations 4.30 and 4.31 gives the following expression:

$$Q = \frac{(P - 0.2S)^2}{P + 0.8S}, \quad P > 0.2S$$
(4.32)

This equation is the basis for estimating the volume of runoff, Q, from a volume of rainfall, P, given the maximum retention, S. Note however that the NRCS method was originally developed as a runoff index for 24-hour rainfall amounts and should therefore be used with caution when attempting to analyze incremental runoff amounts during the course of a storm (Kibler 1982) or the runoff volume for durations other than 24 hours. The curve-number model predicts the same runoff for a given rainfall amount, regardless of duration. This is obviously not correct, because long-duration low-intensity storms will have a smaller runoff amount than short-duration high-intensity storms, when both storms have the same total rainfall. The curve-number model is less accurate when the runoff is less than 10 millimeters and in these cases a different method should be used to determine the runoff (SCS 1986). In application the curve-number model is quite satisfactory when used for its intended purpose, which is to evaluate the effects of land-use changes and conservation practices on direct runoff. Because it was not developed to reproduce individual historical events, only limited success has beeen achieved in using it for that purpose.

Instead of specifying S directly, a curve number (CN) is usually specified, and the CN is related to S by the expression

$$S = \frac{1000}{CN} - 10 \qquad (4.33)$$

where S is given in inches. Equation 4.31 is modified from the original formula, which requires S in inches. Clearly, in the absence of available storage ($S = 0$, impervious surface), the CN is equal to 100, and for an infinite amount of storage, the CN is equal to 0. The CN, therefore varies between 0 and 100. The utilization of the CN in the rainfall–runoff relation is the basis for the naming of the curve-number model. In fact, the NRCS adopted the term "curve number" because the rainfall–runoff equation may be expressed graphically with curves for different values of CN.

In practical applications, the CN is considered a function of several factors, including hydrologic soil group, type of vegetation cover, treatment (management practice), hydrologic conditions, antecedent runoff conditions (ARC), and the extent of impervious area in the catchment. CNs corresponding to a variety of urban land uses are given in Table 4.8. Soils are classified into four hydrologic soil groups: A, B, C, and D; descriptions of these groups are given in Table 4.9. Soils are grouped based on their profile characteristics, including depth, texture, organic matter content, structure, and degree of swelling when saturated. The minimum infiltration rates associated with the hydrologic soil groups are given in Table 4.9, where the rates apply to bare soil after prolonged wetting. The NRCS has classified more than 5000 types of soils into these four groups (Rawls et al. 1982). Local NRCS offices can usually provide information on local soils and their associated soil groups, but the soil characteristics should also be considered. There are a variety of methods for determining the cover type. The most common are field reconnaissance, aerial photographs, and land-use maps. In agricultural practices, treatment is a cover-type modifier used to describe the management of cultivated lands. It includes mechanical practices, such as contouring and terracing, and management practices, such as crop rotation and reduced or no tillage. The term "hydrologic condition" indicates the effects of cover type and treatment on infiltration and runoff and is generally estimated from

TABLE 4.8
Curve Numbers for Various Urban Land Uses

Cover Type and Hydrologic Condition	Curve Numbers of the Hydrologic Soil Group			
	A	B	C	D
Lawns, open spaces, parks, and golf courses				
Good condition: Grass cover on 75% or more of the area	39	61	74	80
Fair condition: Grass cover on 75% of the area	49	69	79	84
Poor condition: Grass cover on 50% or less of the area	68	79	86	89
Paved parking lots, roofs, driveways, and so on	98	98	98	98
Streets and roads				
Paved with curbs and storm sewers	98	98	98	98
Gravel	76	85	89	91
Dirt	72	82	87	93
Paved with open ditches	83	89	82	83
Commercial and business areas (85% impervious[a])	89	92	94	95
Industrial districts (72% impervious[a])	81	88	91	93
Row, town, and residential houses, with plot sizes of 1/8 acre or less (65% impervious[a])	77	85	90	92
Residential house average plot size				
1/8 acre or less (town houses, 65% impervious[a])	77	85	90	92
1/4 acre (38% impervious[a])	61	75	83	87
1/3 acre (30% impervious[a])	57	82	81	86
1/2 acre (25% impervious[a])	54	70	81	86
1 acre (20% impervious[a])	51	68	79	84
2 acre (12% impervious[a])	46	65	77	82

[a] The impervious area is assumed to be directly connected to the drainage system, with the impervious area having a CN of 98; the pervious area is assumed equivalent to open space in good hydrologic condition.

TABLE 4.9
Description of the Four NRCS Soil Groups

Group	Description	Minimum Infiltration Rate (mm/h)
A	Deep sand, deep loess, and aggregated silts	>7.5
B	Shallow loess and sandy loam	3.8–7.6
C	Clay loams, shallow sandy loam, soils low in organic content, and soils usually high in clay	1.3–3.8
D	Soils that swell significantly when wet, heavy plastic clays, and certain saline soils	0–1.3

the density of plant and residue cover. Good hydrologic conditionst indicate that the soil has a low runoff potential for that specific hydrologic soil group, cover type, and treatment. The percentage of impervious area and the means of conveying runoff from impervious areas to drainage systems should generally be considered when estimating the CN for urban areas. An impervious area is considered directly connected if the runoff from the area flows directly into the drainage system. The impervious area is not directly connected if the runoff from the impervious area flows over a pervious area and then into a drainage system. This method is used to obtain some hydrologic properties of soils in arid and semi-arid regions.

The ARC is a measure of the actual available storage relative to the average available storage at the beginning of the rainfall event and is closely related to the antecedent moisture content of the soil. It is grouped into three categories: ARC I, ARC II, and ARC III. The average CN normally cited for a particular land area corresponds to ARC II conditions, and these CNs can be adjusted for drier than normal conditions (ARC I) or wetter than normal conditions (ARC III) using Table 4.10.

TABLE 4.10
Antecedent Runoff Condition Adjustments

CN for ARC II	Corresponding CN for Condition	
	ARC I	ARC III
100	100	100
95	87	99
90	78	98
85	70	97
80	63	94
75	57	91
70	51	87
65	45	83
60	40	79
55	35	75
50	31	70
45	27	65
40	23	60
35	19	55
30	15	50
25	12	45
20	9	39
15	7	33
1	4	26
5	2	17
0	0	0

ARC = antecedent runoff condition, CN = curve number.

The CN adjustments in Table 4.10 can be approximated by the following relations (Chow et al. 1988):

$$CN_I = \frac{CN_{II}}{2.3 - 0.0013CN_{II}}$$ (4.34)

and

$$CN_{III} = \frac{CN_{II}}{0.43 + 0.0057CN_{II}}$$ (4.35)

where CN_I, CN_{II}, and CN_{III} are the curve numbers under ARC I, ARC II, and ARC III conditions, respectively.

The guidelines for selecting the CNs given in Tables 4.8 and 4.10 are useful in cases where site-specific data on the maximum retention, S, are not available or cannot be reasonably estimated. The parameters of the curve-number model given in Table 4.8 have been estimated based on the data obtained from small (<4 hectares) agricultural watersheds in the midwestern United States, which is appropriate for arid and semi-arid regions. Current empirical evidence suggests that hydrological systems are overdesigned when using CNs from Tables 4.8 and 4.10 (Schneider and McCuen 2005). Whenever S is available for a catchment, the CN should be estimated directly using Equation 4.31, and, if necessary, adjusted using Table 4.10. As a precautionary note, the NRCS does not recommend the use of the curve-number model when the CN is less than 40.

4.9.4 Initial Loss Plus Uniform Loss Rate

This is an often used, and generally accepted, simplified rainfall loss method for flood hydrology. This simplified method assumes that the rainfall loss process can be simulated as a two-step procedure as illustrated in Figure 4.7. First, all rainfall is lost to runoff until the accumulated rainfall is equal to the initial loss; second, after the initial loss is satisfied, a portion of all future rainfall is lost at a uniform rate. Two parameters are needed to apply this method: the initial loss (starting loss or STRTL) and the uniform loss rate (constant loss or CNSTL), according to U.S. Hydrologic Engineering Center model (HEC-1) nomenclature.

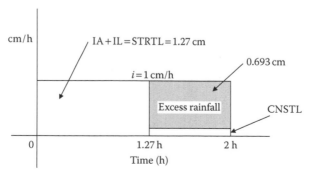

FIGURE 4.7 Initial- and uniform method.

The initial loss (STRTL) is the sum of all losses prior to the onset of runoff and is made up of the surface-retention loss (IA) and the initial amount of infiltration (IL); therefore, STRTL = IA + IL. Values of the infiltration component (IL) of STRTL for bare ground according to soil texture classification are shown in columns 3 through 5 in Table 4.11. These values have been derived from the Green–Ampt infiltration equation, and the relevant parameter values are shown in Table 4.12. The value of IL "dry" should be used for soil that is usually in a state of low moisture content that is at or near the wilting point for vegetation.

Values of IL for bare ground that have been classified according to the hydrologic soil group are shown in Table 4.13. These values within each hydrologic soil group have been derived from the data in Table 4.11 for various texture classifications.

The uniform loss rate (CNSTL) represents the long-term, equilibrium infiltration capacity of the soil. The values of CNSTL shown in column 2 of Table 4.11 for soils grouped according to their texture classification are equivalent to the hydraulic conductivity at natural saturation (KSAT) as determined for the Green–Ampt equation (Table 4.12). The values of CNSTL for soils classified according to the hydrologic soil groups are shown in Table 4.13. These values within each hydrologic soil group have been selected after an inspection of the KSAT values in Table 4.12 for the various soil texture classifications.

TABLE 4.11
Values of Initial Loss Plus Uniform Loss-Rate Parameters for Bare Ground According to Soil Texture Classification

Soil Texture Classification	Uniform Loss Rate (inches/h) CNSTL	Initial Loss (inches) IL		
		Dry	Normal Saturated	Saturated
(1)	(2)	(3)	(4)	(5)
Sand	4.6	1.3	1.3	0
Loamy sand	1.2	0.8	0.8	0
Sandy loam	0.40	0.7	0.6	0
Loam	0.25	0.8	0.7	0
Silty loam	0.15	0.6	0.5	0
Sandy clay loam	0.06	0.6	0.5	0
Clay loam	0.04	0.5	0.4	0
Silty clay loam	0.04	0.6	0.5	0
Sandy clay	0.02	0.4	0.3	0
Silty clay	0.02	0.4	0.3	0
Clay	0.01	0.3	0.2	0

Source: Adapted from Maricopa County Hydrology Manual. 1989. Flood district of Maricopa County, Phoenix, AZ.

TABLE 4.12
Values of the Green–Ampt Loss-Rate Parameters for Bare Ground

Soil Texture Classification	K(SAT) (inches/h)	ψ (inches)	Δθ		
			Dry	Normal	Saturated
(1)	(2)	(3)	(4)	(5)	(6)
Sand	4.6	1.9	0.35	0.30	0
Loamy sand	1.2	2.4	0.35	0.30	0
Sandy loam	0.40	4.3	0.35	0.25	0
Loam	0.25	3.5	0.35	0.25	0
Silt loam	0.15	6.6	0.40	0.25	0
Silt	0.10	7.5	0.35	0.15	0
Sandy clay loam	0.06	8.6	0.25	0.15	0
Clay loam	0.04	8.2	0.25	0.15	0
Silty clay loam	0.04	10.8	0.30	0.15	0
Sandy clay	0.02	9.4	0.20	0.10	0
Silty clay	0.02	11.5	0.20	0.10	0
Clay	0.01	12.4	0.15	0.05	0

Source: Adapted from Maricopa County Hydrology Manual. 1989. Flood district of Maricopa County, Phoenix, AZ.

Dry: Nonirrigated lands, such as deserts and rangelands.

Normal: Irrigated lawn, turf, and permanent pasture.

Saturated: Irrigated agricultural land.

TABLE 4.13
Values of Initial Loss Plus Uniform Loss-Rate Parameters for Bare Ground According to Hydrologic Soil Group

Hydrologic Soil Group	Uniform Loss Rate, CNSTL (inches/h)	Initial Loss, IL (in inches)		
		Dry	Normal	Saturated
(1)	(2)	(3)	(4)	(5)
A	0.40	0.6	0.5	0
B	0.25	0.5	0.3	0
C	0.15	0.5	0.3	0
D	0.05	0.4	0.3	0

EXAMPLE 4.5

Use the same data for the soil given in Example 4.3 for a silty clay soil, $\theta_e = 0.423$, $\psi = 29.22$ centimeters, and $K = 0.05$ centimeters per hour. The initial effective saturation is $s_e = 0.2$. Find the initial loss, IL, and the constant loss, CNSTL. Use the curve-number method to get the same losses. Draw a curve to demonstrate the results, assuming that rainfall is at 1 centimeter per hour for a 2-hour storm. Also show the excess rainfall.

SOLUTION

$\Delta\theta = (1 - s_e)\,\theta_e = (1 - 0.20) \times 0.423 = 0.338$, and $\psi\Delta\theta = 29.22 \times 0.338 = 9.89$ centimeters.

From Table 4.14, for natural land, the average IA = 0.2 inches. From Table 4.12, for silty clay, $K_{sat} = 0.02$ inches per hour, and from Table 4.11:

uniform loss rate $(CNSTL) = 0.02$ inches/h $= 0.0508\,cm/h$

$$IL\,(normal) = 0.3\ inches = 0.762\,cm$$

$$STRTL = IA + IL = 0.2 + 0.3 = 0.5\ inches = 1.27\,cm$$

$$occurring\ in\ 1.27\,h$$

rainfall depth $-STRTL = 2 - 1.27 = 0.73\,cm$

excess rainfall event $= 2 - 1.27 = 0.73\,h$

excess rainfall depth $= 0.73 - K_{sat} \times (0.73\,h) = 0.73 - 0.0508 \times 0.73 = 0.693\,cm$

For demonstration, see Figure 4.7.

Using the curve-number method, the soil was found to belong to group D. When dry, $CN_I = 80$. However, the soil CN should be adjusted against that of the normal group, CN_{II}, which, from Table 4.10, is equal to 90.

TABLE 4.14
Surface-Retention Loss for Various Land Surfaces

Land Use and/or Surface Cover	Surface-Retention Loss, IA (inches)
Natural	
Deserts and rangelands, with flat slopes	0.35
Hillslopes, Sonoran desert	0.15
Mountains, with vegetated surfaces	0.25
Developed (Residential and Commercial)	
Lawn and turf	0.20
Desert landscape	0.10
Pavement	0.05
Agricultural	
Tilled fields and irrigated pastures	0.50

Source: Adapted from Maricopa County Hydrology Manual. 1989. Flood district of Maricopa County, Phoenix, AZ.

Therefore, $\quad s = \dfrac{1000}{90} - 10 = 1.11$ inches

IL ≈ 0.2 $S = 0.222$ inches $= 0.566$ centimeters, a little lower than the loss-rate method.

4.10 MEASUREMENT OF INFILTRATION

Information about the infiltration characteristics of the soil at a given location can be obtained by conducting controlled experiments within small areas. The experimental setup is called an infiltrometer. There are two types of infiltrometers:

1. Flooding-type infiltrometers
2. Rainfall simulators

4.10.1 FLOODING-TYPE INFILTROMETERS

The flooding-type, or ring, infiltrometer consists of a set of two concentric rings (Figure 4.8). The two rings are inserted into the ground and water is maintained on the soil surface to a common fixed level in both the rings. The outer ring provides a water jacket to the infiltrating water of the inner ring and hence prevents the spreading out from the inner ring. Measurements of water volume are taken on the inner ring only. The volume of water added to the inner ring is recorded with respect to time, upon which the infiltration rate can be calculated.

The three main disadvantages of flooding-type infiltrometers are as follows:

1. The raindrop impact effect is not simulated.
2. The embedding of the tube or rings disturbs the soil structure.
3. The results of the infiltrometer depend to some extent on its size, with larger meters giving fewer rates than smaller ones; this is due to the border effect.

4.10.2 RAINFALL SIMULATORS

A small plot of land about 2×4 meters in size is provided with a series of nozzles on the longer side with arrangements to collect and measure the surface runoff rate. The

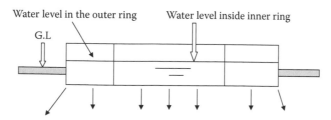

FIGURE 4.8 Double-ring infiltrometer.

specially designed nozzles produce raindrops, which fall from a height of 2 meters and can produce various intensities of rainfall. Experiments are conducted under controlled conditions with various combinations of intensities and durations, and the surface runoff is measured in each case. Using the water-budget equation involving the volume of rainfall, infiltration, and runoff, the infiltration rate and its variation with time is calculated. If the rainfall intensity is higher than the infiltration rate, the infiltration-capacity values are obtained.

Rainfall simulator infiltrometers give lower values than flooding-type infiltrometers due to the effect of the rainfall impact and turbidity of the surface water present in the former type.

4.11 INFILTRATION INDEXES

In hydrological calculations involving floods, it is convenient to use a constant value for the infiltration rate for the duration of the storm. The average infiltration rate is called the infiltration index, and two types of indexes are in common use.

The ϕ index is the average rainfall above which the rainfall volume is equal to the runoff volume. The ϕ index is derived from the rainfall hyetograph with the knowledge of the resulting runoff volume. The initial loss is also considered to be infiltration. The ϕ value is found by treating it as a constant infiltration capacity. If the rainfall intensity is less than ϕ, then the infiltration rate is equal to the rainfall intensity; however, if the rainfall intensity is larger than ϕ, the difference between rainfall and infiltration in an interval of time represents the runoff volume (Figure 4.9). The amount of rainfall in excess of the ϕ index is called the rainfall excess. The ϕ index thus accounts for the total abstraction and enables the estimation of runoff magnitudes for a given rainfall hyetograph.

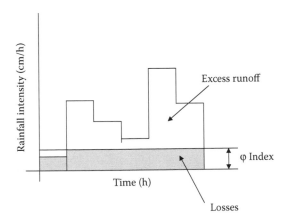

FIGURE 4.9 ϕ index method.

REFERENCES

Chow, V. T. et al. 1988. *Applied Hydrology*. New York: McGraw-Hill.

Horton, R. E. 1939. Analysis of runoff-plat experiments with varying infiltration-capacity. *Transactions of the American Geophysical Union* 20:693–711.

Kibler, D. F. 1982. Desk-top runoff methods. In D.F. Kibler, ed., *Urban Stormwater Hydrology*, 87–135, Washington, DC: American Geophysical Union.

Maricopa County Hydrology Manual. 1989. Flood district of Maricopa County, Phoenix, AZ.

Rawls, W. J. and D. L. Brakensiek. 1982. Estimating soil water retention from soil properties. *Journal of Irrigation and Drainage Engineering* 108(2):166–171.

Rawls, W. J., D. L. Brakensiek, and N. Miller. 1983. *ASCE J Hydraul Eng* 109(1):62–71.

Schneider, L. E. and R. H. McCuen. 2005. Statistical guidelines for curve number generation. *Journal of Irrigation and Drainage Engineering* 131(3):282–290.

Singh, V. P. 1992. *Elementary Hydrology*. Englewood Cliffs, NJ: Prentice Hall.

Tholin, A. L., and G. J. Reefer. 1960. Hydrology of urban runoff. *Trans Am Soc Civ Eng* 125:1308–79.

BIBLIOGRAPHY

Bedlent, P. B., and W. C. Huber. 1988. *Hydrology and Floodolain Analysis*. Reading, MA: Addison-Wesley.

Brakensiek, D. L., and W. J. Rawls. 1983. Hydrologic classification of soils. In *Proceedings of the Natural Resources Modeling Symposium*. Pingree Park, CO: U.S. Department of Agriculture, ARS-30.

Brakensiek, D. L., W. J. Rawls, and G. R. Stephenson. 1984. "Modifying SCS hydrologic soil groups and curve numbers for rangeland soils." American Society of Agricultural Engineers, 1984 Annual Meeting, Kennewick, WA.

Chin, D. A. 2006. *Water Resources Engineering*. 2nd ed. Upper Saddle River, NJ: Pearson Prentice Hall.

Gray, D. M. 1970. *Principles of Hydrology*. Huntington, NY: Water Information Center.

Hicks, W. I. 1944. A method of computing urban runoff. *Trans Am Soc Civ Eng* 109.

Li, R. M., M. A. Stevens, and D. B. Simons. 1976. Solutions to the Green–Ampt infiltration equation. *ASCE J Irrig Drain Div* 102(IR2):239–48.

Linsley, R. K., M. A. A. Kohler, and J. L. H. Paulhus. 1982. *Hydrology for Engineers*. 3rd ed. New York: McGraw-Hill.

Rawls, W. J., and D. L. Brakensiek. 1983. A procedure to predict Green and Ampt infiltration parameters. In *Proceedings of the American Society of Agricultural Engineers, Conference on Advances in Infiltration*, Chicago, IL, 102–12.

Sabol, G. V. 1983. Analysis of the urban hydrology program and data for Academy Acres for the Albuquerque Metropolitan Arroyo Flood Control Authority. Las Cruces, NM: Hydro Science Engineers, Inc., 5–7.

Sabol, G. V., T. J. Ward, and A. D. Seiger. 1982a. Rainfall infiltration of selected soils in the Albuquerque drainage area for the Albuquerque Metropolitan Arroyo Flood Control Authority. Las Cruces, NM: Civil Engineering Department, New Mexico State University, 110.

Sabol, G. V., T. J. Ward, L. Coons, A. D. Seiger, M. K. Wood, and J. Wood. 1982b. Evaluation of rangeland best management practices to control non-point pollution. Las Cruces, NM: Civil Engineering Department, New Mexico State University, 102.

Soil Conservation Service. 1986. Urban hydrology for small watersheds. Technical Release 55. Washington, DC: U.S. Department of Agriculture.

Soil Conservation Service. 1993. *National Engineering Handbook, Section 4: Hydrology.*
    Washington. DC: U.S. Department of Agriculture.
Subramanya, K. 1984. *Engineering Hydrology.* New York: McGraw-Hill.
USACE. 1982. Gila River Basin, Phoenix, Arizona and Vicinity, Hydrology Part 2, Design
    Memorandum No. 2: Los Angeles District.
Viessman, W., Jr. 1967. A linear model for synthesizing hydrographs for small drainage areas.
    Washington, DC: American Geophysical Union.
Weisner, C. J. 1970. *Hydrometeorology.* London: Chapman & Hall.

5 Catchment Characteristics and Runoff

5.1 INTRODUCTION

When a storm occurs, some of the rainfall infiltrates into the ground and some of it evaporates. The rest flows as a thin sheet of water over the land surface, called over-land flow. If there is a relatively impermeable stratum in the subsoil, the infiltrating water moves laterally in the surface soil and joins the stream flow. This is called underflow, subsurface flow, or interflow (Figure 5.1). If there is no impervious layer in the subsoil, the infiltrating water percolates into the ground as deep seepage and builds up the groundwater table (GWT or phreatic surface). The groundwater may also contribute to the stream flow if the GWT is higher than the water surface level of the stream. Surface flow is the overland flow that reaches the stream channel first and contributes to the stream flow. The interflow is slower and reaches the stream channel after a few hours, and the groundwater flow is the slowest and reaches the stream channel some time after surface runoff. The term "direct runoff" is used to include the overland flow and the interflow.

Direct surface flow can be analyzed for relatively large drainage areas using the unit hydrograph method and for smaller areas using overland flow analysis. The direct runoff results from the occurrence of an immediately preceding storm, while the ground water contribution, which takes days or months to reach the stream, has no direct relation with the immediately preceding storm.

When the overland flow starts due to a storm, some flowing water is held in pits and small ponds; this storage of water is called depression storage. The volume of water in transit in the overland flow, which has not yet reached the stream channel, is called surface detention or detention storage. The portion of runoff in a rising flood in a stream, which is absorbed by the permeable banks of the stream above the normal phreatic surface, is called bank storage (Figure 5.2).

5.2 CATCHMENT CHARACTERISTICS

The entire area of stream basin whose surface runoff due to a storm drains into the river is considered a hydrologic unit and is called drainage basin, watershed, or catchment area (Figure 5.3). The boundary line along a topographic ridge, which separates two adjacent drainage basins, is called a drainage divide. The single point

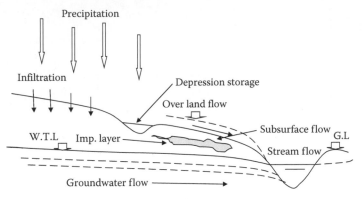

FIGURE 5.1   Disposal of rainwater and runoff.

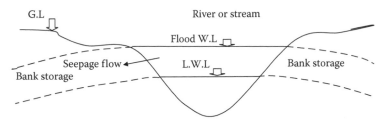

Seepage flow from both sides of the stream stored
as bank storage during the water level fluctuation
between the flood water level and the low water level.

FIGURE 5.2   Bank storage.

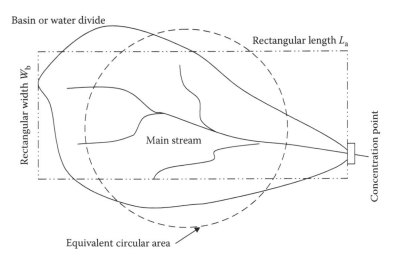

FIGURE 5.3   Drainage basin characteristics.

or location at which all surface drainage from a basin comes together or concentrates as outflow from the basin in the stream channel is called the concentration point or measuring point, since the stream outflow is usually measured at this point.

The time required for the rain falling at the most distant point in a drainage area (i.e., on the fringe of the catchment) to reach the concentration point is called the time of concentration. This is a very significant variable because only storms of a duration greater than the time of concentration will be able to produce runoff from the entire catchment area and cause high-intensity floods.

The characteristics of the drainage basin may be physically described by the following:

- Number of streams
- Stream order
- Length of streams
- Stream density
- Drainage density

5.2.1 STREAM DENSITY

The stream density (D_s) of a drainage basin is expressed as the number of streams per square kilometer as

$$D_s = \frac{N_s}{A} \tag{5.1}$$

where N_s is the number of streams and A is the area of the basin.

5.2.2 DRAINAGE DENSITY

Drainage density (D_d) is expressed as the total length of all stream channels (perennial and intermittent) per unit area of the basin and serves as an index of the areal channel development of the basin:

$$D_d = \frac{L_s}{A} \tag{5.2}$$

where L_s is the total length of all stream channels in the basin.

Drainage density varies inversely with the length of the overland flow and indicates the drainage efficiency of the basin. A high value indicates a well-developed network and torrential runoff causing intense floods, while a low value indicates moderate runoff and high permeability of the terrain.

The boundary line along a topographic ridge, which separates two adjacent basins, is called the drainage divide. The line of the GWT from which the water table slopes downward and away from the line on both sides is called the groundwater divide.

5.2.3 Shape of a Drainage Basin

The shape of a drainage basin can generally be expressed by the following:

  1. From factor F_j
  2. Compactness coefficient (C_c)

$$F_j = \frac{W_b}{L_b} = \frac{A}{L_b^2} \tag{5.3}$$

where $A = W_b \, L_b$ (Figure 5.3), W_b is the axial width of basin, and L_b is the axial length of basin, that is, the distance from the measuring point to the most remote point on the basin.

$$(C_c) = \frac{P_b}{2\sqrt{\pi A}} \tag{5.4}$$

where P_b is the perimeter of the basin and $2\sqrt{\pi A}$ is the circumference of circular area that equals the area of the basin.

If R is the radius of an equivalent circular area, $A = \pi R^2$, $R = \sqrt{\dfrac{A}{\pi}}$

Circumference of the equivalent circular area $= 2\pi \sqrt{\dfrac{A}{\pi}} = 2\sqrt{\pi A}$

A fan-shaped catchment produces a greater flood intensity because all the tributaries are of nearly the same length, hence the time of concentration is nearly the same. In fern-shaped catchments, the time of concentration is more than the fan-shaped catchment and the discharge is distributed over a longer period (Figure 5.4).

Schumm (1956) used an elongation ratio (E_r), defined as the ratio of the diameter of a circle of the same area as the basin to the maximum basin length: the values range from 0.4 to 1.

The drainage basin characteristics influence the time lag of the unit hydrograph and the peak flow (Taylor and Schwarz 1952).

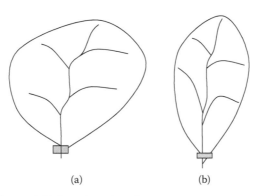

(a)   (b)

FIGURE 5.4   (a) Fan- and (b) fern-shaped catchments.

5.2.4   STREAM ORDER

Horton (1932) suggested a classification of stream order as a measure of the amount of branching within a basin. A first-order stream is a small, unbranched tributary (Figure 5.5). A second-order stream has only first-order tributaries. A third-order stream has only first- and second-order tributaries. The order of a particular drainage basin is determined by the order of the principal stream.

Order is extremely important to the map scale used. A careful study of aerial photographs will often show three or four orders of streams (mostly ephemeral rills and channels) that are not indicated on a standard 1:24,000 scale topographic map. The 1:24,000 scale map shows one or two orders more than does a 1:62,500 scale map. Even standard maps are not consistent in their delineation of streams. Thus, if order is to be used as a comparative parameter, it must be carefully defined. For some uses, it may be desirable to make an adjustment to estimates of order on the basis of a detailed field survey of a few small tributary basins.

5.2.5   CHANNEL SLOPE

The slope of a channel affects the velocity of flow and must play a role in the hydrograph shape. Typical channel profiles (Figure 5.6) are concave upward. In addition, all but the very smallest watersheds contain several channels, each with its own

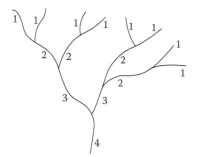

FIGURE 5.5   Definition sketch for stream order.

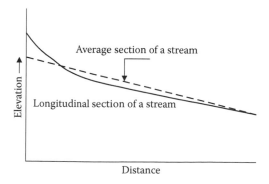

FIGURE 5.6   Longitudinal section of a stream.

profile. Thus, defining an average channel slope of a watershed is difficult. Only the main stream is usually considered in describing the channel slope of a watershed. Taylor and Schwarz (1952) calculated the slope of a uniform channel having the same length and time of flow as the main channel. Since the velocity is proportional to the square root of the slope, the procedure used by Taylor and Schwarz is equivalent to weighting channel segments by the square root of their slope, which gives relatively less weight to the steep upstream reaches of the stream. Thus, if the channel were divided into n equal segments, each of slope s_i, a simple index of slope Sa would be

$$Sa = \left(\sum_{i=1}^{i=n} \frac{\sqrt{s_i}}{n} \right)^2$$

(5.5)

5.2.6 Mean and Median Elevation

The mean elevation is determined as the weighted average of elevations between two adjacent contours. The mean elevation of a drainage basin is given by

$$\bar{Z} = \frac{\sum a_i z_i}{\sum a_i}$$

(5.6)

where $\bar{Z}$ is the mean elevation of the drainage basin, a_1 and a_2 are the areas between the successive contours of the basin, Z_1 and Z_2 are the mean elevations between the two successive contours, and $\sum a_i = A$.

The median elevation is the elevation at 50% area of the catchment and is determined from the area-elevation curve. The area-elevation curve is obtained by plotting the contour elevation against the area or percent of the area, above or below that elevation, as given in Figure 5.7 for one of the subcatchments on the north coast of Egypt. The area-elevation curve is also called the hypsometric curve for the basin.

5.2.7 Hydraulic Characteristics of Streams

Hydraulic geometry describes the characteristics of the channels of a basin, this is, the variation of mean depth, width, and velocity at a particular cross section and between cross sections. These relationships apply to alluvial channels, where the cross section is readily adapted to the flows which occur, but they are less reliable where rock outcrops control the channel characteristics. The basic equations of hydraulic characteristics are (Leopold and Maddock 1953)

$$w = aq^b$$

(5.7)

$$d = cq^f$$

(5.8)

$$v = kq^m$$

(5.9)

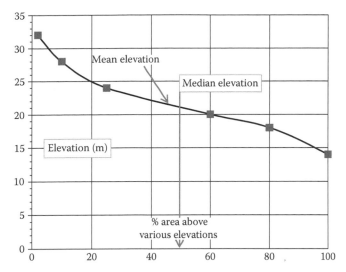

FIGURE 5.7   Area-elevation curve.

TABLE 5.1
Values of the Exponents in Equations 5.7 to 5.9

	At-Station			Between-Station		
	b	f	m	b	f	m
Average, midwestern United States	0.26	0.40	0.34	0.5	0.4	0.1
Ephemeral streams in semi-arid United States	0.29	0.36	0.34	0.5	0.3	0.2
Average, 158 United States stations	0.12	0.45	0.43			
10 stations on the Rhine River (Germany)	0.13	0.41	0.43			
Appalachian streams (USA)				0.55	0.36	0.09
Kaskaskia River, IL (USA)				0.51	0.39	0.14

where q is discharge, w is channel width, d is mean depth, v is mean velocity, and a, b, c, f, k, and m are numerical coefficients. Since $q = wdv$, it follows that $ack = 1$ and $b + f + m = 1$. The equations plot as straight lines on log–log plots, with the exponents representing the slope of the lines and the coefficients the intercept when $q = 1$. Plotting data for a fairly large number of streams indicates marked regional conformity in the values of the exponents, and the regions agree closely enough to suggest the possibility of some degree of universality to the values.

Table 5.1 presents values of exponents b, f, and m as determined by various investigators. At-station values describe the variation of w, d, and v with q at a given cross section, while the between-station values reflect the change in the values of w, d, and v as q increases in the downstream direction along a stream. However, in arid and semi-arid regions, sediment transport should be taken into consideration during the calibration of the stream flow in different geomorphological catchments.

5.2.8 CLASSIFICATION OF STREAMS

Streams may be classifies as follows:

1. Influent and effluent streams
2. Intermittent and perennial streams

5.2.8.1 Influent and Effluent Streams

If the GWT is below the bed of the stream, seepage from the stream feeds the groundwater, resulting in the build up of a water mound (Figure 5.8). Such streams are called influent streams. Irrigation channels function as influent streams, and many rivers that cross desert areas irrigate these areas and are considered influent streams. Such streams will dry up completely during rainless periods and are called ephemeral streams. These streams are generally seen in arid regions and flow only for a few hours after the rainfall. They are of no use for conventional hydropower. However, they can occasionally be used in pure pumped storage schemes, where the actual consumption of water is only marginal.

When the GWT is above the water-surface elevation in the stream, groundwater feeds the stream (Figure 5.9). Such streams are called effluent streams. The base flow of surface streams is the effluent seepage from the drainage basin. Most perennial streams are mainly effluent streams.

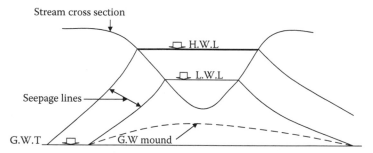

FIGURE 5.8   Influent streams.

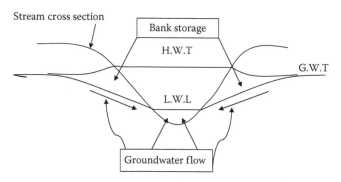

FIGURE 5.9   Effluent streams.

5.2.8.2 Intermittent and Perennial Streams

If the GWT lies above the bed of the stream during the wet season but drops below the bed during the dry season, this means that the stream flows during the wet seasons due to surface runoff and groundwater contributions but becomes dry during the dry seasons. Such streams are called intermittent streams. In the case of perennial streams, the GWT never drops below the bed of the stream even in the most severe droughts, and therefore, these streams flow throughout the year. For power development, a perennial stream is the best; power can also be generated from intermittent streams by providing adequate storage facilities.

5.2.9 TIME OF CONCENTRATION

The time of concentration is often used to characterize the response of a catchment to a rainfall event. Time of concentration is the time to equilibrium of a catchment under steady rainfall excess; it can also be longest travel time that it takes surface runoff to reach the discharge point of a catchment. Most equations for estimating the time of concentration, t_c, express t_c as function of the rainfall intensity, i, catchment length scale, L, average catchment slope, S_o, and a parameter that describes the catchment surface, C. Hence the equations for t_c typically have the following functional form:

$$t_c = f(i, L, S_o, C) \tag{5.10}$$

The time of concentration of a catchment includes the time of overland flow and the travel time in drainage channels leading to the catchment outlet. The most popular equations are described in Sections 5.2.9.1 through 5.2.9.5.

5.2.9.1 Kinematic-Wave Equation

A fundamental expression for the time of concentration in overland flow can be derived by considering the one-dimensional approximation of the surface-runoff process and the law of conservation of mass and by defining the time of concentration (t_c) of a catchment as the time required for a kinematic wave to travel distance L from the catchment boundary to the catchment outlet. The following equation can then easily be developed:

$$t_c = \left(\frac{L}{\alpha \times i_e^{m-1}} \right)^{1/m} \tag{5.11}$$

If the Manning equation is used to relate the runoff rate to other variables, then Equation 5.11 can be written in the form (ASCE 1992):

$$t_c = 6.99 \left(\frac{(nL)^{0.6}}{i_e^{0.4} S_0^{0.3}} \right) \tag{5.12}$$

where t_c is in minutes, i_e in millimeters per hour, L in m, n is the Manning roughness coefficient for overland flow, and S_o is the ground slope. Estimates of the Manning roughness coefficient for overland flow are given in Table 5.2, where the surface types are ordered with increasing roughness. Based on Equation 5.12, the time of concentration for overland flow should be regarded as a function of the rainfall-excess rate (i_e), the catchment-surface roughness (n), the flow length from the catchment boundary to the outlet (L), and the slope of the flow path (S_o).

Equation 5.12 assumes that the surface runoff is described by the Manning equation, which is valid only for turbulent flows (Re > 2000); however, at least a portion of the surface runoff will be in the laminar (Re < 200) and transition regimes (Wong and Chen 1997). This limitation of the Manning equation can be addressed using the Darcy–Weisbach equation, which yields

$$t_c = \left(\frac{0.21\left(3.6\times10^6\,v\right)^k CL^{2-k}}{i_e^{1+k} S_o} \right) \tag{5.13}$$

where t_c is in minutes, v is in square meters per second, L is in meters, and i_e is in millimeters per hour. The values of C were adopted by Wenzel (1970) and others, who

TABLE 5.2
Manning's *n* for Overland Flow

Surface Type	Manning *n*	Range
Smooth concrete	0.011	0.01–0.014
Bare sand	0.01	0.01–0.016
Graveled surface	0.012	0.010–0.018
Asphalt	0.012	0.010–0.018
Bare clay	0.012	0.010–0.016
Smooth earth	0.018	0.015–0.021
Bare clay-loam (eroded)	0.02	0.012–0.033
Bare smooth soil	0.10	—
Range (natural)	0.13	0.01–0.32
Sparse vegetation	0.15	—
Short grass	0.15	0.10–0.25
Light turf	0.20	—
Woods, no underbrush	0.20	0.1–0.3
Dense grass	0.24	0.15–0.35
Lawns	0.25	0.20–0.30ᵣ
Dense turf	0.35	0.30–0.35
Pasture	0.35	0.30–0.40
Dense shrubbery and forest litter	0.40	—
Woods, light underbrush	0.40	0.3–0.5
Bermuda grass	0.41	0.30–0.50
Bluegrass sod	0.45	0.39–0.63
Woods, dense underbrush	0.80	0.6–0.95

Source: Adapted from American Society of Civil Engineers. 1992. *Design and Construction of Urban Stormwater Management Systems.* New York: American Society of Civil Engineers.

indicated that for concrete surfaces, C values of 41.8, 2, and 0.04 are appropriate for laminar, transition, and turbulent flow regimes, respectively. Equation 5.13, called the Chen and Wong formula (Wong 2005), can be used to account for various flow regimes in overland flow and, assuming a single flow regime (laminar, transition, or turbulent), will tend to underestimate the time of concentration (Wong and Chen 1997). Overland flow occurs predominantly in transition regimes and Equation 5.13 may be most applicable using $k \sim 0.5$.

5.2.9.2 NRCS Method

National Resources Conservation Services (NRCS) (SCS 1986) proposed that overland flow consists of sequential flow regimes: sheet flow and shallow concentrated flow. Sheet flow is characterized by runoff that occurs as a continuous sheet of water flowing over land surface, while shallow concentrated flow is characterized by flow in isolated rills and then gullies of increasing proportions. Ultimately, most surface runoff enters open channels and pipes, the third regime of surface runoff included in the time of concentration. In many cases, the time of concentration of a catchment is expressed as the sum of the travel time, that is, as sheet flow plus the travel time as shallow concentrate flow plus the travel time as open-channel flow. The flow characteristics of sheet flow are different enough from the shallow concentrated flow that separate equations for each are recommended. The flow length of the sheet flow regime should generally be less than 100 meters, and the travel time t_f (in hours), over a flow length L (in meters) is estimated by (SCS 1986)

$$t_f = 0.0288 \left(\frac{(nL)^{0.8}}{P_2^{0.5} S_0^{0.4}} \right) \tag{5.14}$$

where n is the Manning roughness coefficient for overland flow (Table 5.2), S_o is the land slope, and P_2 is the 2-year 24-hour rainfall (in centimeters). Equation 5.14 was developed from the kinematic-wave equation (Equation 5.11) by Overton and Meadows (1976) using the following assumptions: (1) the flow is steady and uniform with a depth of about 3 centimeters; (2) the rainfall intensity is uniform over the catchment; (3) the rainfall duration is 24 hours; (4) infiltration is neglected; and (5) the maximum flow length is 100 meters. In considering the validity of these assumptions, note that overland flow depth may be significantly different from 3 centimeters in many areas, the rainfall duration may differ from 24 hours, and the actual travel time can increase if there is a significant amount of infiltration in the catchment. By limiting the maximum flow length to 100 meters, the catchment is necessarily small, and the assumption of a spatially uniform rainfall distribution is reasonable. After the maximum distance of 100 meters, sheet flow usually becomes shallow concentrated flow, and the average velocity, V_{sc}, is taken to be a function of the slope of the flow path and the type of land surface, in accordance with the Manning equation as follows:

$$V_{sc} = \frac{1}{n} R^{2/3} S_o^{1/2} \tag{5.15}$$

where n is the roughness coefficient, R is the hydraulic radius, and S_o is the slope of the flow path. Commonly for unpaved areas, $n = 0.05$ and $R = 12$ centimeters; and for paved areas, $n = 0.025$ and $R = 6$ centimeters. Equation 5.15 can also be expressed in the form:

$$V_{sc} = kS_o^{1/2} \qquad (5.16)$$

where $k \, (= R^{2/3}/n)$ is called the intercept coefficient. Several suggested values of k are given in Table 5.3. In addition to intercept coefficients for shallow concentrated flow, Table 5.3 also gives bulk intercept coefficients for the entire overland flow, including sheet flow and shallow concentrated flow regimes. The average velocity (V_{sc}) can be derived from Equation 5.16 and is then combined with the flow length (L_{5C}) of shallow concentrated flow to yield the flow time (t_{sc}) as

$$t_{sc} = \frac{L_{5C}}{V_{sc}} \qquad (5.17)$$

The total time of concentration (t_c) of overland flow is taken as the sum of the sheet flow time (t_f) given by Equation 5.14 and the shallow concentrated flow time (t_{sc}) given by Equation 5.17. The overland flow time of concentration is added to the channel flow time to obtain the time of concentration of the entire catchment.

5.2.9.3 Kirpich Equation

An empirical time of concentration formula that is especially popular is the Kirpich formula, (Kirpich 1940) given by

$$t_c = 0.19 \frac{L^{0.77}}{S_o^{0.385}} \qquad (5.18)$$

TABLE 5.3
Intercept Coefficient for Overland-Flow Velocity versus Slope

Land Cover and Flow Regime	K(m/s)
Forest with heavy ground litter; hay meadow (overland flow)	0.76
Trash fallow or minimum tillage cultivation; contour or strip cropped; woodland (overland flow)	1.52
Short grass pasture (overland flow)	2.13
Cultivated straight row (overland flow)	2.74
Nearly bare and untilled (overland flow); alluvial fans in western mountain regions	3.05
Grassed waterway (shallow concentrated flow)	4.57
Unpaved (shallow concentrated flow)	4.91
Paved area (shallow concentrated flow); small upland gullies	6.19

Source: Adapted from McCuen, R. H., P. A. Johnson, and R. M. Ragan. 1996. Hydrology. Technical Report HDS-2, U.S. Federal Highway Administration, Washington, DC.

---

where t_c is the time of concentration in minutes, L is the flow length in meters, and S_o is the average slope along the flow path. Equation 5.18 was originally developed and calibrated from NRCS data reported by Ramser (1927) on seven partially wooded agricultural catchments in Tennessee, ranging in size from 0.4 to 45 hectares (ha), with slopes varying from 3 to 10%; it has found widespread use in urban applications to estimate both overland flow and channel flow times. Equation 5.18 is most applicable for natural basins with well-defined channels, bare-earth overland flow, and in-mowed channels, including roadside ditches.

Rossmiller (1980) reviewed field applications of the Kirpich equation and suggested that overland flow on concrete or asphalt surfaces should be multiplied by 0.4; for concrete channels, multiply t_c by 0.2; and for general overland flow and flow in natural grass channels, multiply t_c by 2. According to Prakash (1987), the Kirpich equation yields relatively low estimates of the time of concentration. The Kirpich formula is usually considered applicable to small agricultural watersheds with drainage areas less than 80 hectares.

5.2.9.4   Izzard Equation

The Izzard equation (Izzard 1944, 1946) was derived from laboratory experiments on pavements and turf where overland flow was dominant. The Izzard equation is given by

$$t_c = \frac{530 K L^{1/3}}{i_e^{2/3}} \quad \text{where} \quad i_e < 3.9 \text{ m}^2/\text{h} \tag{5.19}$$

where t_c is the time of concentration in minutes, L is the overland flow distance in meters, i_e is the effective rainfall intensity in millimeters per hour, and K is a constant given by

$$K = \frac{2.8 \times 10^{-6} i_e + c_r}{S_o^{1/3}} \tag{5.20}$$

where c_r is a retardance coefficient that is determined by the catchment surface as given in Table 5.4 and S_o is the catchment slope.

5.2.9.5   Kerby Equation

The Kerby equation (Kerby 1959) is given by

$$t_c = 1.44 \left(\frac{Lr}{\sqrt{S_o}} \right)^{0.467} \tag{5.21}$$

where t_c is the time of concentration in minutes, L is the length of flow in meters, r is a retardance roughness coefficient given in Table 5.5, and S_o is the slope of the catchment. The Kerby equation is an empirical relation developed by Kerby (1959) using published research on airport drainage obtained by Hathaway (1945), consequently, it is sometimes referred to as the Kerby–Hathaway equation. Catchments with areas less than 4 hectares, slopes less than 1%, and retardance coefficients less than 0.8

TABLE 5.4
Values of c_r in the Izzard Equation

Surface	c_r
Very smooth asphalt	0.0070
Tar and sand pavement	0.0075
Crushed-slate roof	0.0082
Concrete	0.012
Closely clipped sod	0.016
Tar and gravel pavement	0.017
Dense bluegrass	0.060

Source: Adapted from Izzard, C. F. 1944. *Trans Am Geophys Union* 25:959–69; and Izzard, C. F. 1946. *Proc Highway Res Board* 26:129–46.

TABLE 5.5
Values of r in the Kerby Equation

Surface	r
Smooth pavements	0.02
Asphalt/concrete	0.05–0.15
Smooth bare packed soil, free of stones	0.10
Light turf	0.20
Poor grass on moderately rough ground	0.20
Average grass	0.40
Dense turf	0.17–080
Dense grass	0.17–0.30
Bermuda grass	0.30–0.48
Deciduous timberland	0.60
Conifer timberland, dense grass	0.60

Sources: Adapted from Kerby, W. S. 1959. *Civ Eng* 29(3):174.

were used to calibrate the Kerby equation, so applications of this equation should also be limited to this range. In addition, since the Kerby equation applies to the overland or sheet flow regime, the length of flow L should be less than 100 meters. Example 5.1 compares the time of concentration equations.

EXAMPLE 5.1

An urban catchment with a concrete surface has an average slope of 0.5% and the distance from the catchment boundary to the outlet is 90 meters. For a 20-minute storm with an effective rainfall rate of 75 millimeters per hour, estimate the times of concentration (t_c) using (a) the kinematic-wave equation, (b) the NRCS method, (c) the Kirpich equation, (d) the Izzard equation, and (e) the Kerby equation.

SOLUTION

If the overland flow is assumed to be fully turbulent, the Manning form of the kinematic-wave equation can be used. From the given data $L = 90$ meters, $i_e = 75$ millimeters, $S_o = 0.005$, and for n concrete surface, Table 5.2 gives $n = 0.012$. The other equations will use the same data. The following table summarizes the t_c values:

Equation	t_c (minute)
Kinematic wave	6
NRCS	10
Kirpich	2
Izzard	6
Kerby	7

Noting that the NRCS method overestimates t_c and that the Kirpich equation would give $t_c = 5$ minutes if the Rosmiller factor of 0.4 were not applied, the overland flow time of concentration of the catchment is around 6 minutes, which coincides with both the Chen and Wong equation and the kinematic-wave equation.

As an accuracy of estimates, Wong (2005) compared nine equations for estimating t_c in experimental plots having concrete and grass surfaces and concluded that t_c equations that do not account for rainfall intensity are valid for only a limited range of rainfall intensities. The equations used in the Wong (2005) study included, among others, the Izzard equation (Equation 5.19), the Kerby equation (Equation 5.21), and the Chen and Wong equation (Equation 5.13). The results of this study showed that the Chen and Wong equation obtained the best agreement with experimental data for both concrete and grass surfaces.

McCuen et al. (1984) compared 11 equations for estimating t_c in 48 urban catchments in the United States. The catchments used in the study all had areas less than 1600 hectares (4000 acres), average impervious areas of approximately 29%, and times of concentration from 0.21 to 6.14 hours with an average time of concentration of 1.5 hours. The equations used in these studies included, among others, the kinematic-wave equation (Equation 5.12), the Kirpich equation (Equation 5.18), and the Kerby equation (Equation 5.21). The results of this study indicated that the error in the estimated value of t_c exceeded 0.5 hours for more than 50% of the catchments, with the standard deviation of the errors ranging from 0.37 to 2.27 hours. These results indicate that relatively large errors can be expected when estimating t_c from commonly used equations. Singh (1989) has noted that this can lead to significant errors in design discharges.

It is good practice to use at least three different methods to estimate the time of concentration and, within the range of these estimates, select the final value by judgment (Prakash 2004). Estimates of travel time in drainage channels are usually much more accurate than estimates of overland flow; therefore, in catchments where flow in drainage channels constitutes a significant portion of the travel time, the time of concentration can be estimated more accurately.

5.2.10 Isochrones

The lines joining all points in a basin of some key time elements in a storm, such as beginning of precipitation, are called isochrones (Figure 5.10). These are the time contours, represent lines of equal travel time (time of concentration), and are helpful in deriving hydrographs.

5.2.11 Factors Affecting Runoff

Low-intensity storms over longer spells contribute to groundwater storage and produce relatively less runoff. A high-intensity storm over a smaller area increases the runoff because losses such as infiltration and evaporation are less. If there is a succession of storms, runoff will increase due to the initial wetness of the soil resulting from antecedent rainfall. Rain during the summer season will produce less runoff; while rain during winter will produce more runoff. Greater humidity decreases evaporation. The pressure distribution in the atmosphere helps the movement of storms. Snow storage and the frozen ground greatly increase the runoff.

Peak runoff (if expressed as cumecs per square kilometer) decreases as the catchment area increases due to a higher time of concentration. A fan-shaped catchment produces a greater flood intensity than a fern-shaped catchment. Steep rocky catchments with less vegetation will produce more runoff compared to flat tracts with more vegetation. If the vegetation is thick, the absorption of water is greater, so there will be less runoff. If the direction of the storm producing rain is downstream from receiving the surface flow, it will produce greater flood discharge than when it is upstream. If the catchment is located on the orographic (windward) side of the mountains, it will receive greater precipitation and hence have greater runoff.

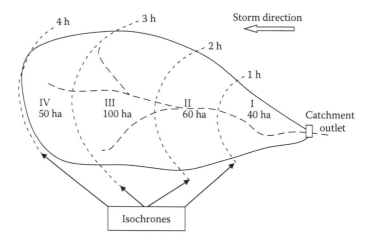

FIGURE 5.10 Isochrones.

If it is on the leeward side, it will receive less precipitation and so have less runoff. Similarly, catchments located at higher altitude will receive more precipitation and yield greater runoff. The land use pattern—arable grass land, forest, or cultivated area—greatly affects runoff.

Storage in channels and depressions (valley storage) will reduce the flood magnitude. Upstream reservoirs, lakes, and ponds moderate the flood magnitudes because of their storage effects. For drainage basins having pervious deposits, a lot of groundwater storage may be created, which may also contribute to the stream flow in the form of delayed runoff.

The various factors that affect the runoff from a drainage basin are as follows:

1. Storm characteristics such as the season, intensity, duration, areal extent (distribution), frequency, and direction of storm movement
2. Meteorological characteristics such as temperature, humidity, wind velocity, and pressure variation
3. Basin characteristics such as size, shape, slope, altitude (elevation), topography, geology (type of soil), land use and vegetation, and orientation
4. Storage characteristics such as depressions, pools and ponds, lakes, stream channels, check dams (in gullies), reservoirs, flood plains, swamps, and groundwater storage in pervious deposits

5.3 ESTIMATION OF RUNOFF

Runoff is the balance of rainwater that flows or runs over the natural ground surface after losses by evaporation, interception, and infiltration.

The yield of a catchment in arid and semi-arid regions is the net quantity of water available for storage, after all losses, for the purposes of water resources utilization and planning.

The runoff from rainfall can be estimated by the following methods:

- Empirical formulas, curves, and tables
- Infiltration method
- Rational method
- Overland flow hydrograph
- Unit hydrograph method

5.3.1 EMPIRICAL FORMULAS, CURVES, AND TABLES

There are several empirical formulas, curves, and tables relating the rainfall and runoff:

$$\text{Usually, } R = aP - b \tag{5.22}$$

$$\text{Sometimes, } R = aP^a \tag{5.23}$$

where R is the runoff, P is the rainfall, and a, b, and n, are constants. Equation 5.22 plots as a straight line plot on natural graph paper. Equation 5.23 plots as a straight line plot on log–log paper, where the constants can be obtained from the straight line plots.

Several empirical formulas have also been developed to estimate the maximum rate of runoff or maximum flood discharge (MFQ) for arid and semi-arid regions. These are given below, based on data from the north coast wadis in Egypt.

$Q = 0.75 \ (P - 8)$ $(p > 8 \, \text{mm})$ (Ball 1937)

$Q = 0.06 \ (P - 1.12) - 0.013t$ (Economides 1968)

$Q = (P - 7.29)^2/(P + 139.64)$ (Sewidan 1978)

$Q = 0.3 \ (P - 10)$ $(p > 10 \, \text{mm})$ (Wakil and Shaker 1989)

$Q = (P - 4.01)^2/(P + 63.22)$ (2 – 5 hours duration) (Zaki 2000)

$Q = (P - 5.40)^2/(P + 39.55)$ (6 – 8 hours duration) (Zaki 2000)

where Q is predicted surface runoff (millimeters), P is storm rainfall (millimeters), and t is the duration of rainfall (in hours).

5.3.2 INFILTRATION METHOD

Runoff from rainfall can be estimated by deducting the infiltration loss, that is, the area under the infiltration curve, from the total precipitation, or by the use of infiltration indices. These methods are largely empirical and the derived values apply only when the rainfall characteristics and the initial soil moisture conditions are identical to those that are derived. The infiltration method may be applied for small watersheds with negligible depression storage, that is, the land surface is almost planed.

5.3.3 RATIONAL METHOD

A rational approach is to obtain the yield of a catchment by assuming a suitable runoff coefficient (Table 5.7):

$$\text{Yield} = CAP \tag{5.24}$$

where A is the area of catchment, P is the precipitation, and C is the runoff coefficient.

The value of the runoff coefficient C varies depending upon the soil type, vegetation, geology, and so on, and Table 5.6 may be taken as a guide.

In the rational method, the drainage area is divided into a number of subareas and the runoff contribution from each area is determined with the known time of concentration for different subareas. The choice of the value of the runoff coefficient C for the different subareas is an important factor in computing the runoff by this method. This method of dividing the area into different zones by drawing lines of time of concentrations contours, that is, isochrones, is illustrated in Example 5.2.

TABLE 5.6
Runoff Coefficients for Various Types of Catchments

Type of Catchment	Value of C
Rocky and impermeable	0.8–1.0
Slightly permeable, bare	0.6–0.8
Cultivated or covered with vegetation	0.4–0.6
Cultivated absorbent soil	0.3–0.4
Sandy soil	0.2–0.3
Heavy forest	0.1–0.2

TABLE 5.7
Discharge Calculation Using Rational Method

Time (hour)	C from Zone I	C from Zone II	C from Zone III	C from Zone IV	Q from Zone I	Q from Zone II	Q from Zone III	Q from Zone IV	QT	T (hour)
1	0.5				2000				2000	1
2	0.6	0.5			2400	3000			5400	2
3	0.7	0.6	0.4		2800	3600	4000		10,400	3
4	0.8	0.7	0.5	0.4	3200	4200	5000	2000	14,400	4
5		0.8	0.6	0.5		4800	6000	2500	13,300	5
6			0.7	0.6			7000	3000	10,000	6
7				0.7				3500	3500	7

EXAMPLE 5.2

A 4-hour rainy storm with an average intensity of 1 centimeter per hour falls over the fern leaf–type catchment as shown in Figure 5.10. The time of concentration from the lines (isochrones) 1, 2, 3, and 4 hours, respectively, to the outlet, are where the discharge measurements are made.

The values of the runoff coefficient C for both zones I and II are 0.5, 0.6, and 0.7 for the first, second, and third hours of rainfall, respectively, and attain a constant value of 0.8 after 3 hours. The values of C for zones III and IV for the same time are 0.4, 0.5, 0.6, and 0.7, respectively. Determine the discharge at the outlet. Plot the discharge versus the time.

SOLUTION

Q per hour for any zone of the catchment = CiA
    For zones I and II

$$Q \text{ for the 1st hour} = 0.5 \times (1/100) \times A \times 10^4 = 50 \text{ Am}^3$$
$$Q \text{ for the 2nd hour} = 0.6 \times (1/100) \times A \times 10^4 = 60 \text{ Am}^3$$
$$Q \text{ for the 3rd hour} = 0.7 \times (1/100) \times A \times 10^4 = 70 \text{ Am}^3$$
$$Q \text{ for the 4th hour} = \qquad\qquad\qquad = 80 \text{ Am}^3$$

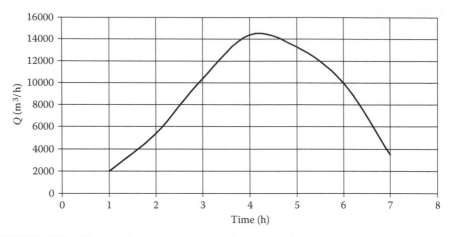

FIGURE 5.11   Flood hydrograph using the rational method.

For zones III and IV

$$Q \text{ for the 1st hour} = 0.4 \times (1/100) \times A \times 10^4 = 40 \text{ Am}^3$$
$$Q \text{ for the 2nd hour} = 0.5 \times (1/100) \times A \times 10^4 = 50 \text{ Am}^3$$
$$Q \text{ for the 3rd hour} = 0.6 \times (1/100) \times A \times 10^4 = 60 \text{ Am}^3$$
$$Q \text{ for the 4th hour} = \qquad\qquad\qquad\quad = 70 \text{ Am}^3$$

The discharge can be easily tabulated for each zone and then added to get the hourly discharges as given in Table 5.6 and Figure 5.11.

The overland flow method and unit hydrograph method are discussed in Chapter 6.

REFERENCES

American Society of Civil Engineers. 1992. *Design and Construction of Urban Stormwater Management Systems*. New York: American Society of Civil Engineers.

Ball, J. 1937. *The Water Supply of Mersa Matruh*. Cairo: Survey and Mines Department.

Economides, P. 1968. Final report, SF: Project Ec/nf. Unpublished internal report of the Pre-Investment Survey of the North Western Coastal Region of Egypt. Alexandria, Egypt.

Hathaway, G. A. 1945. Military airfields: A symposium, design of drainage facilities. *Transactions ASCE* 110: 697–733.

Horton, R. E. 1932. Drainage basin characteristics. *Trans Am Geophys Union* 13:350–61.

Izzard, C. F. 1944. *Trans Am Geophys Union* 25:959–69.

Izzard, C. F. 1946. *Proc Highway Res Board* 26:129–46.

Kerby, W. S. 1959. *Civ Eng* 29(3):174.

Kirpich, P. Z. 1940. Time of concentration of small agricultural watersheds. *Civ Eng* 10(6):362.

Leopold, L. B., and T. G. Maddock. 1953. The hydraulic geometry of stream channels and some physiographic implications. *US Geol Surv Prof Pap* 252.

McCuen, R. H., P. A. Johnson, and R. M. Ragan. 1996. Hydrology. Technical Report HDS-2, U.S. Federal Highway Administration, Washington, DC.

McCuen, R. H., S. L. Wong, and W. J. Rawls. 1984. Estimating urban time of concentration. *J Hydraul Eng* 110(7):887–904.

Overton, D. E. and M. E. Meadows. 1976. *Storm Water Modeling*. New York: Academic Press.

Prakash, A. 2004. *Water Resources Engineering*. New York: ASCE Press.

Ramser, C. E. 1927. Runoff from small agricultural areas. *J Agric Res* 34(9).

Rossmiller, R. L. 1980. The rational formula revisited. In *Proceedings of the International Symposium on Urban Storm Runoff*, July 28–31. Lexington, KY: University of Kentucky.

Schumm, S. A. 1956. Evolution of drainage systems and slopes in badlands at Perth Amboy, New Jersey. *Bull Geol Soc Am* 67:597–646.

Sewidan, A. S. 1978. Water budget analysis for northwestern coastal zone of the Arab Republic of Egypt. PhD thesis, Faculty of Sciences, Cairo University.

Singh, V. P. 1989. *Hydrologic Systems, Rain Fall-Runoff Modeling*, vol. I. Englewood Cliffs, NJ: Prentice Hall.

Soil Conservation Service. 1986. Urban hydrology for small watersheds. Technical Release 55. Washington, DC: Soil Conservation Service, US Department of Agriculture.

Taylor, A. B., and H. E. Schwarz. 1952. Unit-hydrograph lag and peak flow related to drainage basin characteristics. *Trans Am Geophys Union* 33:235–46.

Wakil, M., and M. R. Shaker. 1989. Rainfall-runoff relationship of the Wadi Mehleb experimental basin preliminary study, University of Alexandria, Egypt.

Wenzel, H. G. 1970. The effect of raindrop impact and surface roughness on sheet flow. Water Resources Center Research Report 34, Water Resources Center, University of Illinois, Urbana, IL.

Wong, T. S. W. 2005. Assessment of time of concentration formulas for overland flow. *ASCE J Irrig Drain Eng* 131(4):383–7.

Wong, T. S. W., and C.-N. Chen. 1997. Time of concentration formulas for sheet flow of varying flow regimes. *J Hydrol Eng* 2(3):136–9.

Zaki, M. H. 2000. Assessment of surface water runoff in Mersa Matruh area. Master's thesis, Faculty of Sciences, Alexandria University.

BIBLIOGRAPHY

American Society of Civil Engineers. 1996. *Manual of Practice No. 28: Hydrology Handbook*, 2nd ed. New York: American Society of Civil Engineers.

Bush, L. M., Jr. 1961. Drainage basins, channels and flow characteristics of selected streams in central Pennsylvania. *US Geol Surv Prof Pap* 282-F:145–81.

Chen, C.-N., and T. S. W. Wong. 1993. Critical rainfall duration for maximum discharge from overland plane. *ASCE J Hydraul Eng* 119(9):1040–1045.

Chorley, R. J., D. E. G. Malm, and H. A. Pogorzelski. 1957. A new standard for estimating drainage basin shape. *Am J Sci* 255:138–41.

Hack, J. T. 1957. Studies of longitudinal stream profiles in Virginia and Maryland. *US Geol Surv Prof Pap* 294-B:45–97.

Horton, R. E. 1945. Erosional development of streams. *Geol Soc Am Bull* 56:281–3.

Leopold, L. B. 1953. Downstream change of velocity of rivers. *Am J Sci* 251:606–24.

Leopold, L. B., M. G. Wolman, and J. P. Miller. 1964. *Fluvial Processes in Hydrology*. San Francisco, CA: Freeman.

Stall, J. B., and Y. S. Fok. 1968. Hydraulic geometry of Illinois streams. Water Resources Center Research Report 15, University of Illinois, Urbana, IL.

Wong, T. S. W. 2006. Effect of loss model on evaluation of manning roughness coefficient of experimental concrete catchment. *J Hydrol* 331(1–2):205–18.

Yang, C. T. 1971. Potential energy and stream morphology. *Water Resour Res* 7:311–22.

6 Stream Flow Measurement

6.1 INTRODUCTION

A stream is a flow channel into which the surface runoff from a specified basin drains. Stream flow is historical data and is measured in units of discharge (cubic meters per second) occurring at a specified time. The measurement of discharge in a stream forms an important branch of hydrometry, the science and practice of water measurement. This chapter deals with only the salient stream flow measurement techniques in arid and semi-arid regions.

Continuous measurement of stream discharge is very difficult. As a rule, direct measurement of discharge is a time-consuming and expensive process. Hence, a two-step procedure is followed. First, the discharge in a given stream is related to the elevation of the water surface (stage) through a series of careful measurements. Next, the stage of the stream is observed routinely in a relatively inexpensive manner, and the discharge is estimated by using the previously determined stage–discharge relationship. The observation of the stage is easy and inexpensive, and if desired, continuous readings can also be obtained. This method is adopted universally to determine the stream discharge.

6.2 MEASUREMENT OF STAGES

The stage of a stream is defined as its water-surface elevation measured above a datum. This datum can be the mean sea level (MSL) or any arbitrary datum connected independently to the MSL.

6.2.1 STAFF GAUGES

The simplest of stage measurements are made by noting the elevation of the water surface in contact with a fixed graduated staff. The staff gauge is made of a durable material with a low expansion coefficient with respect to both temperature and moisture. It is fixed rigidly to a structure, such as an abutment, pier, or wall. The staff may be vertical or inclined with clearly and accurately graduated permanent markings. The markings are distinctive, easy to read from a distance, and similar to those on a surveying staff. Figure 6.1 shows a marble staff gauge graduated in centimeters and fixed to a wing wall upstream of a regulator, a common irrigation practice in Egypt to record water levels both upstream and downstream of hydraulic structures. When

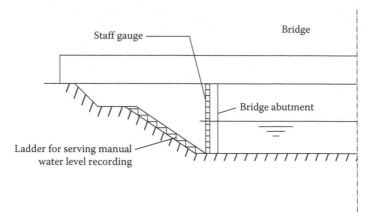

FIGURE 6.1   Staff gauge fixed to a hydraulic structure.

installing sectional gauges, care must be taken to provide an overlap between various gauges and to refer all the sections to the same common datum.

6.2.2 AUTOMATIC STAGE RECORDERS

The staff gauges discussed in Section 6.2.1 are manual gauges. Although they are simple and inexpensive, they have to be read frequently to define the variation of stage with time accurately. Automatic stage recorders overcome this disadvantage and have found considerable use in stream flow measurement practice.

The float-operated stage recorder is the most common type of automatic stage recorder in use. In this stage recorder, a float operating in a stilling well is balanced by a counterweight over the pulley of the recorder. Float displacement due to the rising or lowering of the water-surface elevation causes an angular displacement of the pulley and hence the input shaft of the recorder. Mechanical linkages convert this angular displacement to the linear displacement of a pen to record over a drum driven by clockwork. The pen traverse is continuous with automatic reversing when it reaches the full width of the chart. A clockwork mechanism runs the recorder for a day, week, or fortnight and provides a continuous plot of stage time. A good instrument will have a large float and less friction. An improved model for this basic analog model sends digital signals transmitted directly onto a central data-processing center and is called a telemetry system. They are common in Egypt to record water levels in streams of catchments, canals, and the remote hydraulic structures of the Nile River.

To protect the float from debris and to reduce the water-surface wave effects on the recording, stilling wells are used in all float-type stage recorder installations. Figure 6.2 shows a typical stilling well installation. Note the intake pipes that communicate with the stream and flushing arrangement to flush these intake pipes of the sediment and debris occasionally. The water-stage recorder must be located above the highest expected water level in the stream to prevent it from being inundated during floods. Further, the instrument must be properly housed in a suitable enclosure to protect it from weather elements and vandalism. Installation of the water-stage recorder is usually costly.

6.2.3 STAGE DATA

Stage data is often presented in the form of a plot of stage against chronological time known as a stage hydrograph (Figure 6.3). In addition to its use in the determination of stream discharge, stage data is important for flood warnings and flood-protection works. Reliable long-term stage data corresponding to peak floods can be analyzed statistically to estimate the design peak river stages for use in the design of hydraulic structures such as bridges and weirs. Historic flood stages are invaluable for the indirect estimation of corresponding flood discharges. In view of these multipurpose uses, the stream stage is an important hydrologic parameter for regular observation and recording.

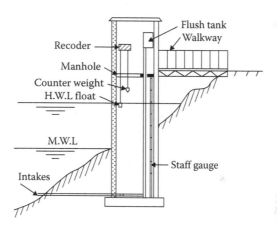

FIGURE 6.2 Stilling well installation.

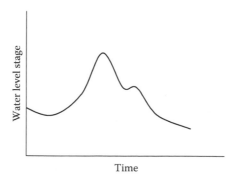

FIGURE 6.3 Water level hydrograph.

6.3 DISCHARGE MEASUREMENT

The discharge (Q) can be measured, either directly or indirectly, by measuring the average velocity (V_a) through the cross-sectional area (A) of a stream and multiplying V_a by A, that is, $Q = V_a \times A$.

Velocity is important for indirect stream flow measurement techniques. Of the many methods for stream flow velocity measurements, the most common is to use current meters. There are different kinds of current meters. To obtain accurate measurements, current meters must be used in moderate and deep channel sections with long flow duration. These conditions are not satisfied in shallow streams with short flow duration, as in the arid and semi-arid regions. For this reason we will not expand upon current meter methods. The method in the next section is suggested for stream flow discharge measurement in arid and semi-arid areas.

6.3.1 VELOCITY MEASUREMENT BY FLOATS

A floating object on the surface of a stream, when timed, can yield the surface velocity by the following equation:

$$v_s = \frac{S}{t} \qquad (6.1)$$

where S is the distance traveled in time t. Although this method is primitive, it still finds applications in special circumstances such as (1) small streams in flood, (2) small streams with a rapidly changing water surface, and (3) preliminary or exploratory surveys. While any floating object can be used, normally specially made leak proof and easily identifiable floats are used (Figure 6.4). A simple float moving

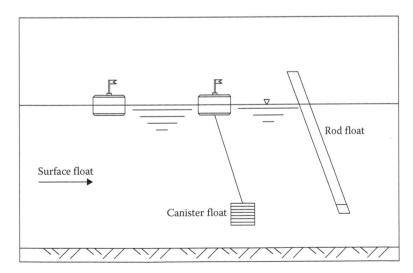

Surface float

Rod float

Canister float

FIGURE 6.4 Velocity measurements by float method.

on a stream surface is called a surface float. However, surface floats are affected by surface winds. To get the average velocity in the vertical direction, special floats in which part of the body is under water are used. Rod floats, in which a cylindrical rod is weighted so that it can float vertically, belong to this category.

When using floats to observe the stream velocity, a large number of easily identifiable floats are released at a uniform spacing across the width of the stream at an upstream section. Two sections on a fairly straight reach are selected, and the time to cross this reach by each float is noted and the surface velocity is then calculated.

6.3.2 Chemical Gauging for Stream Flow Measurement

The chemical method for flow measurement depends upon the continuity principle applied to a tracer that is allowed to mix completely with the flow. The dilution principle is to inject the tracer of concentration C_1 at a constant rate Q_t at section 1. At section 2, the concentration gradually rises from the background value C_0 at time t_1 to a constant value C_2. At the steady state, the continuity equation for the tracer is $Q_t C_1 + Q C_0 = (Q + Q_t) C_2$, that is

$$Q = \frac{Q_t (C_1 - C_2)}{(C_2 - C_0)} \tag{6.2}$$

This technique, in which Q is estimated by knowing C_1, C_2, C_0, and Q_1, is known as the constant rate injection method.

The dilution method of gauging is based on an assumption of steady flow. If the flow is unsteady and the flow rate changes appreciably during gauging, the storage volume in the reach changes and the steady-state continuity equation used to develop Equation 6.2 is not valid. Systematic errors can be expected in such cases.

The tracer used should ideally have the following properties:

1. It should not be absorbed by the sediment, channel boundary, and vegetation. It should not chemically react with any of these surfaces and also should not be lost by evaporation.
2. It should be nontoxic.
3. It should be capable of being detected in a distinctive manner in small concentrations.
4. It should not be very expensive.

The tracers used are of three main types:

1. Chemicals (common salt and sodium dichromate)
2. Fluorescent dyes (rhodamine WT and sulpho-rhodamine B extra)
3. Radioactive materials (such as bromine-82, sodium-24, and iodine-132)

Common salt can be detected with an error of ±1% up to a concentration of 10 parts per million (ppm). Sodium dichromate can be detected up to 0.2 ppm. Fluorescent dyes can be detected at levels of tens of nanograms per liter (~1 in 10^{11}) and hence require very small amounts of solution for injections. Radioactive traces are detectable up to accuracies of tens of picocuries per liter (~1 in 10^{14}) and therefore permit large-scale dilutions. However, they involve the use of very sophisticated instruments and should be handled by trained personnel only. The tracer is chosen based on the availability of detection instrumentation, environmental effects of the trace, and overall cost of the operation.

The length of the reach between the dosing section and the sampling section should be adequate for complete mixing of the tracer with the flow. This length depends upon the geometric dimensions of the channel cross section discharge and turbulence levels.

The length L varies from about 2 kilometers for a mountain stream with a discharge of about 1 cubic meters per second to about 100 kilometers for river in a plain with a discharge of about 300 cubic meters per second. The mixing length becomes very large for large rivers and is one of the major constraints of the dilution method. Artificial mixing of the tracer at the dosing section may prove beneficial for small streams in reducing the mixing length of the reach.

The dilution method has a major advantage: the discharge is estimated directly in an absolute way, making it a particularly attractive method for small turbulent streams, such as those in mountainous areas. Where suitable, it can be used as an occasional method for checking the calibration, stage-discharge curves, and so on obtained by other methods.

EXAMPLE 6.1

A 20 grams per liter solution of a fluorescent tracer was discharged into a stream at a constant rate of 15 cubic centimeters per second. The background concentration of the dye in the stream water was found to be zero. At a downstream section sufficiently far away, the dye was found to reach an equilibrium concentration of 5 parts per billion. Estimate the stream discharge.

SOLUTION

By Equation 6.2 for the constant-rate injection method

$$Q = \frac{Q_t(C_1 - C_2)}{(C_2 - C_0)}$$

$$Q_t = 15\,cm^3/s = 15 \times 10^{-6}\,m^3/s$$

$$C_1 = 0.020,\ C_2 = 5 \times 10^{-9},\ C_0 = 0$$

$$Q = \frac{15 \times 10^{-6}}{5 \times 10^{-9}}(0.020 - 5 \times 10^{-9}) = 60\,m^3/s$$

6.3.3 ELECTROMAGNETIC METHOD

The electromagnetic method is based on the Faraday's principle that an electromagnetic field (emf) is induced in the conductor (water in the present case) when it cuts a normal magnetic field. Large coils buried at the bottom of the channel carry a

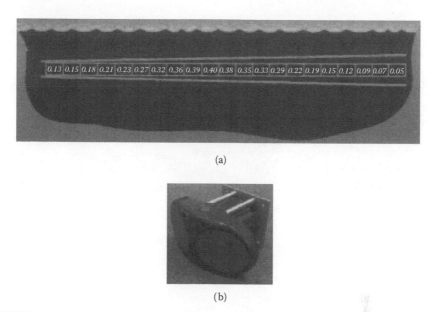

0.13 0.15 0.18 0.21 0.23 0.27 0.32 0.36 0.39 0.40 0.38 0.35 0.33 0.29 0.22 0.19 0.15 0.12 0.09 0.07 0.05

(a)

(b)

FIGURE 6.5 Acoustic Doppler arrangements. (a) The horizontal acoustic Doppler Current Profiler measuring the velocity in a channel. (b) The acoustic Doppler equipment.

current I to produce a controlled vertical magnetic field. Electrodes provided at the sides of the channel section measure the small voltage produced due to the flow of water in the channel.

The method involves sophisticated and expensive instrumentation and has been successfully tried in a number of installations. The fact that this kind of addition gives the total discharge once it has been calibrated makes it specially suited for field situations where the cross-sectional properties can change with time due to weed growth, sedimentation, and so on. Another specific application is in tidal channels where the flow undergoes rapid changes in both magnitude and direction. Present-day commercially available electromagnetic flow meters can measure the discharge to an accuracy of ±3%, the maximum channel width that can be accommodated is 100 meters and the minimum detectable velocity is 0.005 meters per second.

6.3.4 ULTRASONIC METHOD

The ultrasonic method is essentially an area-velocity method in which the average velocity is measured by using ultrasonic signals. Ultrasonic flow meters have frequencies around 500 kHz. Sophisticated electronics are needed to transmit, detect, and evaluate the mean velocity of a flow along the path. In a given installation, a calibration (usually performed by current meter) is needed to determine the system constants. Available commercial systems have been installed successfully in many places and accuracies of about 2% for the single-path method and 1% for the multipath method are reported. The systems are currently available for rivers up to 500 meters wide (Figure 6.5).

The specific advantages of the ultrasonic system of river gauging are as follows:

- It is rapid and has a high accuracy.
- It is suitable for automatic recording of data.
- It can handle rapid changes in the magnitude and direction of flow, as in tidal rivers.
- The installation cost is independent of the size of rivers.

The accuracy of this method is limited by factors that affect the signal velocity and the averaging of flow velocity, such as (1) unstable cross sections, (2) fluctuating weed growth, (3) high loads of suspended solids, (4) air entrainment, (5) salinity, and (6) temperature changes.

6.4 FLOW-MEASURING STRUCTURES

Structures like notches, weirs, flumes, and sluice gates are used for flow measurement in hydraulic laboratories. These conventional structures are used in field conditions also, but their use is limited by the ranges of head, debris, or sediment load of the stream and the backwater effects produced by the installations. To overcome these limitations, a wide variety of flow-measuring structures with specific advantages are in use.

The basic principle governing the use of flow-measuring structures such as weirs and flumes is that these structures produce a unique control section in the flow. In these structures, the discharge Q is a function of the water-surface elevation measured at a specified upstream location:

$$Q = f(H) \qquad (6.3)$$

where H is the water-surface elevation mounted from a specified datum.

The various flow-measuring structures can be broadly classified under three categories:

1. Thin plate structures are usually made from a vertically set metal plate. The V-notch, rectangular full width, and contracted notches are typical examples of this category.
2. Long-base weirs, also known as broad-crested weirs, are made of concrete or masonry and are used with large discharge values.
3. Flumes are made of concrete, masonry, or metal sheets, depending on their use and location. They depend primarily on width constriction to produce a control section.

The structures suitable for streams in arid and semi-arid regions are discussed next.

6.4.1 Weirs

A weir is a small dam that regulates and measures the flow of water in an open channel, lake, reservoir, and so on. A measuring weir is an overflow structure built perpendicular to an open-channel axis to measure the flow rate of water. A properly built and operated weir of a given shape has a unique depth of water at the measuring station in the upstream pool for each discharge. The crest overflow shape governs the variations in the discharge with the head measurement.

When approach conditions allow full contractions at the ends and bottom, the weir is a contracted weir. For full contraction, the ends of the weir should not be closer to the sides and bottom of the approach channel than a specified distance. If the specified distances are not met, then the weir is considered to be partially contracted. When the sides of the flow channel act as the ends of a rectangular weir, no side contraction exists, and the nappe does not contract from the width of the channel. This type of weir is a suppressed weir.

Velocity of approach is important because it can change weir calibrations by effectively reducing the crest length and/or measuring head. In addition, a variable discharge coefficient results as increasing velocity changes the curvature of flow springing from the weir edge.

The recommended types of weirs, which are presented in this chapter, to measure the flow rate of water are

1. Clear over fall weir
2. Standing wave weir

6.4.1.1 Clear over Fall Weir

The most popular type is the Fayum-type weir, a clear over weir type. This type has been calibrated at the Hydraulic Research Institute in Egypt and can be used in streams in arid and semi-arid regions. Figure 6.6 shows a cross section through the weir structure. The discharge of the Fayum-type weir was found to have the following parameters:

$$Q = 1.652 \, LH^{1.54} \quad \text{(for } H \text{ up to } 0.14 \, \text{m)} \quad \text{m}^3/\text{s} \tag{6.4}$$

$$Q = 1.956 \, LH^{1.72} \quad \text{(for } H > 0.14 \, \text{m)} \quad \text{m}^3/\text{s} \tag{6.5}$$

where H is the head above the weir crest and L is the weir length.

Note that the velocity of approach is neglected in this case because it is almost insignificant (Soliman 1980).

EXAMPLE 6.2

Design a clear over fall weir at a certain location of a stream where there is a drop in the water level of 1.3 meters. The maximum stream discharge is 20 cubic meters per second.

SOLUTION

Select the Fayum-type weir. For this weir to be a clear over fall weir, the downstream water level should be lower than the weir crest by a few centimeters.

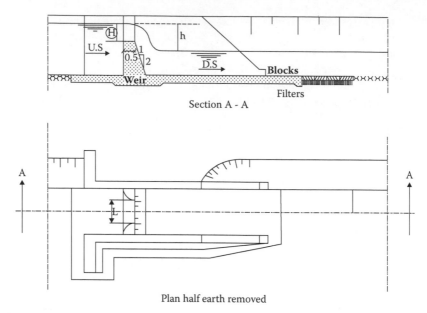

Section A - A

Plan half earth removed

FIGURE 6.6   Fayum-type weir.

Select the weir crest to be higher than the downstream water level by 10 centimeters. Therefore, the maximum head = 1.3 − 0.1 = 1.2 meters. For this condition, Equation 6.5 can be used to find the weir length:

$$20 = 1.956 \times L \times (1.2)^{1.72}$$

Then $L = 7.47$ meters. For construction facilities, select $L = 7.5$ meters; therefore we have to find the exact value of H:

$$20 = 1.956 \times 7.5 \times (H)^{1.72}$$

From which $H = 1.197$ meters. Therefore, the exact crest level should be = 1.3 − 1.197 = 0.103 meters = 10.3 centimeters above the downstream level.

6.4.1.2 Standing Wave Weir

The standing wave weir is considered a modified broad crested weir. The upstream corner is rounded to reduce the vertical control of the nappe. As shown in Figure 6.7, the geometrical shape of the weir helps maintain the critical depth so it occurs over the weir crest even for relatively high discharges, forming a standing wave over the crest.

The main advantage of this type of weir is its ability to pass large discharges with relatively low heads. The effect of the change of the downstream water level does not change the conditions of flow if the submergency h/H is not more than 85%.

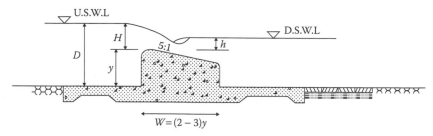

FIGURE 6.7 Standing-wave weir. (Note—submergency h/H not more than 85%; $y = D/3$)

The dimensions for such weirs, which are the type adopted by the Egyptian Ministry of Water Resources and Irrigation (Figure 6.7), are as follows:

$$Y = 0.35D$$
$$W = 2 \text{ to } 3Y$$

where Y is the height of the upstream side, D is the water depth upstream the weir, and W is the width of the weir.

The discharge equation of such weirs is as follows:

$$Q = 2.05LH^{1.6} \text{ (for submergency } h/H \text{ up to 85\%)} \tag{6.6}$$

where H is the upstream head above the crest and h is the downstream head above the crest.

6.4.2 CUT-THROAT FLUMES

The development of flumes started with the Parshall flume and continued in the 1960s due to a group of researchers in the Utah Water Research Laboratories in Logan, Utah, headed by G. Skogerboe. This group proposed changes to the geometrical configurations of the Parshall flume, starting with a suppression of the bottom dip in the throat, thus converting the Parshall flume into a flat, converging–diverging flume with a straight transition. Then, they found that the flow into the outlet transition improved if the throat was deleted, hence the name "cut-throat" for the new flume. This evolved design has straight-sided inlet transitions with a 3:1 taper ratio and an outlet transition with a 6:1 ratio (see Figure 6.8) and is one of the simplest flow measuring devices that can be built in an open channel. It resembles the Parshall flume in some ways; for example, the upstream head h_a is measured at a location near the minimum width section, thus at a point of high flow curvature. The downstream depth is also measured within the flume and is not the same as the tailwater depth in the channel following the flume. This design aspect means that the cut-throat flume cannot be calibrated by theoretical means and the stage–discharge curves must be obtained by recourse to experiment. On the other hand, this flume has an advantage over the Parshall flume of geometrical similarity that

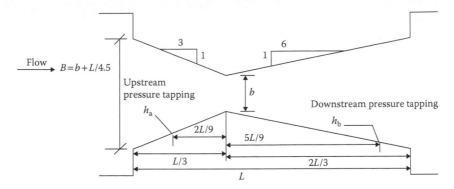

FIGURE 6.8 Definition sketch for the geometry of a cut-throat fume.

permits easier scaling of flow properties based on model studies. The experimental values gave the following equation:

$$Q = C_d \, b \, h_a \sqrt{2gh_a} \qquad (6.7)$$

C_d was obtained by Keller (1984) and ranges from 0.45 for h_a/b up to 1 and 0.5 for higher values of h_a/b. Here h_a is the pressure head measured in the tapping shown in Figure 6.8 and b is the throat width.

These experimental results reveal that little scale effect can be detected even though the throat size changed over a range of 4:1. Keller reported that scale effect could be reduced further by correcting h_a for the boundary layer displacement thickness. Therefore, any flume should be calibrated after its construction, before its first use.

REFERENCES

Keller, R. 1984. Cut-throat flume characteristics. *ASCE J Hydraul Div* 110:1248–63.
Soliman, M. M. 1980. *Irrigation Engineering Design.* Cairo: Ain Shams University Publications.

7 Stream-Flow Hydrographs

7.1 INTRODUCTION

Engineering hydrology is concerned primarily with three characteristics of stream flow: monthly and annual volumes available for storage and use, low flow rates that restrict the stream-based uses of water, and floods. Detailed analysis of flood hydrographs is usually important in flood-damage mitigation, flood forecasting, and establishing design flows for many structures that convey floodwaters.

7.2 CHARACTERISTICS OF THE HYDROGRAPH

The water that constitutes stream flow can reach the stream channel by any of several paths from the point where it first reaches the earth as precipitation. Some water flows over the soil surface as surface runoff and reaches a stream soon after its occurrence as rainfall. Other water filters through the soil surface and flows beneath the surface of the stream. This water moves more slowly than surface runoff and contributes to the sustained flow of the stream during periods of dry weather. In hydrological studies involving rate of flow in streams, these components of total flow must be distinguished. The first step in such studies is to divide the observed hydrographs of stream flow into components before analyzing the relationship between rainfall and runoff, determining the characteristic shape of hydrographs for a basin, or studying drought conditions.

The route followed by a water particle from the time it reaches the ground until it enters a stream channel is devious. Three main routes of travel can be conveniently visualized: overland flow, interflow, and groundwater flow, as discussed in Chapter 5.

Overland flow, or surface runoff, is the water that travels over the ground surface to a channel. The word channel here refers to any depression that may carry a small rivulet of water in turbulent flow during a rain and for a short while after. Such channels are numerous, and the distance water must travel is rarely more than a few hundred meters because overland flow is relatively short. Therefore, overland flow soon reaches a channel; if it occurs in sufficient quantity, is an important element in the formation of floods. The amount of surface runoff may be quite small, however, because surface flow over a permeable soil surface can occur only when the rainfall rate exceeds the filtration capacity, as discussed before. In many small and moderate storms, surface runoff may occur only from impermeable surfaces within the basin or from precipitation that falls directly on the water surface of the basin. Except in urban areas, the total of both impermeable area and water surface is usually only a small part of the basin area. Hence, surface runoff is an important factor in stream flow only during heavy or high-intensity rains, which occur in the catchments of many arid and semi-arid regions.

The distinctions between the three components of flow are arbitrary. Water may start out as surface runoff, infiltrate from a sheet of overland flow, and complete its trip to a stream as an interflow in the permeable area. Certainly, interflow varies from groundwater only in the speed of travel. In limestone terrain, groundwater frequently moves at relatively high velocities as turbulent flow through solution channels and fractures in the limestone. Streams in limestone country often exhibit a high ratio of flood peak flows to the average flow, a condition characteristic of streams having small groundwater contributions. In such terrains, groundwater flow actually has some of the characteristics ascribed to interflow. For convenience, it is customary to consider the total flow to be divided into two parts: storm—or direct—runoff and base flow. The distinction is actually based on the time of arrival in the stream rather than on the path followed. Direct runoff is presumed to consist of overland flow and a substantial portion of the interflow, whereas base flow is considered to be largely groundwater. However, in arid and semi-arid regions, many catchments yield direct runoff because the groundwater table is far below the ground surface.

The hydrograph that results from the direct runoff of an isolated storm is typically a single-peaked skewed distribution of discharge and is known variously as a storm hydrograph, flood hydrograph, or simply a hydrograph. It has three characteristic regions (Figure 7.1): (1) the rising limb AB, joining point A, the starting point of the rising curve, and point B, the point of inflection, (2) the crest segment BC between the two points of inflection, with a peak P in between, and (3) the falling limb or depletion curve CD starting from the second point of inflection C. Thereafter, the recession curve represents withdrawal of water from storage within the basin. The shape of the recession curve is largely independent of the characteristics of the storm that causes the rise.

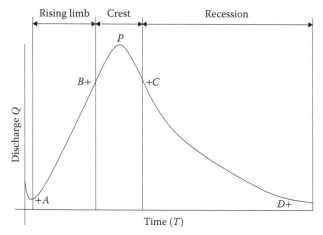

FIGURE 7.1   Characteristics of a hydrograph.

In cases where runoff-producing rainfall occurs over only a part of large basins, the recession may vary from storm to storm, depending on the particular area of runoff generation. If rainfall occurs while the recession from a previous storm is in progress, the recession will naturally be distorted. However, the recession curve for a basin is a useful tool in hydrology.

The recession curve (sometimes called a depletion curve because it represents depletion from storage) is described by the characteristic depletion equation

$$q_1 = q_0 k_r \qquad (7.1)$$

where q_0 is the flow at any time, q_1 is the flow one time unit later, and k is a recession constant that is less than unity. Equation 7.1 can be written in a more general form as

$$q_t = q_0 k_r^t \qquad (7.2)$$

where q_t is the flow at time t units after q_0. The time unit is frequently 24 hours, although a shorter unit may be necessary for small basins. The numerical value of k_r depends on the time unit selected. Integrating Equation 7.2 and remembering that the volume of water discharged during time dt is $q \cdot dt$ and is equal to the decrease in storage, $-dS$, during the same interval, we see that the storage S_t remaining in the basin at time t is

$$S_t = \frac{-q_t}{\ln k_r} \qquad (7.3)$$

Equation 7.2 will plot as a straight line on semilogarithmic paper, with q on the logarithmic scale. If the recession of a stream rise is plotted on semilogarithmic paper, the result is usually not a straight line but a curve with gradually decreasing slope, that is, with increasing values of k_r. This is because the water comes from three different types of storage: stream channels, surface soil, and the groundwater, each having different lag characteristics.

Barnes (1940) suggests that the recession can be approximated by three straight lines on a semilogarithmic plot. The transition from one line to the next is often so gradual that it is difficult to select the points of change in the slope. This is not surprising considering the heterogeneity of the typical watershed. Several aquifers may be contributing groundwater, while influent seepage is simultaneously occurring at other points in the stream. In most areas, runoff occurs in varying amounts over the watershed.

The slope of the last portion of the recession curve should represent the characteristic k_r for groundwater because, presumably, both interflow and surface runoff have ceased. By projecting this slope backward in time and replotting the difference between the projected line and the total hydrograph, a recession curve that consists largely of interflow for a given time period is obtained. With the slope applicable to interflow thus determined, the process can be repeated to establish the recession characteristics of surface runoff. Example 7.1 demonstrates the recession curve analysis.

EXAMPLE 7.1

Table 7.1 shows the flood hydrograph of one of the catchments in one of the semi-arid regions (Figure 7.2). Determine the surface flow–recession coefficient. Neglect the base flow because the water table is far below the stream base.

SOLUTION

From the hydrograph, the point of inflection is at $t = 26$ hours and $q_0 = 100$ m³/s.

TABLE 7.1
Flood Hydrograph Values

Time from Beginning of Storm (h) (1)	Ordinate of Storm (O. S.) Hydrograph (m³/s) (2)
0	10
6	30
12	87.5
18	111.5
24	102.5
30	85
36	71
42	59
48	47.5
54	39
60	31.5
66	26
72	21.5
78	17.5
84	15
90	10
96	2

Total volume $= 758 \times (6 \times 3600)$
$= 1,63,62,000$ m³; excess rainfall $= 4.1$ cm

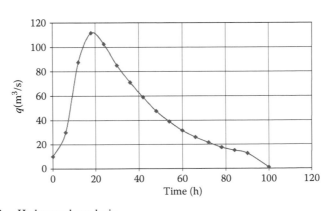

FIGURE 7.2 Hydrograph analysis.

At $t = 80$ hours, $q_t = 19$ m³/s. Recall Equation 7.2: $q_t = q_0 k_r^t$. In log format, Equation 7.2 becomes $t \log k_r = \log q_t/q_0$.

$\log k_r = (1/t) \log q_t/q_0$, with $t = 80 - 26 = 54$ hours and $q_t/q_0 = 19/100 = 0.19$; that is, $\log k_r = 1/54 \log 0.19 = -0.013356$ and $k_r = 10^{-0.013356} = 0.969$.

The logarithmic relation of the regression curve is shown in Figure 7.3. Figure 7.4 shows the total hydrograph drawn on semilog paper. It is interesting to note that part of the rising limb takes a straight-line form until the point of inflection.

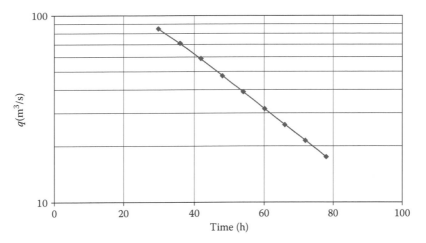

FIGURE 7.3   Regression curve on semilog paper.

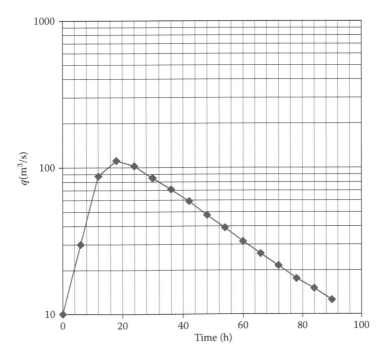

FIGURE 7.4   Hydrograph curve on semilog paper.

7.3 HYDROGRAPH SEPARATION

Although the streams in arid regions where the majority of hydrographs are charted are considered to be direct runoff because the water tables are far below the stream bases, there are some streams in semi-arid regions where the groundwater contributes to the stream hydrograph. Such cases may require hydrograph separation.

There is no real basis for distinguishing direct and groundwater flow in a stream at any instant. Because definitions of these two components are relatively arbitrary, the method of separation is usually equally arbitrary. To apply the unit hydrograph concept, the method of separation should be such that the time base of direct runoff remains relatively constant from storm to storm. This is usually achieved by terminating the direct runoff at a fixed time after the peak of the hydrograph.

There are three methods of base-flow separation in common use.

> *Method 1—Straight-Line Method*: In this method, the separation of the base flow is achieved by joining a straight line with the beginning of the surface runoff to a point on the recession limb representing the end of the direct runoff. In Figure 7.5, point A represents the beginning of the direct runoff. It is usually easy to identify in view of the sharp change in the runoff rate at that point.
>
> In Figure 7.5, point C, marking the end of the direct runoff, is rather difficult to locate exactly. An empirical equation for the time interval N (days) from the peak to the point B is

$$N = 0.8 \, A^{0.2} \tag{7.4}$$

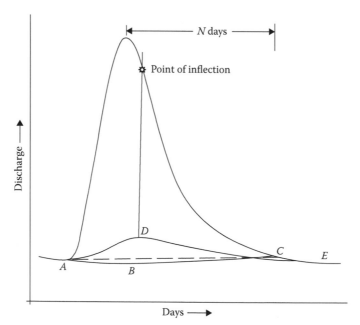

FIGURE 7.5   Hydrograph separation.

where A is the drainage area in square kilometers and N is in days. Points A and B are joined by a straight line to separate the base flow from surface runoff. The value of N obtained in Equation 7.4 is only approximate, and the position of B should be decided by considering a number of hydrographs for the specific catchment. This method of base-flow separation is the simplest of all the three methods.

Method 2—Curve Method: In this method, the base-flow curve existing prior to the commencement of the surface runoff is extended until it intersects the ordinate drawn at the peak (point B in Figure 7.5). This point is joined to point C by a straight line. Segments AB and BC separate the base flow from the surface runoff. This is probably the most widely used base-flow separation procedure.

Method 3—Line Segment Method: A third method of separation is illustrated by line segment ADE (Figure 7.5). This line is constructed by projecting the groundwater recession after the storm back under the hydrograph to a point under the inflection point of the falling limb. An arbitrary rising limb is sketched from the point of rise of the hydrograph to connect with the projected base-flow recession. This type of separation may have some advantages where groundwater is relatively plentiful and reaches the stream fairly rapidly, as in limestone terrain.

7.4 UNIT HYDROGRAPH CONCEPT

Because the physical characteristics of the basin, such as shape and size, are constant, one might expect considerable similarity in the shape of hydrographs from storms with similar rainfall characteristics. This is the essence of the unit hydrograph as proposed by Sherman (1932). The unit hydrograph is a typical hydrograph for the basin. It is called a unit hydrograph because, for convenience, the runoff volume under the hydrograph is commonly adjusted to 1 centimeter (or 1 inch).

The unit hydrograph is viewed as a unit impulse in a linear system. Thus, the principle of superposition applies, and 2 centimeters of runoff would produce a hydrograph with all ordinates twice as large as those of the unit hydrograph, that is, with the hydrographs summed. Mathematically, the unit hydrograph is the kernel function $U(t-T)$ in

$$q(t) = \int i(t)U(t-T)dt \qquad (7.5)$$

where $q(t)$ is the output hydrograph and $i(t)$ is the input hyetograph. The convolution of the unit hydrograph and rainfall excess yields the direct-runoff hydrograph of the storm.

Although the physical characteristics of the basin remain relatively constant, the variable characteristics of storms can cause variations in the shape of the resulting hydrographs. The storm characteristics are rainfall duration, time–intensity pattern, areal distribution of rainfall, and amount of runoff. Their effects are discussed in Sections 7.4.1 through 7.4.4.

7.4.1 DURATION OF RAIN

The unit hydrograph may be employed in two ways. It may be developed for a short duration and all storms, treated by dividing the rainfall excess into similar intervals. The alternate approach is to derive a series of unit hydrographs covering the range of durations experienced on the watershed. Because of a lack of data on hourly rainfall, the second approach was widely used when the unit hydrograph was first introduced. Theoretically, an infinite number of unit hydrographs would be required to cover the range of durations. Actually, the effect of small differences in duration is slight and a tolerance of ±25% in duration is usually acceptable. Thus, only a few unit hydrographs are actually required. Where a computer solution is used, a short-duration unit hydrograph is preferable.

7.4.2 TIME–INTENSITY PATTERN

In order to derive a separate unit hydrograph for each possible time–intensity pattern, an infinite number of unit hydrographs would be required. Practically, unit hydrographs can be based only on an assumption of runoff uniform intensity. However, large variations in rain intensity (and hence in the runoff rate) during a storm are reflected in the shape of the resulting hydrograph. The time scale of critical intensity variations depends mainly on basin size. Rainfall bursts lasting only a few minutes may cause clearly defined peaks in the hydrograph for a basin of a few hectares, whereas for basins with areas of several hundred square kilometers, only changes in the intensity of a storm that last for hours will cause distinguishable effects on the hydrograph. If the unit hydrographs for a basin are applicable to a storm of shorter duration than the critical time for the basin, hydrographs of longer storms can be synthesized quite easily. A basic duration of about one-fourth of the basin lag is generally satisfactory.

7.4.3 AREAL DISTRIBUTION OF RUNOFF

The areal pattern of runoff can cause variations in the hydrograph shape. If the area of high runoff is near the basin outlet, a rapid rise, a sharp peak, and a rapid recession usually result. Higher runoff in the upstream portion of the basin produces a slow rise and recession and a lower, broader peak. Unit hydrographs have been developed for specific runoff patterns, for example, heavy upstream, uniform, or heavy downstream. This is not wholly satisfactory because of the subjectivity of classification. A better solution is to apply the unit hydrograph method only to basins small enough to ensure that the usual areal variations will not be great enough to cause major changes in the hydrograph shape. The limiting basin size is fixed by the accuracy desired and regional climatic characteristics. Generally, however, unit hydrographs should not be used for basins over 5000 square kilometers (2000 square miles) unless reduced accuracy is acceptable. This does not apply to rainfall variations caused by topographic controls. Such rainfall patterns are relatively fixed characteristics of the basin.

7.4.4 Amount of Runoff

It has been assumed, for a linear unit hydrograph, that the ordinates of flow are proportional to the volume of runoff for all storms of a given duration and that the time bases of all such hydrographs are equal. This assumption is obviously not completely valid because, considering the character of recession curves, duration of recession must be a function of peak flow. Moreover, unit hydrographs for storms of the same duration but different magnitudes do not always agree. Peaks of unit hydrographs derived from very small events are commonly lower than those derived from larger storms. This may be because the smaller events contain less surface runoff and relatively more interflow and groundwater than the larger events or because the flow time of the channel is longer at low flow rates.

In the range of floods frequently experienced on a watershed, it is relatively simple to verify the adequacy of the assumption of linearity by comparing the hydrographs from storms of various magnitudes. If important nonlinearity exists, the unit hydrographs derived should be used only for reconstructing events of similar magnitude. A series of unit hydrographs covering the appropriate ranges of magnitude would be required for each period. Much more critical is the extreme event, exceeding any which has occurred on the watershed. There is no way to obtain empirical evidence about the changes in the unit hydrograph peak. Many hydrologists increase the peaks of unit hydrographs derived from ordinary floods by 5–20% before using them for estimation of extreme floods. This increase is based on the belief that channel flow time shortens as flood magnitude increases. However, if extreme floods overflow onto floodplains, the opposite effect might result. Caution should be exercised in using the unit hydrograph to extrapolate extreme events.

The unit hydrograph can be defined as the hydrograph of 1 centimeter (or 1 inch, in a different system) of direct runoff from a storm of specified duration. For a storm of the same duration but with a different amount of runoff, the hydrograph of direct runoff is assumed to have the same time base as the unit hydrograph and ordinates of flow approximately proportional to the runoff volume. The duration assigned to a unit hydrograph should be the duration of rainfall producing significant runoff, as determined by inspection of the hourly rainfall data.

7.5 DERIVATION OF UNIT HYDROGRAPHS FROM SIMPLE HYDROGAPHS

The unit hydrograph is best derived from the hydrograph of a storm of reasonable uniform intensity and with a runoff volume close to or greater than 1 centimeter (or 1 inch). The first step (Figure 7.5) is to separate the base flow from direct runoff. The volume of direct runoff is then determined, and the ordinates of the direct-runoff hydrograph are divided by the observed runoff depth. The adjusted ordinates form a unit hydrograph.

A unit hydrograph derived from a single storm may have errors, so it is desirable to average the unit hydrograph from several storms of the same duration. This should not be an arithmetic average of concurrent coordinates because, if peaks do

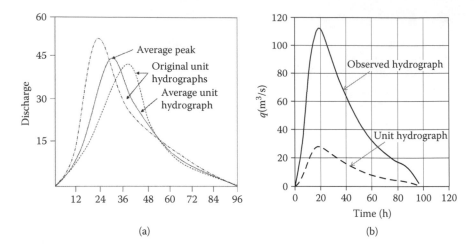

FIGURE 7.6 (a) Construction of an average unit hydrograph and (b) derivation of a unit hydrograph from a simple storm.

not occur at the same time, the average peak will be lower than the individual peaks. The proper procedure is to compute the average peak flow and the time to peak. The average unit hydrograph is sketched with reference to flow and time to peak, then sketched to conform to the shape of the other graphs, passing through the computed average peak, and having a unit volume of 1 inch or 1 cm. The following example demonstrates the aforementioned procedures (Figure 7.6a).

EXAMPLE 7.2

The observed flows for the stream of Example 7.1 are for a storm of a 6-hour duration. The drainage area of the stream is 400 square kilometers. Derive the unit hydrograph.

SOLUTION

From Table 7.2, the excess rainfall (*d*) of the flow hydrograph (as direct runoff) = 4.1 centimeters. Dividing the hydrograph ordinates by (*d*), the unit hydrograph ordinates can be obtained as shown in Table 7.2, column 3.

Dividing the area of the unit hydrograph by the area of the catchment, we get *d* = 1.01 centimeters, which is considered a good approximation for a unit hydrograph runoff depth of 1 centimeter. The total hydrograph and unit hydrograph are also given in Figure 7.6b.

7.6 DERIVATION OF UNIT HYDROGRAPHS FROM COMPLEX STORMS

Because there are no ideal separate storms available on record, it is best to develop the unit hydrograph from a complex storm. If individual bursts of rain in the storm result in well-defined peaks, it is possible to approximately separate the hydrographs of several bursts and use these hydrographs as independent storms. If the resulting

TABLE 7.2
Flood Hydrograph Values and Unit Hydrograph

Time from Beginning of Storm (h) (1)	Ordinate of Storm (O. S.) Hydrograph (m³/s) (2)	Unit Hydrograph Ordinates (3)
0	10	0
6	30	7.5
12	87.5	21.875
18	111.5	27.875
24	102.5	25.625
30	85	21.25
36	71	17.75
42	59	14.75
48	47.5	11.875
54	39	9.75
60	31.5	7.875
66	26	6.5
72	21.5	5.375
78	17.5	4.375
84	15	3.75
90	10	2.5
96	2	0.5
	Total volume = 758 × (6 × 3600) = 1,63,62,000 m³; excess rainfall = 4.1 cm	Excess rainfall = 1.01 cm; O.K.

unit hydrographs are averaged, errors in the separation are minimized. If the hydrograph does not lend itself to separation, the analysis begins with estimates of direct runoff volumes represented by rainfall depths $d_1, d_2, \ldots d_n$ in successive periods during the storm.

The first ordinate $q_1 = d_1 U_1$, and since d_1 is known (or estimated), U_1 can be computed (Figure 7.7). The second ordinate is then

$$q_2 = d_1 U_2 + d_2 U_1 \qquad (7.6)$$

The only unknown in this equation is U_2—this can be computed and all the ordinates can be determined in a similar way. This process can be repeated to cover the number of storms given. A general equation for any ordinate of the total hydrograph q_n in terms of runoff d and unit hydrograph ordinate U is

$$q_n = d_n U_1 + d_n - 1 U_2 + d_n - 2 U_3 + \cdots + d_1 U_n \qquad (7.7)$$

Although the procedure just outlined may seem simple, there are numerous difficulties. Each computation depends on all the preceding computations for values of U. Errors in estimating the runoff increments, errors in the observed stream flow, and variations in the intensity and areal distribution of rainfall during several storm periods can lead to cumulative errors, which may become large in the latter portion of the unit hydrograph. Large negative ordinates sometimes develop.

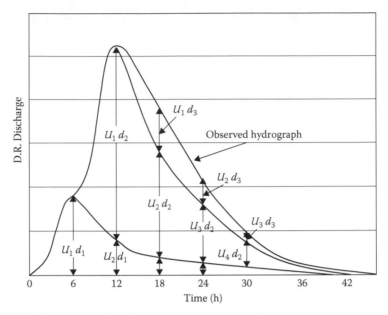

FIGURE 7.7 Derivation of unit hydrographs from a complex storm.

A unit hydrograph can also be developed by successive approximations. A unit hydrograph is assumed and used to reconstruct the storm hydrograph. If the reconstructed hydrograph does not agree with the observed hydrograph, the assumed unit hydrograph is modified and the process repeated until a unit hydrograph that seems to give the best fit is determined. This condition has been applied in this book for the hydrologic model developed by the author, as mentioned in Appendix D.

A more elegant but laborious method is the use of least squares (Snyder 1955), a statistical technique used to find the constants a and b_i in an equation of the form

$$q = a + b_1 d_1 + b_2 d_2 + b_3 d_3 + \cdots + b_i d_i \qquad (7.8)$$

Equations 7.7 and 7.8 are similar. Theoretically for the unit hydrograph, a should be zero and the b should be equivalent to the unit hydrograph ordinates U_i. The least squares technique uses data from a number of flood events, for which values of q and d are established to develop a set of average values of U_i.

7.7 UNIT HYDROGRAPHS FOR VARIOUS DURATIONS

If a unit hydrograph for duration t hours is added to itself lagged by t hours, the resulting hydrograph represents the hydrograph for 2 centimeters of runoff in $2t$ hours. If the ordinates of this graph were divided by 2, the result is a unit hydrograph for duration $2t$ hours. The final graph represents the flow from 1 centimeter of runoff generated at uniform intensity of $1/2t$ centimeters per hour in $2t$ hours.

This simple example illustrates the ease with which a unit hydrograph for a short duration can be converted into a unit hydrograph for any multiple of the original duration.

A more convenient technique for conversion to either a shorter- or longer-duration hydrograph is the S curve—or summation curve—method. The S curve is the hydrograph that would result from an infinite series of runoff increments of 1 centimeter (or 1 inch) in t hours. Thus, each S curve applies to a specific duration within which each centimeter of runoff is generated. The S curve is constructed by adding together a series of unit hydrographs, each lagged t hours with respect to the preceding one (Figure 7.8b). If the time base of the unit hydrograph is T hours, then continuous rainfall producing 1 centimeter of runoff every T hours would develop a constant outflow at the end of T hours. Thus, only T/t unit hydrographs need be combined to produce an S curve that should reach equilibrium at flow q_a.

$$q_a = \frac{2.78A}{t} \, \text{m}^3/\text{s} \tag{7.9}$$

where A is the drainage area in square kilometers and t is the duration in hours.

Commonly, the S curve tends to fluctuate about the equilibrium flow. This means that the initial unit hydrograph does not actually represent the runoff at a uniform rate over time t. If a uniform rate of runoff is applied to a basin, equilibrium flow at the rate given by Equation 7.9 must eventually develop. Thus, the S curve serves as an approximate check on the assumed duration of effective rainfall for the unit hydrograph. A duration that results in minimum fluctuation of the S curve can be found by trial. Note, however, that fluctuation of the S curve can also result from nonuniform runoff generation during the period t hours, unusual areal distribution of rain, or errors in basic data. For this reason, the S curve can indicate only an approximate duration.

Construction of an S curve does not require tabulating and adding hydrographs with successive lags of t hours. Table 7.3 illustrates the computation table for a 6-hour unit hydrograph, starting with an initial unit hydrograph for which $t = 6$ hours. The table also illustrates the development of a 12-hour unit hydrograph, shown in the last column. For the first time unit, the unit hydrograph and S curve are identical (columns 2 and 4). The S-curve additions (column 3) are the ordinates of the S curve set ahead by 6 hours. Because an S-curve ordinate is the sum of all concurrent unit-hydrograph ordinates, combining the S-curve additions with the initial unit hydrograph is the same as adding all previous unit hydrographs.

The difference between two S curves with initial points displaced by t' hours gives a hydrograph for the new duration t' hours. Because the S curve represents runoff production at a rate of 1 cm (or 1 inch) in t hours, the runoff volume represented by this new hydrograph will be t'/t cm (or inches). Thus, the ordinates of the unit hydrograph for t' hours are computed by multiplying the S-curve differences by the ratio t/t', as given in the last column of Table 7.3. Figure 7.8a shows also the two S curves with a displacement of 12 hours to get the unit hydrograph for a 12-hour duration.

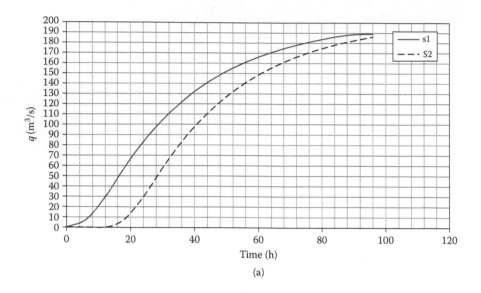

(a)

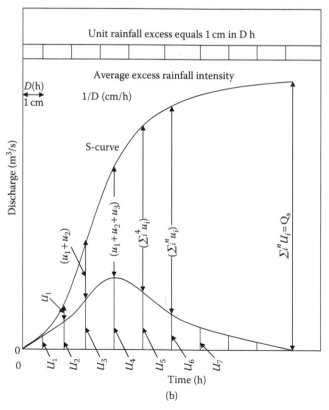

(b)

FIGURE 7.8 (a) Graphical illustration of an S curve; and (b) derivation of a unit hydrograph from an S curve.

TABLE 7.3
Derivation of 12-Hour Unit Hydrograph by the S-Curve Method

Time (h)	Unit Hydrograph Q (m³/s)	S Curve (S1)	S Curve (S2)	12-Hour Hydrograph	12-Hour Unit Hydrograph.
(1)	(2)	(3)	(4)	(5)	(6)
0	0	0	0	0	0
6	7.5	7.5	0	7.5	3.75
12	21.875	29.375	0	29.375	14.6875
18	27.875	57.25	7.5	49.75	24.875
24	25.625	82.875	29.375	53.5	26.75
30	21.25	104.125	57.25	46.875	23.4375
36	17.75	121.875	82.875	39	19.5
42	14.75	136.625	104.125	32.5	16.25
48	11.875	148.5	121.875	26.625	13.3125
54	9.75	158.25	136.625	21.625	10.8125
60	7.875	166.125	148.5	17.625	8.8125
66	6.5	172.625	158.25	14.375	7.1875
72	5.375	178	166.125	11.875	5.9375
78	4.375	182.375	172.625	9.75	4.875
84	3.75	186.125	178	8.125	4.0625
90	2.5	188.625	182.375	6.25	3.125
96	0.5	189.125	186.125	3	1.5
		189.125	188.625	0.5	0.25
		189.125	189.125	0	0

7.8 SYNTHETIC UNIT HYDROGRAPHS

Only a relatively small number of streams are gauged. Unit hydrographs can be derived as described in Section 7.7 only if records are available. Hence, a means of deriving unit hydrographs for ungauged basins is necessary and requires a relation between the physical geometry of the basin and the resulting hydrograph. Three approaches have been used: formulas relating hydrographs to basin characteristics, transposition of unit hydrographs from one basin to another, and storage routing.

Most attempts to derive formulas for the unit hydrograph have tried to determine the time of peak, peak flow, and time base. These items plus the fact that the runoff volume must equal 1 centimeter allow sketching of the complete hydrograph. The key item in most studies has been the basin lag, most frequently defined as the time from the centroid of rainfall to the peak of the hydrograph.

7.8.1 SNYDER METHOD

The first synthetic unit hydrograph procedure was presented by Snyder (1938). His unit hydrograph model is an empirical model that was developed based on studies of

20 watersheds located primarily in the Appalachian Highlands of the United States. Snyder collected rainfall and runoff data from gauged watersheds with areas ranging from 26 to 26,000 square kilometers, derived the unit hydrographs, and related the hydrograph parameters to measurable watershed characteristics. Although Snyder's methodology was developed several decades ago, it is still widely used.

Snyder's observations indicated that the peak runoff per unit (1 centimeter) of rainfall excess, Q_p (cubic meter per second), is given by

$$Q_p = 2.78 \frac{C_p A}{t_1} \tag{7.10}$$

where t_1 is the time lag in hours, A is the catchment area in square kilometer, and C_p is a coefficient. The time lag, t_1, is defined as the time from the centroid of the rainfall excess to the peak runoff. The duration of rainfall excess, t_r, is related to the time lag, t_1, by

$$t_r = \frac{t_1}{5.5} \tag{7.11}$$

Based on Snyder's observations, the time lag t_1 is related empirically to the catchment characteristics by the following equation:

$$t_1 = 0.75 \, C_t (LL_c)^{0.3} \tag{7.12}$$

where C_t is the basin coefficient that accounts for the slope, land use, and associated storage characteristics of the river basin, L is the length of the main stream from the outlet to the catchment boundary in kilometers, and L_c is the length along the main stream from the outlet to a point nearest to the catchment centroid in kilometers. The term $(LL_c)^{0.3}$ is sometimes called the shape factor of the catchment (McCuen 2005). The parameters C_p (in Equation 7.10) and C_t (in Equation 7.12) are best found via calibration, as they are not physically based. Snyder reported that C_t typically ranges from 1.35 to 1.65, with a mean of 1.5; however, values of C_t have been found to vary significantly outside this range in very mountainous or very flat terrains. Bedient and Huber (1992) reported that C_p ranges from 0.4 to 0.8, where larger values of C_p are associated with smaller values of C_t. The variability in C_t and C_p is evident from the range of values reported in Table 7.4. If the duration of the desired hydrograph for the watershed of interest is significantly different from that specified in Equation 7.11, the following relationship can be used to adjust the time lag t_1:

$$t_{1R} = t_1 + 0.25 \, (t_R - t_r) \tag{7.13}$$

where t_R is the desired duration of the rainfall excess, and t_{1R} is the lag of the desired unit hydrograph. For durations other than the standard duration, the peak runoff Q_{pR} is given by

$$Q_{pR} = 2.78 \frac{C_p A}{t_{1R}} \tag{7.14}$$

TABLE 7.4
Variability in Snyder Unit Hydrograph Parameters

C_t	C_p	Location	References
1.01–4.33	0.26–0.67	27 watersheds in Pennsylvania	Miller et al. (1983)
0.40–2.26	0.31–1.22	13 watersheds (1.28–192 km²) in Central Texas	Hudlow and Clark (1969)
0.3–0.7	0.35–0.59	Northwestern United States (Connecticut; Sacramento and lower San Joaquin rivers)	Linsley (1943)
0.4–2.4	0.4–1.1	12 watersheds (0.05–635 km²) in eastern New South Wales, Australia	

The time base of the unit hydrograph, T_B, in hours can be estimated using the following equation:

$$T_B = 72 + 3t_{lR} \qquad (7.15)$$

As a general rule of thumb, values of T_B/t_{lR} should be of the order of five, and if Equation 7.15 yields $T_B/t_{lR} \gg 5$, then it is recommended to take $T_B/t_{lR} = 5$. This restriction is particularly applicable for small watersheds. The U.S. Corps of Engineers developed the following empirical equations to help define the shape of the unit hydrograph (Wurbs and James 2002):

$$W_{50} = \frac{5.87}{\left(\dfrac{Q_{pR}}{A}\right)^{1.08}} \qquad (7.16)$$

$$W_{75} = \frac{3.35}{\left(\dfrac{Q_{pR}}{A}\right)^{1.08}} \qquad (7.17)$$

where W_{50} and W_{75} are the widths in hours of the unit hydrograph at 50 and 75% of Q_{pR}, respectively; here, Q_{pR} is the peak discharge in cubic meters per second, and A is the catchment area in square kilometers. These time widths are proportioned such that one-third lies before the peak and two-thirds after the peak.

Snyder's synthetic unit hydrograph formulas have been tried elsewhere with varying success. The coefficients C_t and C_p vary considerably. Other investigators have proposed formulas for synthetic unit hydrographs. Many have found the product LL_c to be significant with an exponent in the vicinity of 0.3. However, the best way any of the methods can be used is to derive coefficients for gauged streams in the vicinity of the problem basin and apply these to the ungauged stream. Perhaps because of approximations inherent in the unit hydrograph concept, synthetic methods of constructing unit hydrographs seem to be of limited value.

7.8.2　NRCS Dimensionless Unit Hydrograph

The Natural Resources Conservation Service (NRCS; formerly, Soil Conservation Service; SCS 1986) developed a dimensionless unit hydrograph that represents the average shape of a large number of unit hydrographs from small agricultural watersheds throughout the United States. This dimensionless unit hydrograph is frequently applied to urban catchments. The NRCS dimensionless unit hydrograph, illustrated in Figure 7.9, expresses the ratio Q/Q_p as a function of the time ratio, t/T_p, where Q_p is the peak runoff and T_p is the time to peak of the hydrograph from the beginning of the rainfall excess. The coordinates of the NRCS dimensionless unit hydrograph are given in Table 7.5.

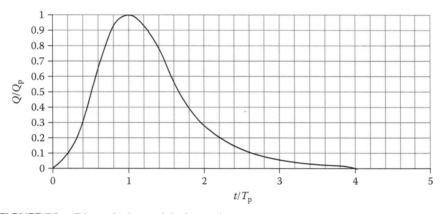

FIGURE 7.9　Dimensionless unit hydrograph.

TABLE 7.5
NRCS Dimensionless Unit Hydrograph

t/T_p	Q/Q_p	t/T_p	Q/Q_p
0.0	0.000	2.6	0.107
0.2	0.100	2.8	0.077
0.4	0.310	3.0	0.055
0.6	0.660	3.2	0.040
0.8	0.930	3.4	0.029
1.0	1.000	3.6	0.021
1.2	0.930	3.8	0.015
1.4	0.780	4.0	0.011
1.6	0.560	4.2	0.008
1.8	0.390	4.4	0.006
2.0	0.280	4.6	0.004
2.2	0.207	4.8	0.002
2.4	0.147	5.0	0.000

The NRCS dimensionless unit hydrograph can be converted to an actual hydrograph by multiplying the abscissa by T_p and the ordinate by Q_p. The time, T_p, is estimated using

$$T_p = \frac{t_r}{2} + t_l \qquad (7.18)$$

where t_r is the duration of the rainfall excess and t_l is the time lag from the centroid of the rainfall excess to the peak of the runoff hydrograph. NRCS recommends that the specified value of t_r not exceed two-tenths of t_c or three-tenths of T_p for the NRCS dimensionless unit hydrograph to be valid. The time lag, t_l, can be estimated by

$$t_l = 0.6t_c \qquad (7.19)$$

The NRCS dimensionless unit hydrograph can be approximated by the triangular unit hydrograph and is illustrated in Figure 7.10. It incorporates the following key properties of the dimensionless unit hydrograph: (1) the total volume under the dimensionless unit hydrograph is the same; (2) the volume under the rising limb is the same; and (3) the peak discharge is the same. Assuming Q_p and T_p to be the same in both the NRCS dimensionless unit hydrograph (Figure 7.9) and the NRCS triangular approximation (Figure 7.10), then the time base of the triangular unit

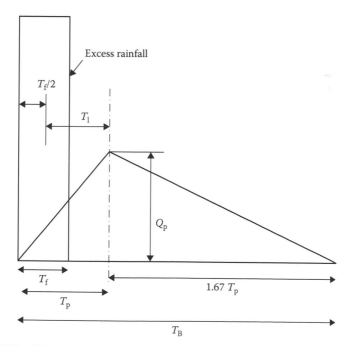

FIGURE 7.10 Triangular SCS unit hydrograph.

hydrograph must be equal to $2.67T_p$. For a runoff depth, d, from a catchment area, A, the triangular unit hydrograph requires that

$$Ad = 0.5 \, Q_p(2.67T_p) \tag{7.20}$$

If the runoff depth, d, is 1 centimeter, A is in square kilometers, Q_p is in (m³/s), and T_p is in hours, then Equation 7.20 yields

$$Q_p = 2.08\left(\frac{A}{T_p}\right) \tag{7.21}$$

This estimate of Q_p provides the peak runoff. The coefficient 2.08 in Equation 7.21 is appropriate for the average rural experimental watersheds used in calibrating the formula, but it should be increased by about 20% for steep mountainous conditions and decreased by about 30% for flat swampy conditions. Flat swampy conditions typically have an average slope less than 0.5%. Adjustment of the coefficient in Equation 7.21 must necessarily be accompanied by an adjustment to the time base of the unit hydrograph to maintain a runoff depth of 1 centimeter.

The triangular unit hydrograph approximation produces sufficiently accurate results for most stormwater management facility designs, including curbs, gutters, storm drains, channels, ditches, and culverts.

7.8.3 Transposing Unit Hydrographs

From Equation 7.12, a general expression for basin lag might be expected to take the form

$$t_p = C_t\left(\frac{LL_c}{\sqrt{s}}\right) \tag{7.22}$$

If known values of lag are plotted against $LL_c/\sqrt{s}$ on logarithmic paper (Figure 7.11), the resulting plot should define a straight line, provided that the values are taken from basins of similar hydrologic characteristics, that is, C_t is constant. A relation such as that shown in Figure 7.11 offers a means of estimating the basin lag. The peak flow and shape of the unit hydrograph can be estimated using a plot relating q_p to t_p. It should be noted that Figure 7.11 was obtained by the U.S. Corps of Engineers from 18 catchments with drainage areas ranging from 6 to 1670.6 square kilometers. The author has checked some lag times obtained from catchments in the Sinai area in Egypt and found that the lag times agree with those obtained from Figure 7.11. This case is also discussed in Chapter 12.

EXAMPLE 7.3

A 2.25-square kilometer urban catchment is estimated to have an average Manning $n = 0.05$, an average slope of 0.5%, and a flow length of 1680 meters

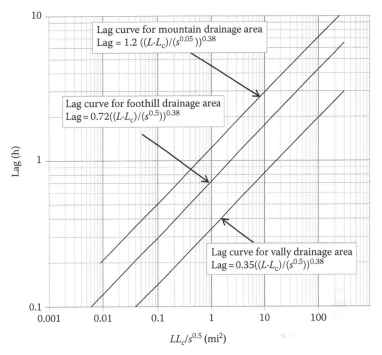

FIGURE 7.11 Relationship between basin lag and basin characteristics.

from the catchment boundary to the outlet. Determine the NRCS unit hydrograph for a 30-minute rainfall excess and verify that it corresponds to a rainfall excess of 1 centimeter.

SOLUTION

Using Equation 5.12, $t_c = 6.99\left(\dfrac{(nL)^{0.6}}{i_e^{0.4} S_0^{0.3}}\right)$

$$t_c = 6.99\left(\frac{(0.05 \times 1680)^{0.6}}{20^{0.4} \times 0.005^{0.3}}\right) = 149.13 \text{ min}$$

$$= 2.49 \text{ h}$$

Take $\qquad t_l = 0.6\, t_c = 1.5 \text{ h}$

$$T_p = \frac{t_r}{2} + t_l = 1.75 \text{ h}$$

$$Q_p = 2.08\left(\frac{A}{T_p}\right) = 2.674 \text{ m}^3/\text{s}$$

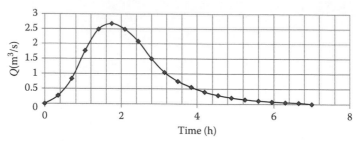

FIGURE 7.12   Derivation of a unit hydrograph from a dimensionless unit hydrograph.

For a triangular relation using Equation 7.20,
Triangular area = 2.67 × 1.75 × (3600) × 2.674/2 = 22492.08 m³,
Rainfall depth = 22,492.08/2.25 × 10⁶ = 0.999648 ≈ 1 cm.
The unit hydrograph is also shown in Figure 7.12 using the dimensionless unit hydrograph relations.

7.9   HYDROGRAPH OF OVERLAND FLOW

In arid and semi-arid regions, the majority of the ground surface in urban areas is impervious. Although the depth of flow in the overland sheet is quite small, the quantity of water temporarily detained in this sheet (called surface detention) is relatively great. It is generally assumed that overland flow is laminar. Hence, from Figure 7.13

$$\rho g(D - y)s = \mu \frac{dv}{dy} \tag{7.23}$$

where ρ is the density, g is the gravity, and μ is the absolute viscosity. The slope is assumed to be so small that the sine and tangent are equal. Since μ/ρ is equal to the kinematic viscosity v,

$$dv = \frac{gs}{v}(D - y)dy \tag{7.24}$$

Integrating and noting that $v = 0$ when $y = 0$, we have

$$V = \frac{gs}{v}\left(yD - \frac{y^2}{2}\right) \tag{7.25}$$

Integrating from $y = 0$ to $y = D$ and dividing by D gives the mean velocity

$$V_m = \frac{gsD^2}{3v} \tag{7.26}$$

and the discharge per unit width $V_m \cdot D$ or

$$q = -bD^3 \tag{7.27}$$

where b is a coefficient involving slope and viscosity.

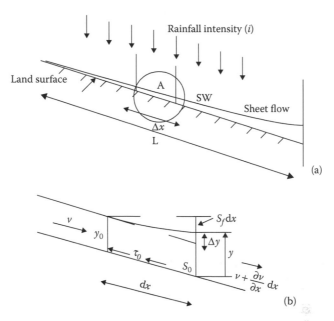

FIGURE 7.13 Definition sketch of sheet flow. (a) Longitudinal section. (b) Detail of A—control volume detail of sheet flow above impervious surface.

The most extensive experiments on overland flow are those of Izzard (1946). His tests on long flumes at various slopes and with various surfaces showed that the time to equilibrium is

$$t_e = \frac{2V_e}{60q_e} \qquad (7.28)$$

where t_e is defined as the time in minutes when flow is 97% of the supply rate and V_e is the volume of water (in cubic feet or cubic meters) in surface detention at equilibrium. From a strip of unit width, the equilibrium flow q_e is

$$q_e = \frac{iL}{3.6 \times 10^6} \qquad (7.29)$$

where i is in millimeters per hour, L is in meters, and q_e is in cubic meters per second. When the average depth on the strip V_e/L is substituted for outflow depth, Equation 7.27 becomes

$$\frac{V_e}{L} = kq_e^{1/3} \qquad (7.30)$$

Substituting Equation 7.29 for q gives

$$V_e = \frac{kL^{4/3}i^{1/3}}{50.2} \qquad (7.31)$$

where V_e is the volume of detention (in cubic meter) on the strip at equilibrium.

Experimentally, k was found to be given by

$$k = \frac{2.8 \times 10^{-5} i + c}{s^{1/3}} \tag{7.32}$$

where s is the surface slope and the retardance coefficient c is given in Table 7.6. Izzard found that the form of the overland flow hydrograph could be presented as a dimensionless graph (Figure 7.14). With t_e and q_e known, the q/q_e curve permits plotting of the rising limb of the overland-flow hydrograph.

The dimensionless recession curve of Figure 7.14 defines the shape of the receding limb. At any time t_a after the end of rain, the factor β is

$$\beta = \frac{60 q_e t_a}{V_a} \tag{7.33}$$

where V_0 is the detention given by Equations 7.30 and 7.32, assuming $i = 0$.

Since Izzard's work, other writers have dealt with the overland flow problem. Relatively few new data have been collected, but they have demonstrated that results

TABLE 7.6

Retardance Coefficient c in Equation 7.29

Surface	Value of c
Very smooth asphalt pavement	0.007
Tar-and-sand pavement	0.0075
Crushed-slate roofing paper	0.0082
Concrete	0.012
Tar-and-gravel pavement	0.017
Closely clipped sod	0.046
Dense bluegrass turf	0.060

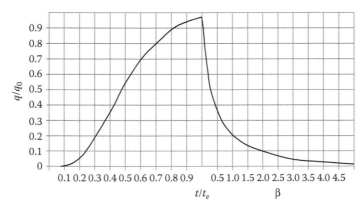

FIGURE 7.14   Dimensionless hydrograph of overland flow.

equivalent to Izzard's can be derived using turbulent-flow equations. Because the experimental data are all from a plane with small roughness elements, the major problem in applying any method to natural watersheds is defining the characteristic of the surface.

The author of this book is in favor of adopting the continuity equation and the equation of momentum to solve this problem (Soliman 2009). This analysis is based on the general continuity equation for a flow system. Having the cross-sectional area of flow A represented by $y \times 1$ for a unit strip of overland flow; $y =$ the depth of flow; $v =$ the mean water velocity; $x =$ the strip length; $t =$ the time; and $q =$ rainfall intensity, i, the continuity equation for this flow system becomes,

$$y\frac{\partial v}{\partial x} + v\frac{\partial y}{\partial x} + \frac{\partial y}{\partial t} = i \tag{7.34}$$

and the momentum equation is

$$v\frac{\partial v}{\partial x} + \frac{g}{y}\frac{\partial \bar{y}y}{\partial x} = g(S_o - S_f) - \frac{v}{y}i \tag{7.35}$$

where S_0 is the strip slope, S_f is the friction slope, and $\bar{y}$ is the distance from the water surface to the centroid of the flow's cross section.

Equations 7.34 and 7.35 are based on the de Saint–Venant equations. Derivations of the equations along with their assumptions may be found in various handbooks on open-channel hydraulics.

It must be noted that the strip's cross-sectional area $A = 1 \times y$ and $\bar{y} = y/2$. Therefore, for a steady-flow condition where equilibrium between rainfall and runoff is reached, Equation 7.34 becomes

$$y\frac{\partial v}{\partial x} + v\frac{\partial y}{\partial x} = i \tag{7.36}$$

$$\text{Since } Q = vy, \quad \text{then}\frac{\partial Q}{\partial x} = y\frac{\partial v}{\partial x} + v\frac{\partial y}{\partial x} \tag{7.37}$$

Therefore, for equilibrium conditions

$$\frac{\partial Q}{\partial x} = i \tag{7.38}$$

By integrating Equation 7.38 using the limits of x from 0 to L and for Q from 0 to Q_T

$$Q_T = iL \tag{7.39}$$

For i in millimeters per hour and L in meters, Equation 7.39 becomes

$$Q_T = \frac{iL}{(3.6)\times 10^6 \, \text{m}^3/\text{s}} \tag{7.40}$$

Equation 7.40 agrees well with that of Izzard's, Equation 7.29.

Since $y \times \bar{y} = y \times y/2 = y^2/2$ by definition and ignoring any eddy losses and momentum correction coefficient, Equation 7.35, which can be applied for both laminar or turbulent stages, becomes

$$v\frac{\partial v}{\partial x} + g\frac{\partial y}{\partial x} = g(S_0 - S_f) - \frac{v}{y}i \tag{7.41}$$

From Equation 7.34, $y\frac{\partial v}{\partial x} = i - v\frac{\partial y}{\partial x}$

Multiplying Equation 7.41 by (y/v), we get

$$y\frac{\partial v}{\partial x} + \frac{gy}{v}\frac{\partial y}{\partial x} = \frac{gy}{v}(S_0 - S_f) - i \tag{7.42}$$

Rearranging Equation 7.42, we get

$$\left(i - v\frac{\partial y}{\partial x}\right) + \frac{gy}{v}\frac{\partial y}{\partial x} = \frac{gy}{v}(S_0 - S_f) - i$$

or

$$\left(2i - v\frac{\partial y}{\partial x}\right) = \frac{gy}{v}\left[(S_0 - S_f) - \frac{\partial y}{\partial x}\right] \tag{7.43}$$

From Figure 7.13, $\Delta y = S_0 \Delta x - S_f \Delta \underline{x}$; therefore

$$\frac{\partial y}{\partial x} = S_0 - S_f$$

Accordingly, Equation 7.43 becomes

$$\left(2i - v\frac{\partial y}{\partial x}\right) = 0 \tag{7.44}$$

Using the Chezy Equation $v = C\sqrt{RS_f}$ for the overland strip of unit width, $R = y$ or

$$v = \left(C\sqrt{S_f}\right)y^{0.5}$$

As a first approximation, take $S_0 = S_f$; therefore, $K \leq C\sqrt{S_0}$.

For $K = \left(C\sqrt{S_0}\right)$, therefore $v = Ky^{0.5}$

From Equation 7.44 we get

$$Ky^{0.5}\frac{\partial y}{\partial x} = 2i \tag{7.45}$$

Integrating Equation 7.45 with the limits from $x = 0$ to $x = L$ and from $y = 0$ to $y = Y_e$

$$(K / 3) (Y_e)^{1.5} = iL$$

or

$$Y_e = \left(\frac{3 \times iL}{K \times 3.6 \times 10^6} \right)^{2/3} \text{m} \tag{7.46}$$

where Y_e is in meters, i is in millimeter per hour, and L is in meters.

Therefore, the detained volume of water over land at equilibrium is $V_e = (1 \times Y_e) \times L$

$$V_e = L \; Y_e = \left(\frac{3i}{K \times 3.6 \times 10^6} \right)^{2/3} L^{5/3} \tag{7.47}$$

To compute the time to equilibrium, Izzard's empirical Equation 7.28 can be used.

To compare the results obtained using the last Equation 7.47 and those of Izzard, use Example 7.4.

EXAMPLE 7.4

A parking lot 50 meters long in the direction of the slope and 100 meters wide has a tar-and-gravel pavement on a slope of 0.005. Assuming a uniform rainfall intensity of 50 millimeters per hour for 60 minutes, find q_e and t_e. Compare the results obtained by the Izzard and Soliman Equations. Find also the section of the gutter that collects the total runoff flow from the parking lot, as given in Figure 7.15.

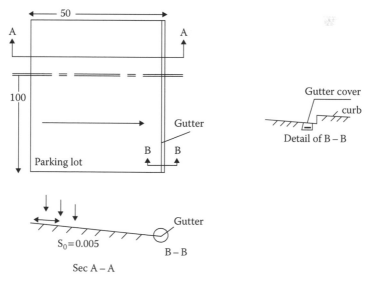

FIGURE 7.15 Overland flow over a parking lot.

SOLUTION

$$q_e = \frac{50 \times 50}{3.6 \times 10^6} = 694.444 \times 10^{-6}\ m^3/s$$

V_e (Izzard) Equation 7.28 = 0.52 m³
For $K = C\ \sqrt{S_0} = 40\ \sqrt{0.005} = 2.83$
V_e (Equation 7.47) = 0.42 m³
$t_e = 2\ V_e/60q_e = 2 \times 0.52/60 \times q_e = 24.76$ min
t_e from Equation 7.44 = 24.76 (0.4/0.52) = 19.2 min

The second value is a little lower than that obtained using the Izzard equation, due to the approximation in S_f. However, this can be corrected by successive trials; on the other hand, Equation 7.47 can be used for both laminar and turbulent flows, whereas the Izzard equation is used only in laminar flow conditions.

7.9.1　DESIGN OF THE SIDE CHANNEL OF THE GUTTER

Maximum discharge q_e through the gutter = $q_e \times L = 694.444 \times 10^{-6} \times 100$
$$= 694.444 \times 10^{-4}\ m^3/s$$

As a first choice, let the maximum velocity through the gutter be = 1 m/s
The cross-sectional area A at the channel outlet = 0.0694 m²
$$= 694\ cm^2$$

For B = bottom width of the channel and Y_0 = the flow depth, $A = B \times Y_0$.

For the best hydraulic section, $B = 2\ Y_0$, $A = 2\ (Y_0)^2 = 694$ cm², $Y_0 = 18.627$ cm, and $B = 37.25$ cm.

To round it off, take $B = 35$ cm and $Y_0 = 20$ cm; $A = 700$ cm².
Therefore, the hydraulic radius $R = 35 \times 20/(35 + 2 \times 20) = 9.33$ cm.
The velocity of flow at the outlet = 0.0694/0.07 = 0.991 m/s.
Using the Manning Equation, the channel slope was found to be 0.0055.

To get the time to equilibrium, t_e, for the channel flow, the routing Equation 7.25 can be used. However, a dimensionless hydrograph as given in Figure 7.16 or a similar one can be adopted for this purpose. Figure 7.16 was developed by the author based on a study made by Brutsaert (1971), who used a numerical model to estimate the discharge of a channel with lateral flow.

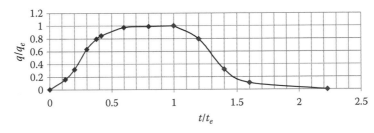

FIGURE 7.16　Dimensionless hydrograph of channel flow.

REFERENCES

Barnes, B. S. 1940. Discussion of analysis of runoff characteristics. *Trans ASCE* 105:106.

Hudlow, M. D. and R. A. Clark. 1969. Hydrological synthesis by digital computer. *Journal of the ASCE Hydraulics Division* 95(3):839–860.

Izzard, C. F. 1946. Hydraulics of runoff from developed surfaces. *Proc High Res Board* 26:129–50.

Linsley, R. K. 1943. Application to the synthetic unit graph in the Western mountain states. *Transactions of the American Geophysical Union* 24:581–587.

McCuen, R. H. 2005. *Hydrologic Analysis and Design.* 3rd edition. Upper Saddle River, NJ: Prentice Hall.

Miller, A. C., S. N. Kerr, and D. J. Spaeder. 1983. Calibration of Snyder coefficients for Pennsylvania. *Water Resources Bulletin* 19(4):625–630.

Sherman, L. K. 1932. Streamflow from rainfall by the unit-graph method. *Eng News Rec* 108:501–5.

Snyder, F. F. 1938. Synthetic unit hydrographs. *Trans Am Geophys Union* 19(1):447–54.

Snyder, W. M. 1955. Hydrograph analysis by method of least squares. *Proc ASCE* 81:793.

Soil Conservation Service. 1986. Urban hydrology for small watersheds. Technical Release 55, Washington, DC: U.S. Department of Agriculture.

BIBLIOGRAPHY

Bedient, P. B., and W. C. Huber. 1992. *Hydrology and Floodplain Analysis.* Upper Saddle River, NJ: Prentice Hall.

Brutsaert, W. 1971. De Saint venant equations experimentally verified. *ASCE J Hydraul Div* 97:8378.

Chow, V. T. 1964. Runoff. In *Handbook of Applied Hydrology,* ed. V. T. Chow, sec. 14. New York: McGraw-Hill.

Commons, G. G. 1942. Flood hydrographs. *Civil Eng* 12:571.

Eagleson, P. S. 1970. *Dynamic Hydrology.* New York: McGraw-Hill.

Henderson, F. M., and R. A. Wooding. 1964. Overland flow and groundwater flow from a steady rainfall of finite duration. *Geophys Res* 69:1531–40.

Langbein, W. B. 1940. Some channel storage and unit hydrograph studies. *Trans Am Geophys Union* 21(2):620–7.

Linsley, R. K., M. A. Kohler, and J. L. H. Paulhus. 1949. *Applied Hydrology.* New York: McGraw-Hill.

McCuen, R. H., and R. E. Beighley. 2003. Seasonal flow frequency analysis. *J Hydrol* 279:43–56.

McDonald, C. C., and W. B. Langbein. 1948. Trends in runoff in the Pacific Northwest. *Trans Am Geophys Union* 29:387–97.

Mockus, V. 1957. *Use of Storm and Watershed Characteristics in Synthetic Hydro-Graph Analysis and Application.* Washington, DC: U.S. Soil Conservation Service.

Morgali, J. R., and R. K. Linsley. 1965. Computer analysis of overland flow. *ASCE J Hydraul Div* 91:81–100.

Sherman, L. K. 1942. The unit hydrograph. In *Hydrology,* ed. O. E. Meinzer, chap. HE. New York: McGraw-Hill, 1942; reprinted New York: Dover, 1949.

Soil Conservation Service. 1986. *Urban Hydrology for Small Watersheds.* Technical Release 55. Washington, DC: Soil Conservation Service, U.S. Department of Agriculture.

Soliman, M. M. 2009. Overland flow process in urban watershed with impervious ground surface. In *Proceedings of Sixth International Conference on Environmental Hydrology,* p. 59. Cairo, Egypt: ASCE.

Williams, H. M. 1945. Discussion of military airfields. *Trans ASCE* 110:820.

Woo, D. C., and E. F. Brater. 1962. Spatially varied flow from controlled rainfall. *Proc ASCE* 88:31–56.

Wurbs, R. A., and W. P. James. 2002. *Water Resources Engineering*. Upper Saddle River, NJ: Prentice Hall.

8 Flood Routing

8.1 INTRODUCTION

Routing is a process used to predict the temporal and spatial variations of a flood hydrograph as it moves through a river reach or reservoir. The effects of storage and flow resistance within a river reach are reflected by changes in both the shape and timing of the hydrograph as the flood wave moves from upstream to downstream. Figure 8.1 shows the major changes that occur to a discharge hydrograph as a flood wave moves downstream. In other words, the hydrograph depicts how a flood wave can be reduced in magnitude and lengthened in time by applying storage between the two points within the reach.

In general, routing techniques can be classified into two categories: hydraulic routing and hydrologic routing. Hydraulic routing techniques are based on the solution of the partial differential equations of unsteady open-channel flow. These equations are often referred to as the St. Venant equations or the dynamic-wave equations. Hydrologic routing uses the continuity equation and an analytical or an empirical relationship between storage within the reach and discharge at the outlet.

8.2 HYDRAULIC ROUTING TECHNIQUES

8.2.1 EQUATIONS OF MOTION

The Saint Venant equations describe one-dimensional (1D) unsteady flow in open channels and consist of the continuity equation, Equation 8.1, and the momentum equation, Equation 8.2. The solution of these equations defines the propagation of a flood wave with reference to distance along the channel and time.

$$A\frac{\partial V}{\partial x} + VB\frac{\partial y}{\partial x} + B\frac{\partial y}{\partial t} = q \tag{8.1}$$

$$S_f = S_0 - \frac{\partial y}{\partial x} - \frac{V}{g}\frac{\partial V}{\partial x} - \frac{1}{g}\frac{\partial V}{\partial t} \tag{8.2}$$

where A is the cross-sectional flow area [L^2]; V is the average velocity of water [LT^{-1}]; x is the distance along channel [L]; B is the water surface width [L]; y is the depth of water [L]; t is the time [T]; q is the lateral inflow per unit length of channel [L^2T^{-1}]; S_f is the friction slope; S_0 is the channel bed slope; and g is the gravitational acceleration [LT^{-2}].

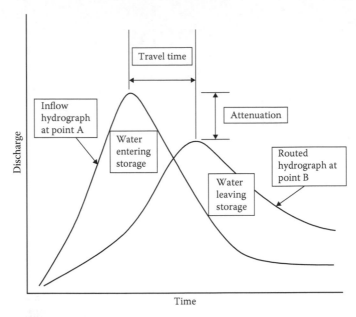

FIGURE 8.1 Discharge hydrograph routing effects.

Solved with the proper boundary conditions, Equations 8.1 and 8.2 constitute the complete dynamic wave equations. The two definitions of the terms of the dynamic wave equations are as follows:

1. Continuity equation:

$$A\frac{\partial V}{\partial x} = \text{prism storage}$$

$$VB\frac{\partial y}{\partial x} = \text{wedge storage}$$

$$B\frac{\partial y}{\partial t} = \text{rate of rise}$$

$$q = \text{lateral inflow per unit length}$$

2. Momentum equation:

$$S_f = \text{friction slope (frictional forces)}$$

$$S_0 = \text{bed slope (gravitational effects)}$$

$$\frac{\partial y}{\partial x} = \text{pressure differential}$$

$$\frac{V}{g}\frac{\partial V}{\partial x} = \text{convective acceleration}$$

$$\frac{1}{g}\frac{\partial V}{\partial t} = \text{local acceleration}$$

Dynamic wave equations are considered to be the most accurate and comprehensive solution to 1D unsteady flow problems in open channels.

8.3 HYDROLOGIC ROUTING TECHNIQUES

Hydrologic routing employs the continuity equation and either an analytical or an empirical relationship between storage within the reach and discharge at the outlet. In its simplest form, the continuity equation can be written as follows: inflow minus outflow equals the rate of change of storage within the reach.

8.3.1 STORAGE EQUATION

Because the methods of flood routing depend upon knowledge of storage in the reach, a way of evaluating the storage must be found. There are two ways of doing this. One is to make a detailed topographical and hydrographical survey of the river reach and the riparian land and thus determine the storage capacity of the channel at different levels. The other is to use the records of past levels of flood waves at the two points and hence deduce the reach's storage capacity, assuming that such storage capacity will not change substantially over time and so may be used to route the passage of large and, more critically, predicted floods. As many data as possible are required for the second method, which is the one generally used, including (1) flow records at the beginning and end of the reach and regarding any tributary streams joining it and (2) rainfall records over any areas contributing direct runoff to it.

8.3.2 CHANNEL ROUTING

The solution to the storage equation in this case is more complicated than that for the simple reservoir because wedge storage is involved. Storage is no longer a function of discharge only. McCarthy (1938), in what has become known as the Muskingum method, proposed that storage should be expressed as a function of both inflow and discharge.

Storage in the reach of a river is divided into two parts, prism and wedge storages, which are illustrated in Figure 8.2. This is because the slope of the surface is not uniform during floods. If the continuity of flow through the reach shown in Figure 8.2 is now considered, it is clear that what enters the reach at point A must emerge at point B or temporarily move into storage as given in Equation 8.3.

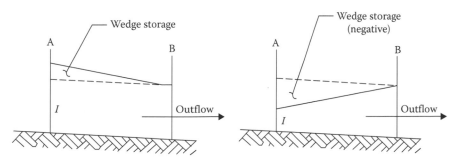

FIGURE 8.2   Prism and wedge storage concepts.

In Figure 8.2, the two cases of flow show the inflow I at the inlet section, the outflow discharge O at the outlet section, and the wedge storage dS in time dt.

The flow equation is approximated for a time interval t by the expression

$$\frac{I_1+I_2}{2}t - \frac{O_1+O_2}{2}t = S_2 - S_1 \qquad (8.3)$$

where subscripts 1 and 2 denote the values at the beginning and end, respectively, of the time t. The time t is called the routing period, and it must be sufficiently short so that the assumption implicit in Equation 8.3 (that is, that the inflow and outflow hydrographs consist of a series of straight lines) does not depart too far from reality. In particular, if t is too long, it is possible to miss the peak of the inflow curve, so the period should be kept shorter than the travel time of the flood-wave crest through the reach. On the other hand, the shorter the routing period, the greater the amount of computation needed.

The two constants can be found as follows. Let Figure 8.3 represent the simultaneous inflow I and outflow D of a river reach. When $I > D$, water is entering the storage water leaving storage in the reach, and when $D > I$, water is leaving.

8.3.3 DEVELOPMENT OF THE MUSKINGUM ROUTING EQUATION

Prism storage is computed as follows: prism storage $= O \times K$, where O is the outflow and K is the travel time through the reach. Wedge storage is computed as follows: wedge storage $= (I - O) \times X \times K$, where $(I - O)$ is the difference between inflow and outflow, X is a weighting coefficient, and K is the travel time. The parameter X is a dimensionless value expressing a weighting of the relative effects of inflow and outflow on the storage S within the reach. Thus, the Muskingum method defines the storage in the reach as a linear function of the weighted inflow and outflow:

$$S = \text{prism storage} + \text{wedge storage}$$
$$S = KO + KX(I-O) \qquad (8.4)$$
$$S = K[XI + (1-X)O]$$

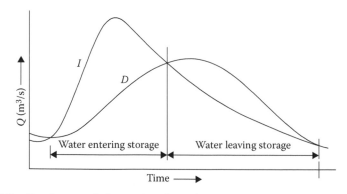

FIGURE 8.3 Simultaneous inflow and outflow of a river reach.

where S is the total storage in the routing reach [L^3]; O = rate of outflow from the routing reach [L^3T^{-1}]; I is the rate of inflow to the routing reach [L^3T^{-1}]; K is the travel time of the flood wave through the reach [T]; and X is the dimensionless weighting factor, ranging from 0 to 0.5.

The quantity within the square brackets of Equation 8.4 is the expression of weighted discharge. When $X = 0$, the equation reduces to $S = KO$, indicating that storage is only a function of outflow, which is equivalent to level-pool reservoir routing with storage as a linear function of outflow. When $X = 0.5$, equal weight is given to both inflow and outflow, and the condition is equivalent to a uniformly progressive wave that does not attenuate. Thus, 0 and 0.5 are the limits of the value of X, and within this range, the value of X determines the degree of attenuation of the flood wave as it passes through the routing reach. A value of 0 produces the maximum attenuation, and 0.5 produces pure translation with no attenuation.

The Muskingum routing equation is obtained by combining Equation 8.4 with the continuity equation, Equation 8.5, and solving for O_2:

$$\frac{O_1 + O_2}{2} = \frac{I_1 + I_2}{2} - \frac{S_2 - S_1}{\Delta t} \tag{8.5}$$

$$O_2 = C_1 I_2 + C_2 I_1 + C_3 O_1 \tag{8.6}$$

The subscripts 1 and 2 in these equations indicate the beginning and end, respectively, of a time interval t. The routing coefficients C_1, C_2, and C_3 are defined in terms of t, K, and X as follows:

$$C_1 = \frac{\Delta t - 2KX}{2K(1-X) + \Delta t} \tag{8.7}$$

$$C_2 = \frac{\Delta t + 2KX}{2K(1-X) + \Delta t} \tag{8.8}$$

$$C_3 = \frac{2K(1-X) - \Delta t}{2K(1-X) + \Delta t} \tag{8.9}$$

Given an inflow hydrograph, a selected computation interval t, and estimates for the parameters K and X, the outflow hydrograph can be calculated.

The following concept from Carter and Godfrey (1960) concisely sums up the choice of values for X and K: The factor X is chosen so that the indicated storage volume is the same whether the stage is rising or falling. For spillway discharges from a reservoir, X may be zero because the reservoir stage, and hence the storage, are uniquely defined by the outflow; hence, the rate of inflow has a negligible influence on the storage in the reservoir at any time. For uniformly progressive flow, X equals 0.50 and both the inflow and outflow are equal in weight. In this wave, no change in shape occurs, and the peak discharge remains unaffected. Thus, the value of X will range from 0 to 0.50, with an average value of 0.25 for river reaches.

The factor K has the dimension of time and is the slope of the storage-weighted discharge relation, which in most flood problems approaches a straight line. Analyses of many flood waves indicate that the time required for the center of the mass of the flood wave to pass from the upstream end of the reach to the downstream end is equal to the factor K. The time between peaks only approximates the factor K. Ordinarily, the value of K can be determined with much greater ease and certainty than that of X.

Now assume a value of X, for example $X = 0.1$, and compute the value of the expression $0.1I + 0.9O$ for various times. Plot these against the corresponding S values taken from the mass curve of storage for the reach. The resulting plot, known as a storage loop, is shown in Figure 8.4a; clearly, there is no linear relationship. Take further values of X (for example, 0.2, 0.3, and so on) until a linear relationship is established, as in Figure 8.4c, when the particular value of X may be adopted. Now obtain K by measuring the slope of the line.

Knowledge about units is required. It is often helpful to work in somewhat unusual units, both to save computation and to keep numbers small. For example, storage S is conveniently expressed in cubic meters per second per day; such a unit is that quantity obtained from 1 cubic meter per second flowing for 1 day equal to 86.4×10^3 m^3. If S is expressed in cubic meters per second per day and the ordinate of Figure 8.4 is in cubic meters per second, then K is in days.

8.3.4 Muskingum–Cunge Channel Routing

The Muskingum–Cunge channel-routing technique (Cunge 1969) is a nonlinear coefficient method that accounts for hydrograph diffusion based on the physical properties of the channel and the inflowing hydrograph.

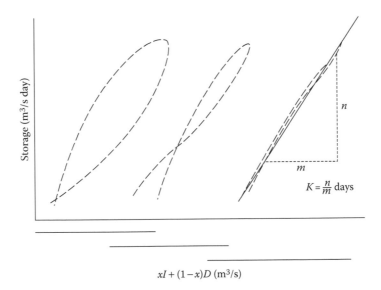

$$xI + (1-x)D \ (\text{m}^3/\text{s})$$

FIGURE 8.4   River-routing storage loops.

8.3.4.1 Development of Equations

The basic formulation of the equations is derived from the continuity equation, Equation 8.10, and the diffusion form of the momentum, Equation 8.11.

$$\frac{\partial A}{\partial t} + \frac{\partial Q}{\partial x} = q_L \tag{8.10}$$

$$S_f = S_0 - \frac{\partial y}{\partial x} \tag{8.11}$$

By combining Equations 8.10 and 8.11 and further linearizing them, the following convective diffusion equation, Equation 8.12, is formulated, which is the basis for the Muskingum–Cunge method:

$$\frac{\partial Q}{\partial t} + c\frac{\partial Q}{\partial x} = \mu\frac{\partial^2 Q}{\partial x^2} + cq_L \tag{8.12}$$

where Q is the discharge [L^3T^{-1}]; A is the flow area [L^2]; t is the time [T]; x is the distance along the channel [L]; Y is the depth of flow [L]; q_L is the lateral inflow per unit of channel length [L^2T^{-1}]; S_f is the friction slope; and S_0 is the bed slope.

The wave celerity (c) and the hydraulic diffusivity (μ) are expressed as follows:

$$c = \frac{dQ}{dA} \tag{8.13}$$

$$\mu = \frac{Q}{2BS_0} \tag{8.14}$$

where B is the top width of the water surface [L].

8.3.4.2 Data Requirements

Data required for the Muskingum–Cunge method are the following:

- Cross section of the representative channel
- Reach length, L
- Manning roughness coefficients, n (for the main channel and overbanks)
- Friction slope (S_f) or channel bed slope (S_0)

This method can be used with a simple (i.e., trapezoid, rectangle, square, triangle, or circular pipe) or more detailed cross section (i.e., cross sections with a left overbank, main channel, and a right overbank). The cross section is assumed to be representative of the entire routing reach. If this assumption is not adequate, the routing reach should be broken up into smaller subreaches, with representative cross sections for each. Reach lengths are measured directly from topographic maps.

Roughness coefficients (Manning's n) must be estimated for the main channels and overbank areas. If information is available to estimate an approximate energy grade line slope (friction slope, S_f), that slope should be used instead of the bed slope. However, if no information is available to estimate the same, the channel bed slope should be used.

EXAMPLE 8.1

This examples outlines routing in a stream channel using the Muskingum method. Given the inflow and outflow hydrographs provided in Table 8.1, derive the constants X and K for the reach.

SOLUTION

In columns 1 and 2, the given hydrographs are listed at a routing period interval, taken as 6 hours. The storage units are (0.25 cubic meters per second per day) because the routing period is 0.25 days. Columns 4–6 are simply tabular statements of the processes of calculating the storage.

A value of X is then chosen, in the first instance 0.2, and the value inside the square brackets of Equation 8.4 is then evaluated in columns 7–9. Columns 6 and 9 are plotted in Figure 8.5, which produces the loop in the Example 8.1, part 2 of the outflow, left-hand side of the Figure 8.5.

A second value of $X = 0.25$ is then tried (refer to columns 10–12), and the resulting plot is the central one of Figure 8.5. A third value, $x = 0.3$, is also tabulated and plotted on the right of the Figure 8.5. By inspection, the central value of $x = 0.25$ most closely approximates a straight line; so, this is chosen as the value of x. K is determined by measuring the slope of the median line, as shown in Figure 8.6, and is found to be 1.5 days:

$$X = 0.25 \quad \text{and} \quad K = 1.5 \text{ days}$$

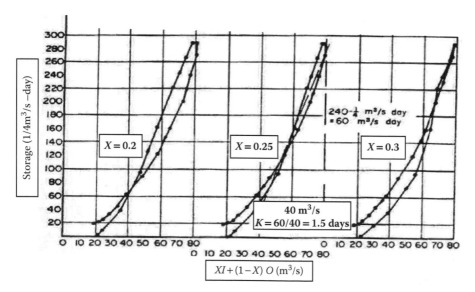

FIGURE 8.5   Storage loops for the reach.

TABLE 8.1
Calculation of Storage Loops

1	2	3	4	5	6	7	8	9	10	11	12	13	14	15
							x = 0.2			x = 0.25			x = 0.3	
	Inflow	Outflow	I−O	Mean Storage	Cumulative Storage									
Hour	I (m³/s)	O (m³/s)	(m³/s)	(¼ m³/s/day)	(¼ m³/s/day)	0.2I	0.8D	Total	0.25I	0.75D	Total	0.3I	0.7D	Total
0	22	22	0	0	0	4	17	21	5	16	21	7	15	22
6	23	21	2	1	1	5	17	22	6	16	22	7	15	22
12	35	21	14	8	9	7	17	24	9	16	25	10	15	25
18	71	26	45	29	38	14	21	35	18	19	37	21	18	39
24	103	34	69	57	95	20	27	47	26	25	51	31	24	55
30	111	44	67	68	163	22	35	57	28	33	61	33	31	64
36	109	55	54	60	223	22	44	66	27	41	68	33	38	71
42	100	66	34	44	267	20	53	73	25	49	74	30	46	76
48	86	75	11	22	289	17	60	77	21	56	77	26	52	78
54	71	82	−11	0	289	14	66	80	18	61	79	21	57	78
60	59	85	−26	−18	271	12	68	80	15	64	79	18	59	77
66	47	84	−37	−31	240	9	67	76	12	63	75	14	59	73
72	39	80	−41	−39	201	8	64	72	10	60	70	11	56	67
78	32	73	−41	−41	160	6	58	64	8	55	63	10	51	61
84	28	64	−36	−38	122	6	51	57	7	48	55	8	45	53
90	24	54	−30	−33	89	5	43	48	6	40	46	7	38	45
96	22	44	−22	−26	63	4	35	39	5	33	38	7	31	38
102	21	36	−15	−18	45	4	29	33	5	27	32	6	25	31
108	20	30	−10	−12	33	4	24	28	5	22	27	6	21	27
114	19	25	−6	−8	25	4	20	24	5	19	24	6	17	23
120	19	22	−3	−4	21	4	18	22	5	16	21	6	15	21
126	18	19	−1	−2	19	4	15	19	4	14	18	5	13	18

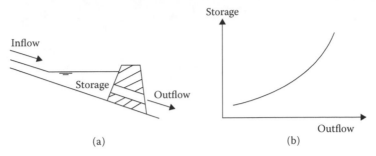

FIGURE 8.6 Storage versus water level in the reservoir: (a) a sketch showing outflow from the reservoir and (b) the relation between storage and outflow.

8.3.5 RESERVOIR ROUTING

If Equation 8.1 is now arranged so that all known terms are on one side, the expression becomes

$$\frac{1}{2}(I_1 + I_2)t + \left(S_1 - \frac{1}{2}O_1 t\right) = \left(S_2 + \frac{1}{2}O_2 t\right) \tag{8.15}$$

The routing process consists of (1) inserting the known values to obtain $S_2 + 1/2\ O_2 t$ and (2) deducing the corresponding value of O_2 from the relationship connecting storage and discharge. This method was first developed by L.G. Puls of the U.S. Army Corps of Engineers.

The simplest case is that of a reservoir receiving inflow at one end and discharging through a spillway at the other. In such a reservoir, it is assumed that there is no wedge storage and that the discharge is a function of the surface elevation, provided that the spillway arrangements are either free-overflow or gated with fixed gate openings. Reservoirs with sluices can be treated also as simple reservoirs if the sluices are opened to defined openings at specified surface-water levels, so that an elevation-discharge curve can be drawn. The other data required are the elevation-storage curve of the reservoir and the inflow hydrograph.

8.3.6 MODIFIED PULS RESERVOIR ROUTING

One of the simplest routing applications is analysis of a flood wave that passes through an unregulated reservoir. The inflow hydrograph is known, and the outflow hydrograph from the reservoir must be computed (Figure 8.7a). Assuming that all gate and spillway openings are fixed, a unique relationship between storage and outflow can be developed, as shown in (Figure 8.7b). The equation defining the storage routing, based on the principle of conservation of mass, can be written in approximate form for a routing interval Δt. Assuming the subscripts 1 and 2 denote the beginning and end of the routing interval, respectively, the equation is written as follows:

$$\frac{O_1 + O_2}{2} = \frac{I_1 + I_2}{2} - \frac{S_2 - S_1}{\Delta t} \tag{8.16}$$

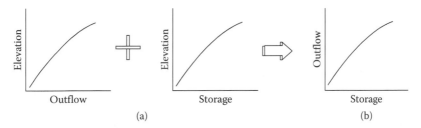

FIGURE 8.7 Reservoir storage routing: (a) the outflow hydrograph must be computed and (b) the relationship between storage and outflow.

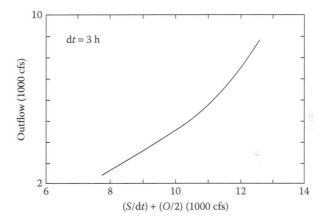

FIGURE 8.8 Reservoir storage outflow curve.

The known values in this equation are the inflow hydrograph and the quantities of storage and discharge at the beginning of the routing interval. The unknown values are the storage and discharge at the end of the routing interval. With two unknowns (O_2 and S_2) remaining, another relationship is required to obtain a solution.

The storage–outflow relationship is normally used as the second equation. How that relationship is derived is what distinguishes various storage routing methods. For an uncontrolled reservoir, outflow and water in storage are both uniquely a function of lake elevation. The two functions can be combined to develop a storage–outflow relationship, as shown in Figure 8.8.

Elevation–discharge relationships can be derived directly from hydraulic equations, whereas elevation–storage relationships are derived using topographic maps. Elevation–area relationships are computed first, and then either the average end-area or conic methods are used to compute volumes.

The storage–outflow relationship provides the outflow for any storage level. For a nearly empty reservoir, the outflow capability would be minimal. If the inflow is less than the outflow capability, the water will flow through. During a flood, the inflow increases and eventually exceeds the outflow capability. The difference between inflow and outflow produces a change in storage, with the difference (on the rising side of the outflow hydrograph) representing the volume of water entering storage.

As water enters storage, the outflow capability increases because the pool level increases; therefore, the outflow increases. This increasing outflow with increasing water in storage continues until the reservoir reaches a maximum level. This occurs the moment the outflow equals the inflow. Once the outflow becomes greater than the inflow, the storage level begins to drop. The difference between the outflow and the inflow hydrographs on the recession side reflects the water withdrawn from storage.

The modified Puls method, as applied to reservoirs, consists of a repetitive solution of the continuity equation. Assume that the surface of the reservoir water remains horizontal, and therefore, outflow is a unique function of reservoir storage. The continuity equation can be manipulated to get both the unknown variables on the left-hand side of the equation.

$$\left[\frac{S_2}{\Delta t} + \frac{O_2}{2}\right] = \left[\frac{S_1}{\Delta t} + \frac{O_1}{2}\right] - O_1 + \frac{I_1 + I_2}{2} \tag{8.17}$$

Since I is known for all time steps and O_1 and S_1 are known for the first time step, the right-hand side of the equation can be calculated. The left-hand side of the equation can be solved by trial and error by assuming a value for either S_2 or O_2, obtaining the corresponding value from the storage–outflow relationship, and then iterating until Equation 8.17 is satisfied. Rather than resort to this iterative procedure, a value of Δt is selected and points on the storage–outflow curve are replotted as the storage-indication curve, as shown in Figure 8.8. This graph allows for a direct determination of the outflow (O_2) once a value of storage indication ($S_2/\Delta t + O_2/2$) has been calculated from Equation 8.17.

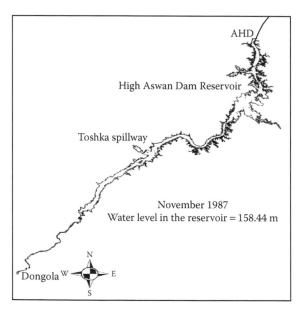

FIGURE 8.9 Storage-indication curve.

8.4   CASE STUDY: FLOOD ROUTING FOR THE HIGH ASWAN DAM RESERVOIR

The flood-routing process is nearly the same for both small and big reservoirs. The same process can also be followed for reservoirs in either arid or wet zones. The following subsections describe the development of a hydrological routing model for the operation of the High Aswan Dam (HAD), along with the results thereof.

The HAD is one of the most important projects in the history of Egypt. The dam, completed in 1968 and located 7 kilometers south of Aswan City, has provided full control of the Nile River flooding. The Nile flow is allowed to pass only through the open-cut channel at the eastern side of the dam, where six tunnel inlets provided with steel gates were constructed for discharge control and water supply to power plants. An escape (spillway) is also provided at the western side of the dam to permit discharge of excess water and to control the maximum level of the upstream reservoir.

The HAD Reservoir (HADR) extends for 500 kilometers along the Nile River upstream of HAD (Figure 8.10) and covers an area of 6000 square kilometers. The reservoir is surrounded by rocky desert terrain. To the west is the great Sahara Desert, and the Eastern Desert on the east side extends to the Red Sea. Another, uncontrolled spillway was constructed at the end of Khor Tushka (on the western side of Lake Nasser, about 256 kilometers upstream of the dam). This spillway is connected to the Tushka Depression by a canal, through which the excess flood can be diverted to the depression.

8.4.1   MODEL DEVELOPMENT

The model (El Mostafa 2007) is based on the mass balance equation of the reservoir, which can be described as the summation of the inflow equal the summation of the outflow and can be represented in the following form:

$$\text{The change in storage} = (\text{Inflow} - \text{Releases} - \text{Tushka-spillway overflow}$$
$$- \text{Evaporation losses})\Delta t \tag{8.18}$$

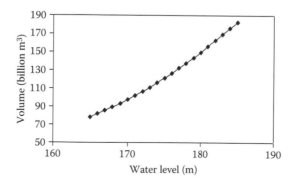

FIGURE 8.10   Boundaries of the High Aswan Dam lake obtained using a land remote-sensing satellite system (LANDSAT) image acquired in November 1987.

As the results of the hydrodynamic model show, the water level in the reservoir is almost constant upstream from the HAD to the Tushka spillway; the release over the weir can be directly calculated from the water level. The Tushka spillway overflow can be calculated from the water level using the following equation (Hydraulic Research Institute 1998):

$$Q_{Tushka} = 2.012B \ (WL - Crest \ level)^{1.569} \qquad (8.19)$$

where B is the width of the spillway crest in meters, and WL is the water level in the reservoir.

Note from Equation 8.19 that the spillway starts to release water when the water level is above its crest level.

The evaporation losses, the Tushka overflow, and storage can be calculated as a function of the water level, as shown in Figure 8.11a.

$$S = 2 \times 10^{-15} \times (WL)^{7.4647} \qquad (8.20)$$

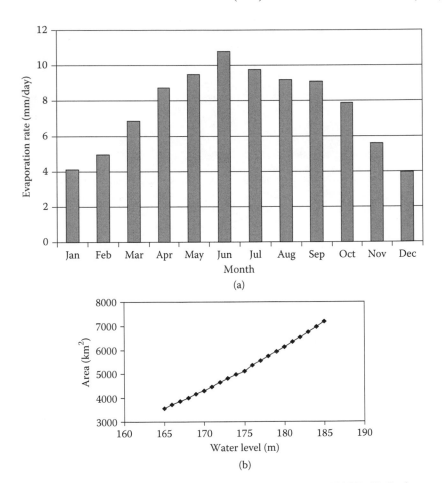

FIGURE 8.11 (a) Estimates of monthly evaporation rates from the HADR. (b) Surface area versus water level in the reservoir.

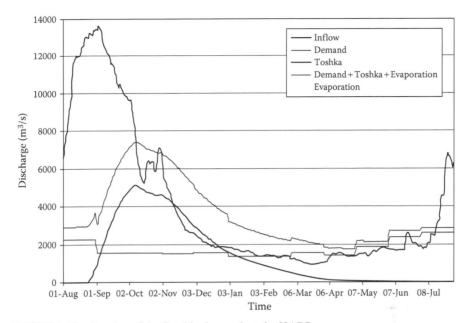

FIGURE 8.12   Routing of the flood hydrograph at the HADR.

where S is the volume of stored water in billion cubic meters and WL is the water level in the reservoir.

Figure 8.11b illustrates the reservoir surface area versus the water level (El Mostafa 2002), from which a relationship between the surface area of the lake and the water level can be derived (see Equation 8.21). The evaporation losses can then be calculated in cubic meters per second by multiplying the surface area by the evaporation rate.

$$A = 1 \times 10^{-10} \times (\text{WL})^{6.0889} \qquad (8.21)$$

where A is the surface area in square kilometers and WL is the water level in the reservoir.

The model has developed interface with options to run for predefined releases. It may also be used to maximize the water level and the hydropower. Figure 8.12 shows the output of the model for the flood hydrograph routing at HADR. The flowchart in Figure 8.13 describes the model parts.

8.4.1.1   Model Constraints

The model constraints can be summarized as follows:

- Maximum release from HAD was assumed to be 280 million cubic meters per day, which is the maximum capacity of the cross sections of the river downstream the dam, in order to prevent scouring of the bed downstream the Nile Barrages.
- Minimum release from HAD was assumed to be 60 million cubic meters per day for reasons of navigation, hydropower, and water quality.

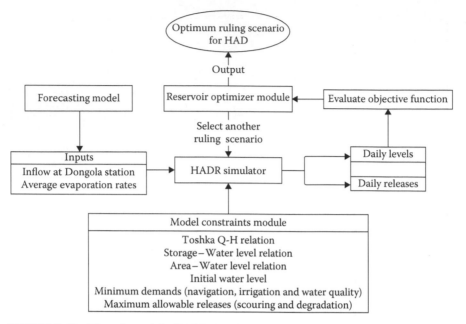

FIGURE 8.13 Flow chart of the HADR simulation model.

- The released volume should not exceed Egypt's share of the Nile water, which, according to the Nile Basin Countries' Agreement, is 55.5 billion cubic meters per year.
- The starting water level upstream of HAD before the beginning of the year and the ending water level are obtained from a long-term forecasting model.
- Relationships derived between (1) water level and area; and (2) water level and storage.
- The relationship between the water level and Tushka spillway discharge was taken into consideration.

8.4.1.2 Model Results

8.4.1.2.1 *Effect of Varying HAD Releases*

The management of the reservoir flood concerns the control of the HAD releases in a way that maximizes the objective functions of the model, such as the water level and the hydropower. To test this effect, the developed model was used to calculate the water level with two release scenarios, maximum and minimum, previously proposed by the HADA, and assuming the maximum recorded hydrograph at the Donqola gagging station, which records the upstream inflow.

Figure 8.14 shows the results of these two scenarios. Note that the proposed release scenarios affect the water level from the beginning of November to the end of April. During this period, the incoming flood is at its peak, whereas the demand is at its lowest, as shown in Figure 8.15.

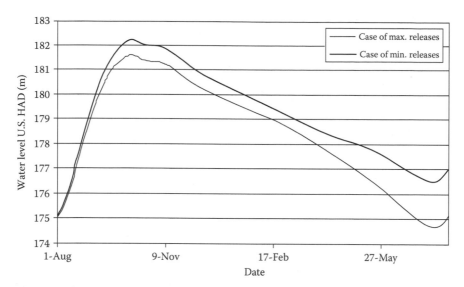

FIGURE 8.14 Water levels resulting during the proposed maximum and minimum proposed release scenarios.

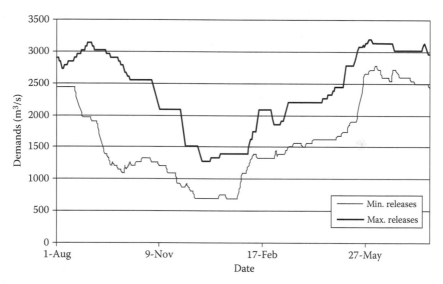

FIGURE 8.15 Maximum and minimum demand scenarios, previously proposed by the HAD Authority (1990–2003).

8.4.1.2.2 Effect of Varying the Initial Water Level

The water level in the reservoir August 1 of a given year, which is defined in the model as the initial water level, begins the simulation. Various simulations are carried out to test the effects of varying this level, with maximum releases to the Nile River downstream of HAD and during a flood year and maximum expected flood

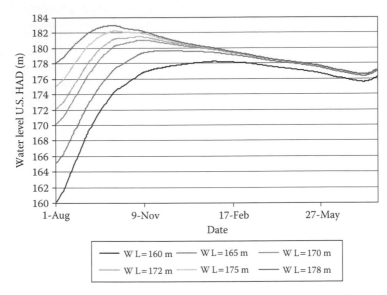

FIGURE 8.16   Water levels of the U.S. HAD during a flood year and with a minimum release scenario.

recorded at the Donqula station. The resultant water levels upstream of HAD during those simulations are presented in Figure 8.16.

From Figure 8.16 note that, in the case of a flood year, the initial starting water level will not affect the ending water level. As the lake continues to fill and the change in water level due to change in storage becomes slightly noticeable—because the top widths of the reservoir cross sections become very wide—any small change in water level will cause great variation in the total storage of the reservoir.

Almost all the proposed scenarios do not exceed the maximum designed level of the HAD, which is 182 meters above mean sea level, except if we start with an initial water level of 178 meters. In almost all cases the ending water level will reach levels somewhere between 176 and 177 meters, which will cause threats to the dam if two successive extreme floods are expected in two successive years.

Another set of simulations are carried out to test the effects of varying the initial water level in the reservoir with minimum releases to the Nile River downstream of HAD and during a drought year. The resulting water levels upstream of HAD during those simulations are presented in Figure 8.17.

From Figure 8.17 note that, in the case of a drought year, the initial starting water level affects the ending water level noticeably. As the HADR releases minimum demands downstream, the change in water level due to the change in storage becomes noticeable and any small change in the storage will cause great variations in the water level upstream of the dam.

The Table 8.2 summarizes the routing techniques and the criteria to be considered when choosing the appropriate technique for routing problems. For more details, refer to Soliman and Ashraf et al. (2007).

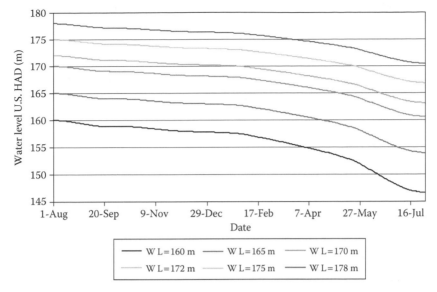

FIGURE 8.17 Water levels of the U.S. HAD during a drought year and with the minimum release scenario.

TABLE 8.2
Selecting the Appropriate Routing Technique

Factors to Consider in the Selection of a Routing Technique	Methods That Are Appropriate for This Specific Factor	Methods That Are Not Appropriate for This Factor
1. No observed hydrograph data available for calibration	Full Dynamic Wave Diffusion Wave Kinematic Wave Muskingum–Cunge	Modified Puls Muskingum Working R & D
2. Significant backwater that will influence discharge hydrograph	Full Dynamic Wave Diffusion Wave Modified Puls Working R & D	Kinematic Wave Muskingum Muskingum–Cunge
3. Flood wave will go out of bank into the flood plains	All hydraulic and hydrologic methods that calculate hydraulic properties of the main channel separately from those of the overbanks	Muskingum
4. Channel slope > 10 feet/mile and $\dfrac{TS_0 u_0}{d_0} \geq 171$	All methods presented	None
5. Channel slopes from 10 to 2 feet/mile and $\dfrac{TS_0 u_0}{d_0} < 171$	Full Dynamic Wave Diffusion Wave Musking–Cunge Modified Puls Muskingum Working R & D	Kinematic Wave

(Continued)

TABLE 8.2 (*Continued*)
Selecting the Appropriate Routing Technique

Factors to Consider in the Selection of a Routing Technique	Methods That Are Appropriate for This Specific Factor	Methods That Are Not Appropriate for This Factor
6. Channel slope < 2 feet/mile and $TS_0\left(\dfrac{g}{d_0}\right)^{1/2} \geq 30$	Full Dynamic Wave Diffusion Wave Muskingum–Cunge	Kinematic Wave Modified Puls Muskingum Working R & D
7. Channel slope < 2 feet/mile and $TS_0\left(\dfrac{g}{d_0}\right)^{1/2} < 30$	Full Dynamic Wave	All others

REFERENCES

Carter, R. W., and R. G. Godfrey. 1960. Storage and flood routing. US Geological Survey Water Supply Paper 1543-B.

Cunge, J. A. 1969. On the subject of a flood propagation method. *J Hydrol Res* 7(2):205–30.

El Mostafa, A. M. et al. 2002. *Impact of High Dam on Nile Flood Wave Routing*, Master's thesis, Ain Shams University, Cairo, Egypt.

El Mostafa, A. M. et al. 2007. *Flood Routing Trough Lake Nasser Using 2D Model*, Doctoral thesis, Ain Shams University, Cairo, Egypt.

Hydraulic Research Institute. 1998. Design of Tushka spillway. Report presented to the Egyptian Ministry of Irrigation, Cairo, Egypt.

McCarthy, G. T. 1938. The unit hydrograph and flood routing. Paper presented at the Conference of the North Atlantic Division, Corps of Engineers, U.S. Army, New London, Connecticut, 24. Providence, Rhode Island: U.S. Engineering Office.

Soliman, M. M., M. Gad, and A. Elmostafa. 2007. 2-D hydrodynamic model for HAD reservoir. In *Proceedings of VI International Symposium on Environmental Hydrology*, p. 40. Cairo: Ain Shams University.

U.S. Army Corps of Engineers. 1994. *Flood Runoff Analysis*. U.S. Army Corps of Engineers manual.

BIBLIOGRAPHY

Chow, V. T. 1964. *Handbook of Applied Hydrology*. Section 14. New York: McGraw-Hill.

Clark, C. O. 1945. Storage and the unit hydrograph. *Trans ASCE* 110:1419.

Gleick, P. H. 1993. *Water in Crisis*. New York: Oxford University Press.

Kohler, M. A. 1958. Mechanical analogs aid graphical flood routing. *J Hydraul Div* 84.

Lawler, E. A. 1964. Flood routing. In *Handbook of Applied Hydrology*, ed. V. T. Chow, section 25-II. New York: McGraw-Hill.

Linsley, R. K., M. A. Kohler, and J. L. H. Paulhus. 1975. *Hydrology for Engineers*. New York: McGraw-Hill.

Price, R. K. 1973. Flood routing methods for British rivers. *Proc Inst Civ Eng* 55:913–30.

Wilson, W. T. 1941. A graphical flood routing method. *Trans Am Geophys Union* 21(3):893.

9 Groundwater Hydrology

9.1 INTRODUCTION

Groundwater and surface water commonly form a linked system. The flow can be in any direction, and the rate of flow varies geographically and seasonally. The interchange is not significant for some aquifers. However, groundwater supplies an estimated 30% of the total flow in surface streams, and seepage from streams is a principal source of flow to some aquifers. Water drawn from wells situated along a bank of an alluvial stream can affect an appreciable reduction in the surface flow and the diversion of surface flow can reduce groundwater recharge. The supply for groundwater and surface water cannot be evaluated independently unless the interchange is minimal. There are two distinct types of circumstances concerning the development and management of groundwater supplies. Groundwater is considered a renewable source, with optimal use restricted to the average rate of recharge; mining of groundwater, however, is sometimes carried out with fixed-term objectives. Average annual recharge can, in extreme cases, be relatively insignificant, as in the major regional Nubian aquifers in northeast Africa.

The environmental consequences of groundwater withdrawal without a good management plan in place include land subsidence, seawater intrusion in coastal aquifers during overdraft situations, or waterlogging of an area during an excess recharge situation.

9.2 DISTRIBUTION OF SUBSURFACE WATER

Water occurs underground in two zones (aeration and saturation zones) that are separated by the water table. The occurrence and movements of water in these two zones are markedly different. The water table exists only in water-bearing formations that contain openings of sufficient size to permit appreciable movement of water. The water table is the lower surface of the zone of suspended water at which the pressure is equal to the atmospheric pressure. The saturation zone extends down as far as there are interconnected openings. The lower boundary may be an impervious layer. The upper saturation zone is called an unconfined aquifer. Sometimes, the saturation zone is bounded on top by another impervious layer and is then known as the confined aquifer. If one of the impervious layers leaks inward or outward, this aquifer is called a leaky aquifer.

Subsoil water is limited to the soil belt and is reached by the roots of plants. Pellicular water adheres to rock surfaces throughout the zone of aeration and is not moved by gravity; however, it may be abstracted by evaporation and transpiration. Gravity or vadose water moves downward by the force of gravity throughout the

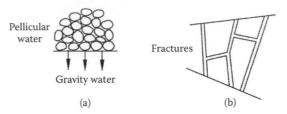

FIGURE 9.1   Pellicular water (a) in granular material and (b) rock fractures.

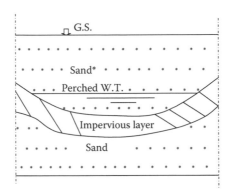

FIGURE 9.2   Perched water table.

saturation zone. Perched water occurs locally above an impervious barrier. Capillary water occurs only in the capillary fringe above the water table. Free water is water that moves by gravity in the unconfined aquifer (see Figures 9.1 through 9.3a and b).

The vadose zone can be more difficult to characterize due to more complex localized flow conditions than those found in the saturation zone below the water table. However, because this zone is nearer to the land surface, remedial actions may not require complete characterization of the vadose-zone flow system for certain site conditions and contaminants if the majority of the affected soils will be treated on site or removed.

9.3   GROUNDWATER FLOW THEORIES

Groundwater flow is generally treated as the flow of fluid in porous media. The classical hydrodynamics of viscous flow are applied. In addition to the equation of continuity and the equation of state, the equation of motion should be considered in any hydrodynamic problem. However, for the flow of fluid through porous media, the equation of motion is replaced by Darcy's law to obtain the groundwater flow equation in a simple manner. Darcy's law for flow through a medium is as follows:

$$V = -K \frac{\partial h}{\partial x_i} \qquad (9.1)$$

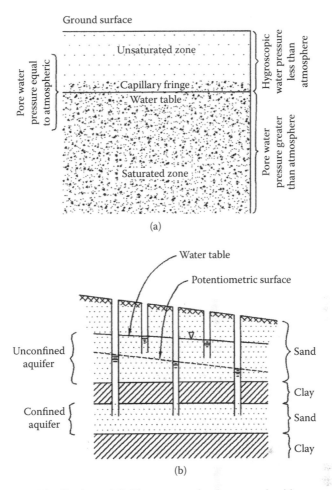

FIGURE 9.3 (a) Distribution of fluid pressures in the ground with respect to water. (b) Unconfined aquifer and its water table; confined aquifer and its potentiometric surface.

where V is the groundwater flow and K is the hydraulic conductivity. However, the actual fluid motion can be subdivided based on the components of flow parallel to the three principal axes:

$$
\begin{aligned}
u &= -K_x \left.\frac{\delta h}{\delta l}\right| \\
v &= -K_y \left.\frac{\delta h}{\delta y}\right| \\
w &= -K_z \left.\frac{\delta h}{\delta z}\right|
\end{aligned}
\tag{9.2}
$$

where h (the hydraulic head) equals $hp + z$, where hp is the pressure head and z is the potential head, K_x, K_y, and K_z are the hydraulic conductivities of the components in the

x, y, and z directions, respectively, u, v, and w are the velocities of flow in the x, y, and z directions, respectively, and dh/dx_i is the hydraulic gradient ($x_i = x$, y, and z directions).

The negative sign (−ve) means that water is flowing in the direction opposite to the increasing hydraulic potentials.

Combining Equation 9.2 with both the continuity equation and the equation of state, the groundwater flow equation, in general form, can be given as follows:

$$\frac{\partial}{\partial x}\left(K_x \frac{\partial h}{\partial x}\right) + \frac{\partial}{\partial y}\left(K_y \frac{\partial h}{\partial y}\right) + \frac{\partial}{\partial z}\left(K_z \frac{\partial h}{\partial z}\right) = S_s \frac{\partial h}{\partial t}$$

or

$$K_x \frac{\partial^2 h}{\partial x^2} + K_y \frac{\partial^2 h}{\partial y^2} + K_z \frac{\partial^2 h}{\partial z^2} = S_s \frac{\partial h}{\partial t} \tag{9.3}$$

where S_s represents a specific storage, defined as the volume of water released from a storage of unit volume of saturated aquifer for a unit decline in hydraulic head, per unit depth. For two-dimensional flows in aquifers of uniform thickness B, the following equation applies:

$$T_x \frac{\partial^2 h}{\partial x^2} + T_y \frac{\partial^2 h}{\partial y^2} = S \frac{\partial h}{\partial t} \tag{9.4}$$

where T_x and T_y, are the transmissivities in the x and y directions, respectively, and $T = BK$ (in square meters per second), where B is the aquifer thickness. $S = S_s \times B$, which is the storage coefficient, defined as the volume of water released from or taken into storage per unit cross-sectional area of the vertical column of aquifer per unit change in head. For the unconfined aquifer, S is taken as the specific yield (S_y).

For well problems with radial flow, Equation 9.4, in cylindrical coordinates, becomes

$$\frac{\partial^2 h}{\partial r^2} + \frac{1}{r}\frac{\partial h}{\partial r} = \frac{S}{T}\frac{\partial h}{\partial t} \tag{9.5}$$

The coefficients S and T may be regarded as empirical values to be determined principally by the pumping-test technique.

Two types of flow are considered for groundwater flow toward wells drilled in aquifers: steady- and unsteady-state flows.

9.3.1 Steady-State Groundwater Flow in Aquifers

Dupuit (1863) was the first person to combine Darcy's law with the continuity equation to derive an equation for well discharge. Dupuit assumed complete axial symmetry and a steady flow through an infinitely extending aquifer. He deduced the following equation for a confined aquifer (Figure 9.4a):

$$Q = \frac{2\pi k m (h_2 - h_w)}{l_n \frac{r_2}{r_w}} \tag{9.6}$$

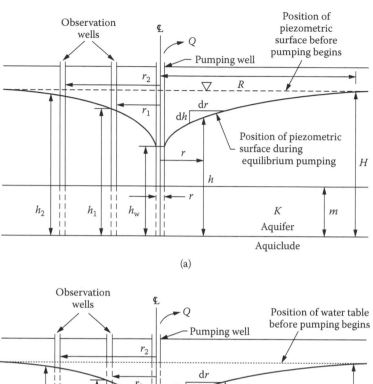

FIGURE 9.4 Drawdown curve around a well in (a) confined aquifer and (b) unconfined aquifer. (Modified from Walton, W. C. 1970. *Groundwater Resource Evaluation.* New York: McGraw-Hill.)

He deduced the following equation for an unconfined aquifer (Figure 9.4b):

$$Q = \frac{\pi k \left(h_2^2 - h_w^2 \right)}{l_n \left. r_2 \middle/ r_w \right.}$$

(9.7)

where Q is the discharge, K is the hydraulic conductivity, and h_2 and h_w are the head levels above the impervious bed at radial distances r_2 and r_w, respectively.

9.3.2   Unsteady-State Groundwater Flow in Confined Aquifers

If there is no replenishment, the area of influence of a well increases, and the piezometric head declines in such a way that the water released from storage equals the well discharge. The differential equation governing such unsteady flow to an axially symmetrical well is shown in Equation 9.5.

For the artesian case, the water is released by the consolidation and compression effects associated with release of pressure. In the case of the water table, the water originates from the recession of the water table and S equals I the specific yield. Theis (1935) applied a solution to Equation 9.5 to the case of constant discharge from an infinitely extending artesian aquifer. For this case (Figure 9.5), the drawdown D and h_o at any radial distance r and time of pumping t is given by

$$D = \frac{Q}{4\pi T} \int_u^\infty \frac{e^{-u}}{u} du \qquad (9.8)$$

where $u = \dfrac{r^2 S}{4Tt}$

Take

$$w(u) = \int_u^\infty \frac{e^{-u}}{u} du \qquad (9.9)$$

Therefore

$$D = \frac{Q}{4\pi T} W(u) \qquad (9.10)$$

The graphical method also can be used to solve Equation 9.10. This is called the matching procedure. A curve such as that shown in Figure 9.6 can be used. This figure

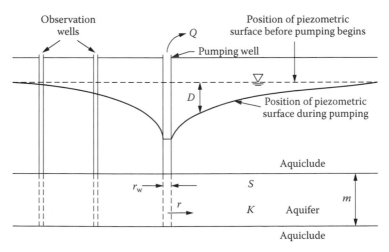

FIGURE 9.5   Transient flow toward a well in a confined aquifer.

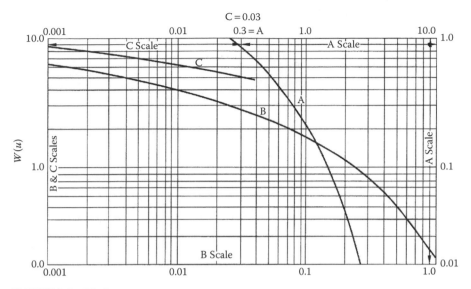

FIGURE 9.6   Theis curve.

shows $W(u)$ plotted as a function of u. Note that the three curves shown in the figure were originally one curve, and was divided into three sections to fit the page size.

From Equations 9.8 and 9.9, for any specific well test, u is proportional to r^2/t and D to $W(u)$. By plotting D as the ordinate and r^2/t as the abscissa on transparent paper to the same scale as the Theis-type curve (Figure 9.6), a field data curve similar to the type curve is obtained. A portion of the field data curve is superimposed and matched to the type curve, keeping the coordinates parallel when matching. Then a specific match point on the matching portion of the curve is chosen and the values of u, $W(u)$, D, and r^2/t for this point are recorded. Substitute these values for D and $W(u)$ into Equation 9.10 and solve for T. Using this value of T and substituting the values for u and r^2/t into Equation 9.9, solve for S.

Theis' nonequilibrium equation is generally applicable to artesian wells tapping confined aquifers and also to wells tapping the water-table aquifer if the drawdown is a small percentage of the saturation thickness of the aquifer. In a number of special cases, however, the equation can be greatly simplified with negligible loss in accuracy. Probably the most important of the modified forms of Theis' nonequilibrium equation are the basic modified equation, the rate of drawdown equation, and the recovery equation.

9.3.2.1   Basic Modified Equation

The basic modified equation by Jacob (1954) is the least modified. It is the basic modification because nearly all other modifications are derived through it. For its derivation and limitations, consider Figure 9.7, which is the same as Figure 9.6, except that the ordinate scale is linear. The solution to Equation 9.8 is an infinite series that can be approximated in the form

$$W(u) = (-1.00)(\ln u) + a = (-2.30)(\log u) + a \tag{9.11}$$

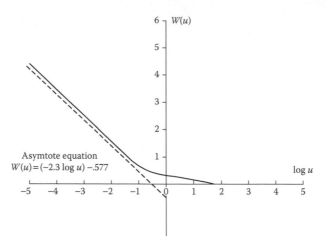

FIGURE 9.7   Semilog-type curve.

where a is nearly a constant for small values of u, approaching (-0.5772) as u approaches zero. The type curve is therefore asymptotic to the straight line.

$$W(u) = (-2.30[\log u]) - 0.57722 \tag{9.12}$$

For small values of u, the type curve nearly parallels the asymptote, with slope equal to -2.30. Thus, the type-curve equation, written in terms of two points, becomes

$$(y_2 - y_1) = m(x_2 - x_1)$$

which becomes

$$W(u_2) - W(u_1) = -2.3(\log u_2 - \log u_1) = 2.3 \log \frac{u_1}{u_2}$$

Replacing u and $W(u)$ with their equivalents, according to Equations 9.8 and 9.9, assuming $S_1 = S_2$, and simplifying, we get

$$D_2 - D_1 = \frac{2.3Q}{4\pi T} \log \frac{r_1^2/t_1}{r_2^2/t_2} \tag{9.13}$$

Equation 9.11 is the basic modified equation. The choice between the two numbering systems is a matter of convenience.

At a constant radial distance r, the rate of drawdown is given by the following equation:

$$\Delta D = D_2 - D_1 = \frac{2.3Q}{4\pi T} \log \frac{t_2}{t_1} \tag{9.14}$$

The values of drawdowns D_1 and D_2 are taken per log cycle of time t:

$$\log\frac{t_2}{t_1} = 1$$

Thus, Equation 9.14 becomes

$$\Delta D = \frac{2.3Q}{4\pi T} \tag{9.15}$$

where ΔD is the drawdown per log cycle of time (such as, $t_1 = 2$, $t_2 = 20$) when D is plotted versus t on semilog paper, forming a straight-line relation; T can be calculated from Equation 9.15. By projecting this line to meet the horizontal axis where $D = 0$ and $t = t_0$, S can be calculated using Equation 9.12 as follows:

$$S = \frac{2.25Tt_0}{r^2} \tag{9.16}$$

9.3.2.2 Adjustment of the Modified Equations for Free-Aquifer Conditions

In the case of free-water table conditions, the saturation thickness B is reduced by the drawdown D, so that $T = KB$ is replaced by

$$K(B - D_{ave}) = Ky_{ave} = \frac{K(y_1 + y_2)}{2}$$

where y_1 and y_2 are the two drawdown curve ordinates corresponding to the two respective times and/or radial distances and D_{ave} is the average drawdown. Then, because $D_2 - D_1 = y_1 - y_2$, the quantity $(D_2 - D_1)\,T$ is replaced by

$$\frac{(y_1 - y_2)Kx(y_1 + y_2)}{2} = \left(\frac{(y_1^2 - y_2^2)}{2}\right)K$$

For example, the basic modified Equation 9.13 then becomes

$$y_1^2 - y_2^2 = \frac{2.3Q}{2\pi k}\log\frac{r_1^2/t_1}{r_2^2/t_2} \tag{9.17}$$

Care must be used in applying a free-aquifer equation that has been derived from artesian well equations by replacing KB by $K(y_{ave})$ as illustrated. The artesian equations are derived on the basis that all streamlines are horizontal, so that the hydraulic gradient in the Darcy equation is equal to $dy/dL = dy/dx$ and the equipotential surfaces representing the areas in the Darcy equation are vertical cylinders. In the case of free-aquifer flow, the upper streamlines slope downward toward the well so that their head losses vary with the sloping flow distance, and the hydraulic gradient is not equal to dy/dx. Furthermore, in the case of a free-aquifer well, the equipotential surfaces representing the areas in the Darcy equation are not vertical cylinders but semicylindrical surfaces that curve inward at the top, as indicated by the curvature of the equipotential lines. No adjustment is made for these factors, so that the resulting free-aquifer equations are applicable only to computations where y represents the height of the hydraulic grade line $(P/w) + Z$ along nearly horizontal streamlines representative of the main flow. This includes all bottom streamlines, a large part of the flow at intermediate elevations where the flow is nearly horizontal, and water-table streamlines at radial distances such that the water table is nearly horizontal. Such equations are applicable also in terms of the water level in the pumping well, because the point (r_e, d_w), where r_e is the effective radius or radius of the borehole, d_w is the drawdown in the well, lies on the hydraulic grade line for all streamlines except those entering the well along the seepage surface.

If the pumping drawdown in the pumping wells is large, the free-aquifer equations may be inapplicable in terms of the free-surface drawdown curve for radial distances as great as 1.5B or 2B. In such cases, the values of water-table drawdown closer to the well can be computed based on empirical relationships.

As Q is proportional to T,

$$\frac{Q_{\text{artesian}}}{Q_{\text{free}}} = \frac{KB}{K(B - D_{\text{ave}})} = \frac{B}{B - D_{\text{ave}}}$$

where B is the saturation thickness.

Thus, failure to adjust modified equations such as Equations 9.14 and 9.8 for free-aquifer conditions by replacing KB with $K(B - D_{\text{ave}})$ gives

$$Q_{\text{computed}} = \frac{B}{B - D_{\text{ave}}} Q_{\text{true}}$$

The computed values of K and $(D_2 - D_1)$ are in error to a similar degree.

9.3.2.3   Recovery Equation

Suppose, as shown in Figure 9.8, the pumping rate Q is suddenly changed to a new rate Q'. The additional drawdown Z caused by the additional discharge $(Q' - Q)$ may be expressed in terms of the Theis nonequilibrium equation as follows:

$$Z = \frac{(Q' - Q)}{4\pi T} W(u') \tag{9.18}$$

where

$$u' = \frac{r^2 S'}{4Tt'} \tag{9.19}$$

S' is the new storage coefficient and t' is the time with reference to the start of Q'. However, in the recovery condition, the drawdown is called the residual drawdown, and Q' equals 0; therefore, Equation 9.18 becomes

$$Z' = \frac{Q}{4\pi T} W(u') \tag{9.20}$$

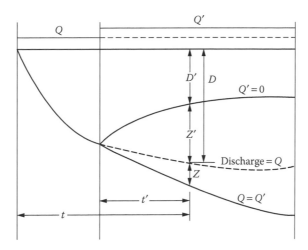

FIGURE 9.8 Hydrographs illustrating drawdowns during changes in rate and recovery conditions.

and the residual drawdown D' becomes

$$D' = D - Z = \frac{Q}{4\pi T}(W(u) - W(u'))$$ (9.21)

For small values of u and u', the modified condition can be used as

$$D' = \frac{Q}{4\pi T}((-2.3\log u + a) - (-2.3\log u' + a))$$ (9.22)

or

$$D' = \frac{2.3Q}{4\pi T}\left(\log\frac{t}{t'} - \log\frac{S}{S'}\right)$$ (9.23)

If S/S' is taken as a constant, D' and t/t' can be plotted on semilog paper, giving a straight-line relation. The line need not pass through the origin where $t/t' = 1$, as it will do so only when $S' = S$. Often $S' \ll S$ owing to imperfect elastic recovery and, in the case of free aquifers, to air pockets and capillary log or land subsidence.

9.3.2.4 Drawdown Equation for Water-Table Conditions

Boulton (1963) was able to derive an equation to solve for the hydrologic factors in the case of a water-table aquifer with a fully penetrating well. In his solution, the gravity drainage to the water table was assumed to be due to the lowering of the water level (dz) between the times T and $T + dT$ because the commenced pumping consisted of the following two parts:

1. A volume (S dz) of water instantaneously released from storage per unit horizontal area, at any time from the start of pumping
2. A delayed yield from storage, at $t(t \geq T)$ from the start of pumping

$$dz \propto S'e^{-\alpha(t-T)} \quad \text{and} \quad n = 1 + S'/S$$

where α is an empirical constant and S' is the total volume of delayed yield from a storage per unit drawdown per unit horizontal area.

With these assumptions, Boulton derived the following equations:

$$D = \frac{Q}{4\pi T} W\left(u_{a,b}, \frac{r}{\beta}\right) \tag{9.24}$$

$$U_a = \frac{Sr^2}{4Tt} = \frac{1}{\Phi} \tag{9.25}$$

$$u_b = \frac{S'r^2}{4Tt} = \frac{1}{\Phi'} = \frac{(r/\beta)^2}{4\alpha t} \tag{9.26}$$

The values of $W(u_{a,b}, r/\beta)$ are plotted against the values of I/U_a and I/u_b on logarithmic paper to construct the type curves, as shown in Figure 9.9. The type curves that lie to the left of the values r/β are termed type A curves, and those to the right of the values are termed type B curves.

The method of using type curves can be briefly described as follows: the observed values of drawdown, D, at a given distance, r, from the pumped well are plotted against the values of time, t, on the same logarithmic scale as that used for the type curves to prepare a graph designated as time-drawdown field-data curve. Placing the time-drawdown curve (on transparent paper) first over the type-A curves and then over the type-B curves and keeping the respective coordinate axes parallel, a value

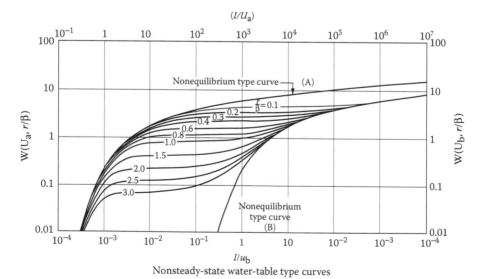

Nonsteady-state water-table type curves

FIGURE 9.9   Delayed-yield type curves.

of r/β is determined from the type curve that gives the best fit. The two following cases may arise:

1. If the time-drawdown curve becomes horizontal after the early-time drawdown, a "match point" is chosen on this segment and, with the time-drawdown curve fitted to the appropriate type-A curve, corresponding values of D, $W(u)$, t, and $1/u_a$ are read off at the match point. The time-drawdown curve is then fitted to the appropriate type-B curve and the value of $1/u_b$ noted for the match point. (Being on the horizontal segment of the type curve, the match point will give the same value of $W[u_a, u_b]$ as before.) The formation constants T, S, S', t, and α are subsequently calculated from Equations 9.24 through 9.26.

2. If the early time–drawdown curve never becomes horizontal, the early time–drawdown time segments of the time-drawdown curve may be fitted, respectively, to the type-A curve and type-B curve having an r/β that gives the best fit. Choosing a match point in each of these segments, the values are read from the curves and substituted into Equations 9.24 to 9.25 to compute the hydrogeologic characteristics T, S, and α.

For case 1, it is evident that type-A and type-B curves, though strictly for situations in which n equals infinity, are applicable when n has a large finite value because they both have zero slope at their intersection. For case 2, however, type-B curve for finite n is generally required. However, if the intermediate slope of the time-drawdown curve is not large, the complete type curve is obtained with sufficient accuracy by joining the appropriate type-A and type-B curves (plotted on $1/u_a$ and $1/u_b$ base) by a straight line tangential to both curves. In this case, the match points on the type-A and type-B curves must be chosen so that they lie on curve segments that are clear for the sloping tangent. If the value of S is not required, the constants T, S', and α may be directly obtained from the type-B curves, in which case, the type-A curves are not needed.

After a relatively long period of pumping, the effects of delayed yield are negligible, and the aquifer's characteristics may be computed using the Theis solution or its approximation as mentioned earlier. Boulton developed a curve that can be used to estimate the time, t_0, when the effects of delayed yield become negligible (Figure 9.10). The figure gives $(\alpha\, t_0)$ as a function of r/β, where α is an empirical coefficient, $\alpha = T/\beta^2 S'$, and r is the radial distance from the pumping well.

Neuman (1973a, 1973b) gave another solution for the average drawdown, D_{av}, in an observation well at a distance r at time t after pumping from a fully penetrating well in an unconfined aquifer with saturation thickness m:

$$D_{av} = \frac{Q}{4\pi T}(Wt_s, \sigma, \beta) \tag{9.27}$$

where $W(t_s, \sigma, \beta)$ is the new well function

$$t_s = \frac{Tt}{Sr^2} \tag{9.28}$$

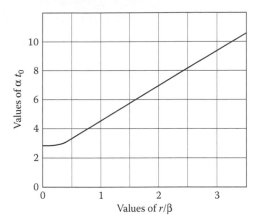

FIGURE 9.10 Curve for estimating the time when delayed yield ceases to influence the drawdown.

$$\sigma = \frac{S}{S'} \tag{9.29}$$

$$\beta = \frac{r^2 k_z}{m^2 k_r} \tag{9.30}$$

K_z and K_r represent the vertical and horizontal hydraulic conductivities, respectively.

On the basis of Equations 9.29 and 9.30, the aquifer properties K_r, S', and K_z can be obtained simply by plotting the drawdown D versus time t. The procedure is as follows:

1. Plot D versus $\log t$.
2. Fit a straight line to the last portion of the data. The intersection of this line with the horizontal axis at $D = 0$ is denoted by t_0. The slope of this line is the change in drawdown over one log cycle, denoted by ΔD.
3. Equation 9.27 can be approximated on the assumption that $S \ll S'$ and $\sigma \approx 0$:

$$D = \frac{Q}{4\pi T} W(t_y, \beta) = \frac{Q}{4\pi T} (2.3 \log 2.25 t_y) \tag{9.31}$$

and

$$T = \frac{2.3Q}{4\pi \Delta D} \tag{9.32}$$

Using this equation to solve for transmissivity (T), the horizontal hydraulic conductivity can be computed as $K_r = T/m$.

4. The specific yield S' can be computed using

$$S' = \frac{2.25 T t_0}{r^2} \tag{9.33}$$

5. Using the computed values of T and S', solve for the dimensionless time t_y from the equation

$$t_y = \frac{Tt}{S'r^2} \tag{9.34}$$

Equation 9.35 (Neuman 1975) can give β, provided that $4 \le t_y \le 100$ (for other cases, Figure 9.11 may be used):

$$\beta = \frac{0.195}{(t_y)1.1053} \tag{9.35}$$

Equation 9.28 can be used to solve for K_z as follows:

$$K_z = \frac{\beta K_r m^2}{r^2}$$

Because the value of S, the storativity of the early time of pumping, is of no importance, only S', K_r, and K_z will be considered.

9.3.2.5 Unsteady-State Flow in Semiconfined Aquifers

The drawdown in a semiconfined aquifer can be described by the Huntush and Jacob formula as follows:

$$D = \frac{Q}{4\pi T} \int_u^\infty \frac{1}{u} e^{-(y)} du \tag{9.36}$$

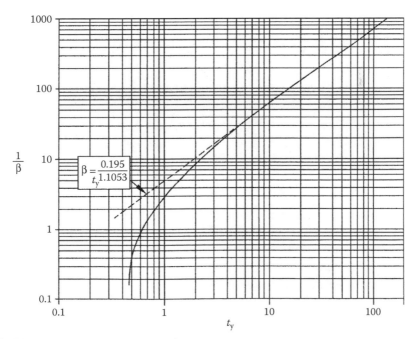

FIGURE 9.11   Logarithmic plot of $1/\beta$ versus t_y.

or

$$D = \frac{Q}{4\pi T} W(u, r/L)$$

where

$$u = \frac{r^2 S}{4Tt} \tag{9.37}$$

and

$$y = \frac{u + r^2}{4L^2 u} \tag{9.38}$$

$$L = \sqrt{TC} \tag{9.39}$$

where $C = D'/K'$, and D' and K' are the thickness and hydraulic conductivity, respectively, of a semipervious aquifer.

Walton (1970) developed a group of curves defining the value of r/L for the application of both methods (Figure 9.12). Lai and Su (1974) gave the following solution for the drawdown in a leaky aquifer for large wells:

$$D = \frac{Q}{4\pi T} F\left(u, \alpha, \frac{r_w}{\beta}, P\right) \tag{9.40}$$

where

$$u = \frac{r^2 S}{4Tt} \tag{9.41}$$

$$\alpha = \frac{r_w^2 S}{r_c^2} \tag{9.42}$$

where r_w is the effective radius of the well bore or open hole and r_c is the radius of the pumping well casing within the range of the water-level fluctuation.

$$r_w\beta = \frac{r_w}{\sqrt{T/D'/K'}} \tag{9.43}$$

Because many assumptions need to be satisfied for a solution with regard to the last condition of Equation 9.38, Equation 9.36 may be considered for its simplicity and the approximate solution obtained for field problems.

9.3.3 EFFECTS OF PARTIAL PENETRATION OF A WELL

Muskat (1937) discussed the problems of partial penetration of a well in detail and presented various methods for determining the flow pattern. He succeeded also in deducing a satisfactory approximate formula for the discharge, however, this formula

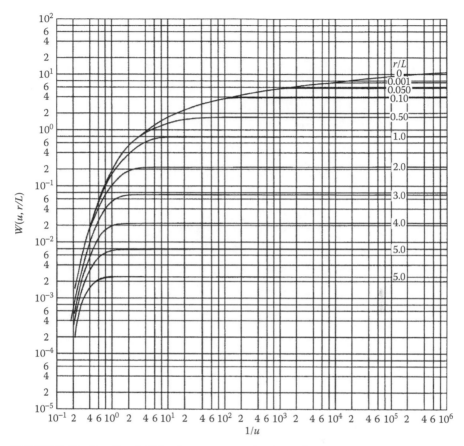

FIGURE 9.12   Family of Walton's-type curves for $W(u, r/L)$ versus I/u and the different values of r/L.

is too complicated for practical application. Muskat later suggested another formula, obtained by Kozenys, as an approximation to his own for steady-state conditions in a confined aquifer:

$$Q = \frac{2\pi D}{\ln} \frac{kmm'}{(r_e/r_w)}\left(1 + 7\sqrt{\frac{r_w}{2mm'}} \cdot \cos\frac{\pi m'}{2}\right) \tag{9.44}$$

where m' is the ratio of the depth penetrated by the well to the thickness of the aquifer, m.

Note that the effects of partial penetration are only apparent in drawdown data collected within an approximate radial distance r of the pumping well:

$$r < 1.5m\sqrt{K_r/K_z}$$

where m is the thickness of aquifer, K_r is the horizontal hydraulic conductivity, and K_z is the vertical hydraulic conductivity (beyond this distance, groundwater flow is essentially horizontal).

In cases where the flow is unsteady, Huntush (1964) gave an equation as follows for the drawdown (D) at any point in an observation well:

$$D = \frac{Q}{4\pi K_r m}\left[W(u) + f(u,x,d/m,1/m,z/m)\right] \tag{9.45}$$

where

$$u = \frac{r^2 S}{4K_r mt} \tag{9.46}$$

$W(u)$ = Theis well function, and

$$f = \left[\frac{2m}{\pi(l-d)}\right]\sum_{n=1}^{\infty}\left(\frac{1}{n}\right)\left[\sin\left(\frac{n\pi l}{m}\right) - \sin\left(\frac{n\pi d}{m}\right)\right]\cos\left(\frac{n\pi z}{m}\right)W(u,x) \tag{9.47}$$

where

$$W_{(u,x)} = \int_u^\infty \frac{e^c}{y} \tag{9.48}$$

$$c = \left(\frac{-y - x^2}{4y}\right) \tag{9.49}$$

$$x = \frac{r}{m}\sqrt{K_z/K_r} \tag{9.50}$$

The remaining variables are as shown in Figure 9.13.

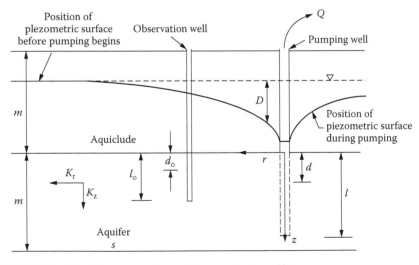

FIGURE 9.13   Partially penetrating well in a confined aquifer.

9.4 HYDRAULICS OF THE WELL AND ITS DESIGN

This chapter so far has discussed the flow of fluid through an aquifer under an energy gradient created by a well. The water must also be transferred through the screen and casing or pump column to the point of discharge. Under some circumstances, the energy expended in moving the water through the well structure may exceed that used in moving it through the aquifer. A better understanding of hydraulic principles involved in moving water through the aquifer should lead to improved well design. Some of the considerations are outlined in this section.

9.4.1 SPECIFIC CAPACITY

Engineers have designated the specific capacity as the ratio of discharge to drawdown. If the hydraulic head losses through the screen and casing are zero and the time effect of storage depletion are ignored, the discharge of an artesian well is expected to be directly proportional to the drawdown. This leads to a constant value of specific capacity corresponding to all values of discharg'e of an artesian well—a condition that is usually assumed. For wells in unconfined aquifers, an increase in drawdown decreases the effective thickness of the aquifer. Thus, even discounting energy losses at the well, the specific capacity would decrease with discharge in the case of the water table.

The hydraulic losses through the well cause further nonlinearity in the relationship of discharge to drawdown. As mentioned by Jacob (1944), flow through the screen and casing usually occurs in the turbulent regime, and the resulting head losses are thus proportional to Q^2. Aquifer losses under conditions of laminar flow should be proportional to Q for an artesian well. Thus, one may write

$$D = BQ + CQ^2 \tag{9.51}$$

where D is the total drawdown in the well and B and C are constants.

Equation 9.51 can be evaluated approximately by performing pumping tests at two different discharge rates, Q_1 and Q_2, measuring the respective values of drawdown, D_1 and D_2; substitution successively into Equation 9.51 provides simultaneous equations in B and C. The difficulty with this procedure is that it is based on an assumption of steady flow and does not take into account the effect of depletion of storage with time.

9.4.2 EFFECTIVE RADIUS

The effective radius of a well, as applied in the formulas of flow, may not be the same as the radius of the screen or hole, especially for wells in unconsolidated sediments. Development of the well or use of gravel envelopes increases the permeability of the formation immediately surrounding the casing. This effect is the same as increasing the radius. The effective radius is defined by Jacob as the distance, measured radially from the axis of the well, at which the theoretical drawdown based on the logarithmic head distribution equals the actual drawdown just outside the screen. Jacob (1944) gives a procedure for determining this quantity using the results of field tests.

9.4.3 WELL SCREENS

A large part of the energy imparted through a well is frequently expended in transferring the water through the screen and pump. For this reason, attention should be given to the hydraulic performance of the well structure. Although considerable progress has been made, collection of data, especially in the field, is difficult because of rapidly changing flow conditions near the wells. Laboratory experiments that simulate field conditions are expensive and arduous. Nevertheless, more attention should be given to this important aspect of the problem of well hydraulics.

The screen is an important part of the well structure. Screens are always required, except in consolidated sediments. They may range from rough, haphazard perforations in a steel casing to highly engineered and carefully manufactured screens made of specially selected materials. The function of a screen is to exclude natural sediments while simultaneously allowing the greatest possible flow of water into the well. The factor of longevity influences the choice of a screen.

The hydraulic performance of the well screen was ably treated by Peterson et al. (Luthin 1957). Water enters the interior of a screen in the form of a radial jet at relatively high velocities. The energy of these jets is rapidly dissipated, and the flow accelerated in the axial direction. From a theoretical consideration of the mechanics involved, these investigators deduced that

$$\frac{\Delta h}{v^2/2g} = \frac{\cosh\,(CL/D+1)}{\cosh\,(CL/D-1)} \tag{9.52}$$

$$r_{\mathrm{w}} = \sqrt{r_{\mathrm{i}}^2(1-n)+nr_{\mathrm{o}}^2}$$

where Δh is the hydraulic head loss involved in the screen and v is the first average velocity along the screen axis (Q/A, where Q is the well discharge and A is the cross-sectional area of the screen), L is the axial length of the screen, and D is the screen diameter. C is defined as the screen coefficient:

$$C = 11.31C_{\mathrm{c}}A_{\mathrm{p}} \tag{9.53}$$

In Equation 9.53, C_{c} is the orifice coefficient of discharge applicable to the screen opening and A_{p} is the fractional ratio of screen opening to total screen surface.

The loss coefficient $\left(\dfrac{\Delta h}{v^2/2g}\right)$ in Equation 9.52 approaches two for values of CL/D exceeding approximately six.

9.4.4 VELOCITY DISTRIBUTION

The velocity distribution around the well screen was assumed constant along its length. This was later proved by Soliman (1965) to be curvilinear and to have the following relationship:

$$U = U_{\mathrm{o}}e^{KL/D} \tag{9.54}$$

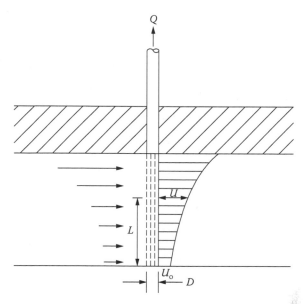

FIGURE 9.14   Velocity distribution around the well screen.

where K is a constant depending on the well screen and the remaining values are as shown in Figure 9.14.

These relations should be considered during the designing of the screen length.

9.5   SLUG TESTS

The slug test considered here is a test for determining the hydraulic conductivity (K) of unconfined or leaky aquifers connected to completely or partially penetrating wells (Figure 9.15). For other cases of confined aquifers, data collected from the pumping tests can be used to determine the hydraulic conductivity and storage coefficient.

Equations describing flow are based on a modified form of the Thiem (Dupuit) equation (Equation 9.6). The rate of groundwater flow, Q, from a well screen between the depths d and L for a specified water level in the well, H_w, is

$$Q = \frac{2\pi k(l-d)H_w}{\ln(R/r_w)}$$

(9.55)

where R is the radius of influence of the injection well and r_w is the effective radius of the well bore; also

$$r_w = \sqrt{r_i^2(1-n)+nr_o^2}$$

if the water level is decreasing within the screen length of the well, the hydraulic conductivity of the filter material or developed zone is much larger than that of the

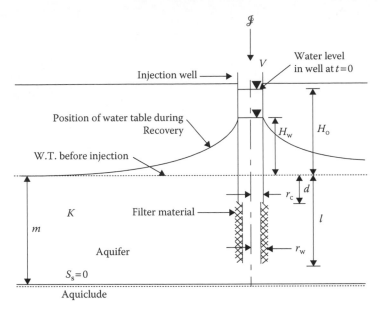

FIGURE 9.15   Slug test in an unconfined aquifer.

aquifer, and n is the porosity of the filter. r_i is the inside radius of the well screen, r_o is the outside radius of the filter material. r_c, in Figure 9.15, is the effective radius of the well casing over which the water level in the well changes.

To develop a simple equation, the following assumptions are made:

- The aquifer is homogeneous and isotropic.
- A volume of water, V, is injected instantaneously at time $t = 0$.
- Head losses through the well screen, filter material, and developed zone (if present) are negligible.

The rate of decrease of the water level in the well is equal to the flow rate divided by the effective cross-sectional area of the well casing as follows:

$$\frac{dH_w}{dt} = -\frac{Q}{\pi r_c^2} \tag{9.56}$$

Combining Equations 8.55 and 8.56 and integrating, K can be given as follows (refer to Figure 9.15 for limits):

$$K = \frac{r_c^2 \ln(R/r_w)\ln(H_o/H_w)}{2(l-d)t} \tag{9.57}$$

Bouwer and Rice (1976) determined the radius of influence, R, for different values of r_w, $(1-d)$, H_w, and m while using measurements made with an electrical resistance analog model. From their experiments, the following empirical equation for estimating R was developed:

$$\ln(R/r_w) = \left[\frac{1.1}{\ln(1/r_w)} + \frac{A + B\ln[(m-1)/r_w]}{2(l-d)/r_w}\right]^{-1} \qquad (9.58)$$

where A and B are dimensionless coefficients that are functions of $(1-d)/r_w$, as shown in Figure 9.16.

If $(\ln(m-1)/r_w) > 6$, then Equation 9.58 becomes

$$\ln(R/r_w) = \left[\frac{1.1}{\ln(1/r_w)} + \frac{A + 6B}{2(l-d)/r_w}\right]^{-1} \qquad (9.59)$$

In addition, if the injection well fully penetrates the aquifer, the following equation is used:

$$\ln(R/r_w) = \left[\frac{1.1}{\ln(1/r_w)} + \frac{C}{2(l-d)/r_w}\right]^{-1} \qquad (9.60)$$

where C can be interpolated from Figure 9.16.

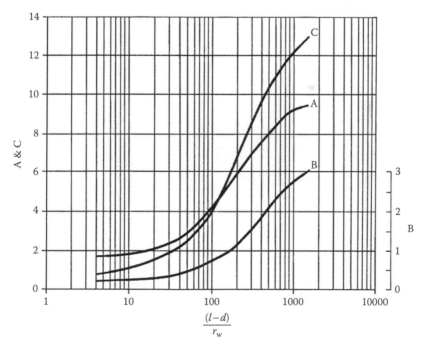

FIGURE 9.16 Values of the coefficients A, B, and C for use in estimating the radius of influence, R. (Adapted from Bouwer, H., and R. C. Rice. 1976. *Water Resour Res* 12(3): 423–8)

9.6 GROUNDWATER RECHARGE

Groundwater recharge may be obtained by artificial or natural means. Artificial groundwater recharge is a planned operation for transferring water from the ground surface into aquifers. Natural groundwater recharge is a phenomenon in which water reaches aquifers without human intervention from surface sources such as streams, natural lakes, or ponds.

The factors affecting natural groundwater recharge are the thickness and properties of soil formation and stratification; surface topography; vegetative cover; land use, soil-moisture content, and depth to the water table; duration, intensity, and seasonal distribution of rainfall; temperature of air and other meteorological factors (humidity, wind, and so on); and influent and effluent streams.

Groundwater recharge (Lerner et al. 1990) may occur by infiltration, injection, or induction. The infiltration process is the entry of water into the saturation zone at the water table surface (Figure 9.17). The injection method is the introduction of water into confined or unconfined aquifers via injection wells (Figure 9.18). Recharge by induction is the entry of water into aquifers from surface-water bodies due to extraction of groundwater (Figure 9.19). Groundwater recharge by infiltration could be natural or artificial, whereas recharge by injection or induction is artificial.

The objectives of artificial groundwater recharge may be listed as follows:

- To serve as a water-conservation mechanism using subsurface storage for local or imported surface waters, to supplement the quantity of groundwater available, and to reduce the cost of pumping.
- To prevent, reduce, and correct adverse conditions such as seawater intrusion, lowering of the water table, land subsidence, and unfavorable salt balance (Figure 9.20).
- To allow heat exchange through the ground by diffusion either to conserve or extract heat energy.

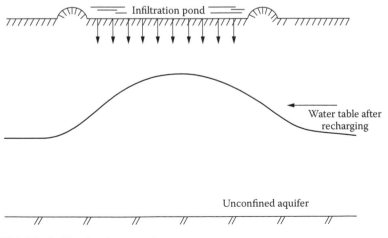

FIGURE 9.17   Infiltration from ponds.

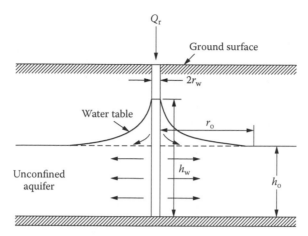

FIGURE 9.18   Water injection through a recharge well.

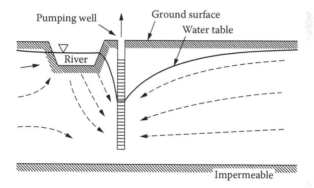

FIGURE 9.19   Induced recharge resulting from a well extraction.

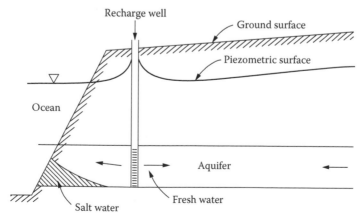

FIGURE 9.20   Control of seawater intrusion by a recharge well.

- To obtain removal of suspended solids by infiltration through the ground and to store reclaimed wastewater for subsequent use. The sources of water for groundwater recharge may be storm runoff collected in ditches, basins, or reservoirs; distant surface water that might be imported into a region by pipeline or aqueduct; or treated wastewater.

Depending on the source and quality of water, type of aquifer, type of soil, topographical and geological conditions, and economic considerations, there are various artificial groundwater recharge methods. These include water-spreading methods—basins, stream channels, ditch furrows, flooding, and irrigation (Figure 9.21); the pit method (Figure 9.22); and the recharge-well method (Figure 9.23).

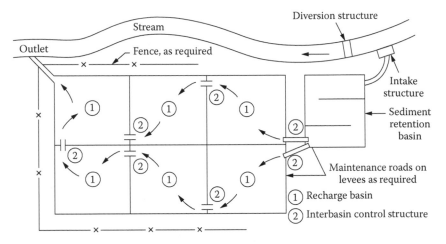

FIGURE 9.21   Multibasin recharge method.

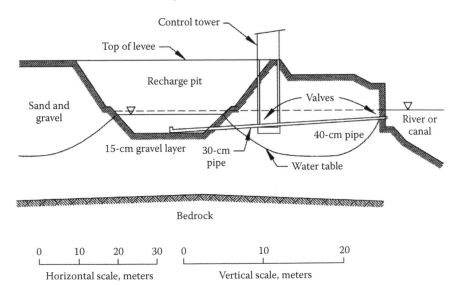

FIGURE 9.22   Cross section through a recharge pit.

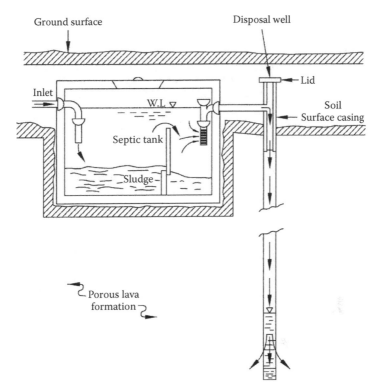

FIGURE 9.23 Recharge well for disposal of septic-tank effluents into a lava formation.

Recent interest has been focused on the reuse of municipal wastewater to recharge groundwater aquifers. This method is used mainly for nonpotable applications, for example, irrigation or industrial purposes, because of questionable health effects. Recharge of wastewater (usually after secondary treatment) improves its quality by removing physical, biological, and some chemical constituents. Storage is provided until subsequent reuse reduces the variations in seasonal temperature and dilutes the recharged water with native groundwater. Land-application practices involve irrigation, spreading overland flow, and recharge wells. Selection of a given system is governed by the soil and subsurface conditions, climate, availability of land, and intended reuse of the wastewater.

Groundwater recharge by infiltration from ponds depends on the rate of infiltration, which in turn depends on the soil characteristics. The infiltration rate of any soil can be measured by a double-ring infiltrometer. The total volume of water infiltrating the soil per unit surface area can be determined by integrating the following equation:

$$f = f_c + (f_0 - f_c)e^{-Kt} \tag{9.61}$$

and by integrating Equation 8.61 the total volume of water (F) infiltrating the soil until time t is as follows:

$$F = f_c t + (1/K)(f_0 - f_c)(1 - e^{-Kt}) \tag{9.62}$$

where f is the rate of infiltration of water into the soil at time t, f_0 is the initial infiltration rate, f_c is the final infiltration rate, and K is a rate constant, f_0, f_c, and K can be derived for any soil from the infiltration tests.

On the other hand, groundwater recharge by injection or induction depends on the hydraulic conductivity of the aquifer to be recharged. Hydraulic conductivity of aquifers can be determined by pumping tests, as mentioned earlier, or by slug tests.

9.7 APPLICATION

9.7.1 UNSTEADY-STATE WELL FORMULAS

9.7.1.1 Confined Aquifer

Because the Theis method is quite clear, Example 9.1 uses Jacob's method to get the different hydrogeological factors.

EXAMPLE 9.1

Drawdown values for a test well 30 meters distant from the productive well are given in the following table:

T_{min}	1	4	10	18	27	33	48	80	139	300
D_{min}	0.23	0.45	0.6	0.68	0.742	0.753	0.79	0.855	0.92	0.99

The discharge is 788 cubic meters per day. Find both S and T'.

SOLUTIONS

1. Theis method

By matching the curves, one point was selected (note that any point can be selected on the curve) to give

$$W(u) = I, \; I/u = 10, \; D = 0.15 \text{ meters, and } t/r^2 = 1.5 \times 10^{-3}$$

$$T = \frac{788}{4 \times 3.14 \times 0.15} \times 1 = 418 \text{ m}^2/\text{day}$$

$$S = 4 \times 418 \times \frac{1.5 \times 10^{-3}}{1440} \times \frac{1}{10} = 1.7 \times 10^{-4}$$

2. Modified Jacob's method

The equation of the asymptote is

$$D = \frac{2.3Q}{4\pi T} \log \frac{2.25Tt}{r^2 S}$$

Therefore, a plot of drawdown D versus $\log t$ forms a straight line. This line is extended until it intersects the time axis at $D = 0$; furthermore, $t = t_0$. Substitution of these values into the last equation gives

$$0 = \frac{2.3Q}{4\pi T}\log\frac{2.25Tt}{r^2 S}$$

Therefore

$$S = \frac{2.25KDt_0}{r^2}$$

Using the data from the aforementioned relation

$$T = \frac{2.3\times788}{4\times3.14\times0.36} = 401 \text{ m}^2/\text{day}$$

and

$$S = \frac{2.25\times401\times0.25}{30^2\times1440} = 1.7\times10^{-4}$$

These values are almost the same as those of the Theis method

3. Recovery method
For demonstration, the recovery data of the last piezometer from this problem will be given here, as shown in the following table:

t (minutes)	1	5	10	20	30	120	180	300
t/t'	831	166	84	42	29	7.9	5.6	3.8
D'(meter)	0.97	0.85	0.76	0.65	0.58	0.36	0.3	0.23

The values of the residual drawdown D' are plotted against the corresponding values of t/t' on semilog paper. The straight line fitted through the plotted points gives a residual drawdown difference per log cycle of t/t' equal to 0.4 meters, that is,

$$T = \frac{2.3Q}{4\pi D'} = \frac{2.3\times788}{4\times\pi\times0.4} = 361 \text{ m}^2/\text{day}$$

The storativity during recovery S' can be related to the storativity during pumping using the following formula:

$$D' = \frac{2.3Q}{4\pi T}\left(\log\left(\frac{t}{t'}\right) - \log\left(\frac{s}{s'}\right)\right)$$

When $D' = 0$, this equation becomes

$$\log\left(\frac{s}{s'}\right) = \log\left(\frac{t}{t'}\right) = 0$$

from which $S = S'$, that is, no change in storativity is recorded during the recovery period.

9.7.1.2 Semiconfined Aquifer

Several methods can be used to obtain the leakance of a semiconfined aquifer.

However, Walton's method will be used here for its simplicity and reliable results. For a semiconfined aquifer, the following set of equations is used.

$$D = \frac{Q}{4\pi T} W(u, r/L)$$

and

$$U = \frac{r^2 S}{4Tt}$$

$m = \sqrt{T \times \dfrac{D'}{K'}} L$ is the leakage factor, and, D'/K' is the hydraulic resistance of the semipervious layer in days, if K' is in meters per day.

Walton was able to plot a group of curves on the log-log scale describing the relation $W(u, L/r)$ versus $1/u$ for different values of r/L.

Walton's method is similar to the Theis graphical solution, which is carried out by plotting the drawdown values D versus the time t (if one observation is used) on the same log-log scale of the Walton graph. Select one of the curves that suits the drawdown time curve from the pumping test. By selecting one point on the two graphs, the values of $W(u, r/L)$, $1/u$, D, and t can be obtained, from which L can be calculated.

EXAMPLE 9.2

The drawdown values in an observation well 90 meters from the productive well and inside a layer overlying the productive aquifer are shown in the following table.

D (meters)	0.077	0.091	0.1	0.12	0.136	0.142
t (hours)	0.73	1.12	1.62	3	5	7

If the discharge is 761 meters per day, find the leakage factor.

SOLUTION

A comparison with the Walton family (Figure 9.12) of type curves shows that the plotted points fall along the curve for $r/L = 0.1$. The point where $W(u, r/L) = 1$ is chosen as the match point.

On the observed data sheet, $D = 0.035$ meters and $t = 5.28$ hours = 0.22 days.

$$T = \frac{Q}{4\pi D} W(u, r/L) = \frac{761 \times 1}{4 \times 3.14 \times .035} = 1729 \text{ m}^2/\text{d}$$

$$S = \frac{4Tt}{r^2} u = \frac{4 \times 1729 \times 0.22}{90^2} \times \frac{1}{10^2} = 1.9 \times 10^{-3}$$

and

$$L = \frac{r}{0.1} = \frac{90}{0.1} = 900 \text{ m}$$

$$\frac{D'}{K'} = \frac{L^2}{T} = \frac{(900)^2}{1729} = 468 \text{ days}$$

9.7.1.3   Water-Table Condition

Boulton's method is used here to obtain the formation constants for Example 9.3.

EXAMPLE 9.3

The following table gives the data of an observation well 90 meters from a pumped well

Time (minutes)	1.17	1.7	2.5	5	7.5	14	26	51	85	175	300	1340
D (meters)	0.004	0.015	0.03	0.054	0.068	0.09	0.11	0.133	0.146	0.161	0.173	0.2

The discharge of the pumped well was 873 meters per day.

SOLUTION

By matching the plotted data, D versus t, the following values were obtained. From the left-hand portion of the curve, a match point is chosen giving $1/u = \theta = 10$, $w(U_A, r/B) = 1$, $D = 0.07$ meters, and $t = 16$ minutes $= 1.11 \times 10^{-2}$ days.

$$T = \frac{Q}{4\pi D} W(u, r/B) = \frac{873}{4 \times 3.14 \times 0.07} \times 1 = 990 \text{ m}^2/\text{d}$$

$$S_A = \frac{u \times 4Tt}{r^2} = \frac{10^{-1} \times 4 \times 990 \times 1.11 \times 10^{-2}}{90^2} = 5.4 \times 10^{-4}$$

Another point is selected on the right-hand portion of the curve along the line $r/B = 0.6$, which fitted the data curve giving: $\theta' = 1$, $W(\theta', r/B) = 1$, $D = 105$ m, and $t = 250$ minutes $= 174 \times 10^{-1}$ days.

$$T = \frac{Q}{4\pi D} W(u, r/B) = \frac{873}{4 \times 3.14 \times 0.105} \times 1 = 660 \text{ m}^2/\text{d}$$

$$S' = \frac{\theta' 4Tt}{r^2} = \frac{1 \times 4 \times 660 \times 1.74 \times 10^{-1}}{90^2} = 5.7 \times 10^{-2}$$

$$B = \frac{90}{0.6} = 150 \text{ m}$$

$$\alpha = \frac{T}{S'B^2} = 0.51 \text{ day}^{-1}$$

For the delay index, $t_0 = 3.6/\alpha = 7$ days, where the delayed yield ceases after 7 days.

9.7.2 GROUNDWATER RECHARGE APPLICATION

An example can be applied to find out the control of the water table by pumping the groundwater from an unconfined sandy aquifer that is recharged by irrigation water. This is very useful when the water table is very near the ground surface and any additional recharged quantity can be removed by a well system, which in this event is called a drainage well system. The mathematical analysis of the problem is given next.

From the definition sketch (Figure 9.24), the following notations are used:

K = hydraulic conductivity
h_w = depth of water inside the well
h_a = saturation thickness at radius r_a
h = saturation thickness at radius r
i_n = rate of infiltration from irrigation water.
Q_p = well discharge

At a certain radius r from the center line of the well

$$q(\text{at } r) = 2\pi rh \, k \, dh/dr \tag{9.63}$$

and

$$q \, dr/r = 2\pi kh \, dh \tag{9.64}$$

but

$$q = Q_p - \pi r^2 i_n \tag{9.65}$$

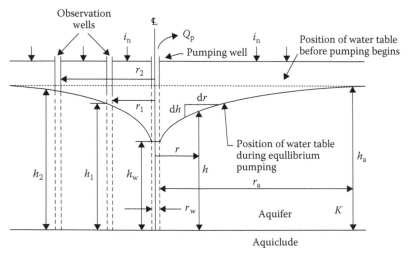

FIGURE 9.24 Pumping well in an unconfined aquifer recharged by irrigation water.

$$\left(Q_{\mathrm{p}} - \pi r^2 i_{\mathrm{n}}\right)\frac{dr}{r} = 2\pi kh \ dh$$

$$\left(\frac{Q_{\mathrm{p}}}{r}\right) dr - i_{\mathrm{n}}\pi r \ dr = 2\pi kh \ dh$$

Integrating from r_{w} to r_{a}, we get

$$Q_{\mathrm{p}}\left(\ln\frac{r_{\mathrm{a}}}{r_{\mathrm{w}}}\right) - i_{\mathrm{n}}\pi\frac{\left(r_{\mathrm{a}}^2 - r_{\mathrm{w}}^2\right)}{2} = \pi k\left(h_{\mathrm{a}}^2 - h_{\mathrm{w}}^2\right)$$

but

$$Q_{\mathrm{p}} = i_{\mathrm{n}}\pi\left(r_{\mathrm{a}}^2 - r_{\mathrm{w}}^2\right)$$

$$\approx i_{\mathrm{n}}\pi r_{\mathrm{a}}^2 \text{ since } r_{\mathrm{a}} \gg r_{\mathrm{w}}$$

$$Q_{\mathrm{p}}\left(\ln\frac{r_{\mathrm{a}}}{r_{\mathrm{w}}}\right) - \left(\frac{Q_{\mathrm{p}}}{2}\right) = \pi K\left(h_{\mathrm{a}}^2 - h_{\mathrm{w}}^2\right)$$

or

$$Q_{\mathrm{p}}\left(\ln\frac{r_{\mathrm{a}}}{r_{\mathrm{w}}} - \frac{1}{2}\right) = \pi K\left(h_{\mathrm{a}}^2 - h_{\mathrm{w}}^2\right) \tag{9.66}$$

or

$$Q_{\mathrm{p}} = \frac{K\left(h_{\mathrm{a}}^2 - h_{\mathrm{w}}^2\right)}{\left(\ln\dfrac{r_{\mathrm{a}}}{r_{\mathrm{w}}} - \dfrac{1}{2}\right)}$$

Arrange Equation 9.66 to obtain

$$\frac{Q_{\mathrm{p}}}{\pi i_{\mathrm{n}}} r_{\mathrm{w}}^2 = 2 \frac{\left(\dfrac{K}{i_{\mathrm{n}}}\right)\left(\dfrac{h_{\mathrm{a}}^2}{r_{\mathrm{w}}^2} - \dfrac{h_{\mathrm{w}}^2}{r_{\mathrm{w}}^2}\right)}{\left(\ln\dfrac{r_{\mathrm{a}}^2}{r_{\mathrm{w}}^2} - 1\right)}$$

$$\frac{Q_{\mathrm{p}}}{\pi i_{\mathrm{n}}} r_{\mathrm{w}}^2 = 2 \frac{\dfrac{K}{i_{\mathrm{n}}}\left(\dfrac{h_{\mathrm{a}}^2}{r_{\mathrm{w}}^2} - \dfrac{h_{\mathrm{w}}^2}{r_{\mathrm{w}}^2}\right)}{\left[\ln\left(\dfrac{Q_{\mathrm{p}}}{\pi i_{\mathrm{n}} r_{\mathrm{w}}^2}\right) - 1\right]} \tag{9.67}$$

For simplicity, let

$$A = \frac{Q_p}{\pi i_n r_w^2}, \quad B = \frac{K}{i_n}, \quad \text{and} \quad C = \frac{h_a^2}{r_w^2}$$

$$A = 2B \frac{\left(\dfrac{(C - (hw/rw)^2)}{r_w^2} \right)}{(\ln A - 1)} \tag{9.68}$$

Changing ln to log, Equation 9.68 becomes

$$A = 2B \frac{\left(\dfrac{(C - (hw/rw)^2)}{r_w^2} \right)}{(2.3 \log A - 1)}$$

or

$$A(2.3 \log A - 1) = 2B \left(\frac{(C - (hw/rw)^2)}{r_w^2} \right) \tag{9.69}$$

Knowing h_w, $\Delta h = h_a - h_w$ is known.

Equation 9.69 can solve any problem using the spreadsheet method. In addition, the spacings of the well network can be found if the aquifer charateristics and the recharge values are known in order to prevent any water table rise. In conclusion, the well network is useful in water-table aquifers to prevent drainage problems.

EXAMPLE 9.4

For an aquifer having h_a = 50 meters, K = 1 meter per hour, i_n = 0.05 meter per hour, and r_w = 0.15 meters, plot a graph to show the relationship between Q and Δh. From this relationship, find the discharge for a drawdown of 4 meters and deduce the well spacings in the well network

SOLUTION

For further arrangement, let Equation 9.69 have the following form:

$$D = A(2.3 \log A - 1)$$

$$= 2B \left(\frac{(C - (hw/rw)^2)}{r_w^2} \right)$$

Therefore, $h_w^2 / r_w^2 = C - D/2B = E$, from which $h_w = r_w \sqrt{E}$.

The calculations of the different values for the aforementioned terms were tabulated with the help of the worksheet given in Table 9.1, covering a wide range of discharges and drawdowns. The relationship between the drawdown Δh and the discharge Q are also plotted, from which the drawdown for any Q for this problem can be selected (Figure 9.25).

TABLE 9.1

Values of the Different Terms in Equation 9.69 to Give the Relationship between Δh and Q

No	Q (m³/h)	A ($Q_p/\pi\, i_n r_w^2$)	B (K/i_n)	$D = A$ (2.3log A −1)	$(C-$ $(hw/rw)^2)$	E $(C-D/2B)$	h_w (m)	$\Delta h = h_a -$ h_w (m)
1	100	141,471.0	100	1,534,470.7	111,111.11	103,438.76	48.2428445	1.7571555
2	200	282,942.1	100	3,264,841.7	111,111.11	94,786.90	46.1812224	3.8187776
3	400	565,884.2	100	6,921,484.1	111,111.11	76,503.69	41.4889508	8.5110492
4	600	848,826.3	100	10,726,009.2	111,111.11	57,481.07	35.9628137	14.0371863
5	800	1,131,768.4	100	14,626,569.5	111,111.11	37,978.26	29.2320189	20.7679811
6	1000	1,414,710.4	100	18,598,541.0	111,111.11	18,118.41	20.1906944	29.8093056

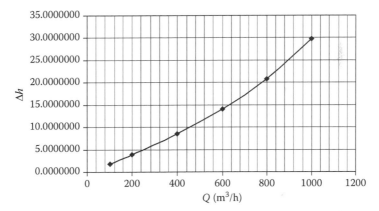

FIGURE 9.25 Relationship between the drawdown in a drainage well and its discharge.

Thus, for $\Delta h = 4$ meters, the discharge from the figure was found to be 210 cubic meters per hour. Note that for a drainage well network, the spacing between the wells depends on the recharge value and aquifer specifications. It is easy to derive the spacing between wells using Equation 9.69.

For well networks, it was found that the triangular well network is economical, that is, the locations of the wells are at the vertices of an equilateral triangle. Therefore, the spacing S between each two wells is $S = 1.732\ r_a$, where r_a can be found from Equation 9.66.

9.8 GROUNDWATER POLLUTION

The movement of contaminants in groundwater is a particularly active area of research. Models have been developed to study saltwater intrusion and leachate migration from waste-disposal sites. Groundwater pollutants can be categorized as bacteria, viruses, nitrogen, phosphorus, metals, organics, pesticides, and radioactive materials. This section discusses information on general subsurface transport.

9.8.1 MIGRATION OF POLLUTANTS IN AQUIFERS

Movement of contaminants in groundwater occurs not only by advection but also by dispersion. Advection, also referred to as convection, is the transport of a solute at a velocity equivalent to that of groundwater movement. It is the movement of the solute at a rate equal to the average pore water velocity due to the hydraulic gradient, which has the form

$$F_c = q \cdot c \qquad (9.70)$$

where F_c is the mass flux (M/T/L^2), q is the average pore water velocity (L/T), and c is the concentration (M/L').

Dispersion refers to the mixing and spreading caused in part by molecular diffusion and in part by the variations in velocity within the porous medium. For many field problems, dispersion caused by molecular diffusion and by flow around grains in the porous medium is negligible compared to dispersion caused by large-scale heterogeneities within the aquifer. Dispersion can simply be defined as the movement of solutes due to varying velocities from pore to pore at high velocities and the subsequent dispersive mass flux in the x direction.

$$F_x = -D_L \frac{\partial c}{\partial x} \qquad (9.71)$$

where F_x is the dispersive flux in the x direction (M/T/L^2), and D is the dispersion coefficient in the longitudinal direction; it has the dimension (L^2/T).

Furthermore, the dispersive mass flux in the y direction becomes

$$F_y = -D_T \frac{\partial c}{\partial y} \qquad (9.72)$$

where F_y is the dispersive flux in the y direction (M/T/L^2) and D_T is the dispersion coefficient in the transverse direction with dimension (L^2/T).

Experiments have demonstrated that, in an isotropic medium, the longitudinal and transverse components of dispersion (Equations 9.71 and 9.72) are linearly dependent on the average speed of groundwater flow. For a uniform flow field with an average linear velocity V_x,

$$D_L = a_L V_x \qquad (9.73)$$

and

$$D_T = a_T V_x \qquad (9.74)$$

where a_L is the dispersivity in the longitudinal direction (L) and a_T is the dispersivity in the transverse direction (T).

Equations 9.71 and 9.72 are equivalent to Pick's law. The solute transport governing the equations can be obtained in the same way as the governing equation of groundwater flow. The equation governing solute transport can be developed by applying the conservation-of-mass approach and Pick's law of dispersion. The equation in statement form is as follows:

Net rate of change in mass of solute within = flux of solute out of the element − flux of solute into the element ± loss or gain of solute mass due to reactions.

The one-dimensional form of the equation for a nonreactive, dissolved constituent in a homogeneous, isotropic aquifer under steady-state, uniform flow is as follows:

$$D\frac{\partial^2 c}{\partial x^2} - u\frac{\partial c}{\partial x} = \frac{\partial c}{\partial t} \tag{9.75}$$

where D is the coefficient of dispersion in the x direction (L^2/T), u is the average linear groundwater velocity (L/T), and c is the concentration (M/L^3).

The two-dimensional equation for the flow of a nonreactive, dissolved chemical species in groundwater becomes

$$\frac{\partial}{\partial x}\left(D_L\frac{\partial c}{\partial x}\right) + \frac{\partial}{\partial y}\left(D_L\frac{\partial c}{\partial y}\right) - q_x\frac{\partial c}{\partial x} - q_y\frac{\partial c}{\partial y} \pm Qc' = \frac{\partial c}{\partial t} \tag{9.76}$$

where D_L and D_T are the hydrodynamic dispersion coefficients in the x and y directions, respectively, c' is the concentration of the solute in a source or sink of strength Q (assumed to be known), q_x, q_y are effective pore-water velocities in the x and y directions, respectively. Generally,

$$q = \frac{Vi}{n_e}$$

where Vi is the flow per unit area and n_e is the effective porosity.

For the case in which the solute transport problem for a soil column is a one-dimensional flow, Equation 9.75 is used, whereas for any two-dimensional problem, Equation 9.76 may be used. For more details in this context, refer to Soliman et al. (1998).

REFERENCES

Boulton, N. S. 1963. Analysis of data from nonequilibrium pumping tests allowing for delayed yield from storage. *Proc Inst Civ Eng* 26(6693):469–82.

Bouwer, H., and R. C. Rice. 1976. A slug test for determining hydraulic conductivity of unconfined aquifers with completely or partially penetrating wells. *Water Resour Res* 12(3):423–8.

Dupuit, J. 1863. *Etudes, theoriques, et pratiques sur le movement des eaux.* 2nd ed. Paris: Dunod.

Huntush, M. S. 1964. Hydraulics of wells. In *Advanced Hydroscience*, vol. 1, pp. 281–431. New York: Elsevier.

Jacob, C. E. 1944. Notes on determining permeability by pumping tests under water table conditions. US Geological Survey, Open File Report.

Jacob, C. E. 1954. On the flow water in an elastic artesian aquifer. *Trans Am Geophys Union* 33:559–569.

Lai, R. Y. S., and C. W. Su. 1974. Nonsteady flow to a large well in a leaky aquifer. *J Hydrol* 22:333–45.

Lerner, D. N., S. I. Arie, and S. Ian. 1990. *Groundwater Recharge*, Vol. 8. International Association of Hydrogeologists.

Luthin, I. 1957. *Drainage of Agricultural Lands*. Madison, WI: American Society of Agriculture.

Muskat, M. 1937. *The Flow of Homogeneous Fluids through Porous Media*. New York: McGraw-Hill.

Neuman, S. P. 1973a. Supplementary comments on theory of flow in unconfined aquifers considering delayed response of the water table. *Water Resour Res* 9(4):1102–3.

Neuman, S. P. 1973b. Calibration of distributed parameter groundwater flow models viewed as a multiple-objective decision process under uncertainty. *Water Resour Res* 9(4):1006–21.

Neuman, S. P. 1975. Analysis of pumping test data from anisotropic unconfined aquifers considering delayed gravity response. *Water Resour Res* 11(2):329–42.

Soliman, M. M. 1965. Boundary flow consideration in the design of wells. *ASCE J Irrig Drain* 91(1):159.

Soliman, M. M. et al. 1998. *Environmental Hydrogeology*. Boca Raton, FL: CRC Press.

Theis, C. V. 1935. The relation between the lowering of piezometric surface and the duration of discharge of a well using groundwater storage. *Trans Am Geophys Union* 16:520.

Walton, W. C. 1970. *Groundwater Resource Evaluation*. New York: McGraw-Hill.

BIBLIOGRAPHY

Boulton, C. A. 1954. The drawdown of the water table under non-steady conditions near a pumped well in an unconfined formation. *Proc Inst Civ Eng* 3(3):564–79.

Boulton, N. S., and T. D. Streltsova. 1975. New equations for determining the formation constant of an aquifer from pumping test data. *Water Resour Res* 11(1):148–53.

Boulton, N. S., and T. D. Streltsova. 1976. The drawdown near an abstraction well of large diameter under non-steady conditions in an unconfined aquifer. *J Hydrol* 30:29–265.

De Wiest, R. J. 1965. *Geohydrology*. New York: John Wiley.

Freeze, R. A., and J. A. Cherry. 1979. *Groundwater*. Englewood Cliffs, NJ: Prentice Hall.

Huntush, M. 1956. Nonsteady flow to well partially penetrating an infinite leaky aquifer. In *Proceedings of Iraqi Science Society*, Baghdad, Iraq.

Neuman, S. P. 1972. Theory of flow in unconfined aquifers considering delayed response of the water table. *Water Resour Res* 8(4):1031–45.

Neuman, S. P. 1979. Perspective on delayed yield. *Water Resour Res* 15(4):899–908.

Soliman, M. M. 1984. *Groundwater Management in Arid Regions*. Vol. 1. Cairo: Ain Shams University.

Soliman, M. M. 1990. Environmental effects on the arid coastal water sheds in Egypt. *International Symposium of Arid Region Hydrology*. San Diego, CA: ASCE.

Todd, D. K. 1959. *Groundwater Hydrology*. New York: John Wiley.

Wenzel, L. K. 1940. Local overdevelopment of groundwater supplies, with special reference to conditions at Grand Island, Nebraska. USGS Water Supply Paper 836, Washington, DC.

10 Sediment Yield from Watersheds

10.1 INTRODUCTION

Arid and semi-arid regions have potential for generating and transporting large quantities of sediment, mainly due to torrential rainfall, excessive weathering, and almost total lack of natural protection against erosion of soil by runoff. Erosion occurs due to the impact of raindrops and surface shear. The bulk of eroded material is deposited at intermediate locations if the surface runoff cannot sustain transport.

Designing tools for evaluating the sediment volume at different hydraulic structures is one of the most important contemporary projects. Considering the present state of knowledge concerning system analysis, geographic information system (GIS)–based techniques offer a unique opportunity to appraise complex interactions of erosion and sedimentation factors (rainfall, runoff, watershed, and land-use characteristics), in addition to enabling the study of contributions of the upland watershed to the sediment load. Sediment sources within the study basin can be grouped broadly into hillsides and channel sources. The undercutting of slopes during flood flows introduces considerable volumes of alluvial material into dams and other hydraulic structures.

Soil is an erodent material, meaning that, with the impact of an erosive agent such as water or wind, erosion occurs. Sediment yield is an important parameter that must be considered during the planning and design stages of flood-relief measures and flood-protection hydraulic structures. The main objective is to predict the amount of sediment yield from soil erosion so that no harm occurs to any hydraulic structure.

10.2 SEDIMENT-YIELD THEORIES

The average annual sediment production from a watershed depends on many factors, such as climate, soil type, land use, topography, and the presence of reservoirs. Adequate data for a complete analysis of all factors are difficult to obtain. Langbein and Schumm (1958) used data from a number of watersheds to construct the curve shown in Figure 10.1, which relates the average annual sediment production per unit area to the mean annual precipitation. Maximum sediment production rates occur at about 305 millimeters of mean annual precipitation because such areas usually have little protective vegetal cover. With higher rainfalls, vegetal cover reduces the erosion, and erosion also decreases with lesser rainfalls.

Fleming (1969) used data from over 250 catchments around the world to derive the relationships (Equation 10.1 and Table 10.1) for the mean annual suspended

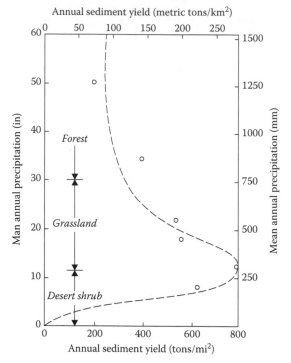

FIGURE 10.1   Sediment yield as a function of the mean annual precipitation.

TABLE 10.1
Values of *a* and *n* in Equation 10.1 for Various Cover Types

Vegetal Cover	*n*	*a*	
		Q (Tons)	Q (Metric Tons)
Mixed broadleaf and coniferous	1.02	117	106
Coniferous forest and tall grassland	0.82	3523	3196
Short grassland and scrub	0.65	19,260	17,472
Desert and scrub	0.72	37,730	34,228

load Q_s (in tons) as a function of the mean annual discharge in cubic feet per second for various vegetal covers:

$$Q = aQ^n \tag{10.1}$$

Errors of ± 50% may be expected from these relations.

   For watersheds without sediment records, the relationships presented above offer an order-of-magnitude estimate of sediment yield. If possible, such estimates should be compared with sediment data on similar watersheds in the same region.

Sediment-yield prediction techniques can be classified under three headings: (1) regional regression equations, (2) physically-based simulation models, and (3) regression models. Based on the type of erosion in the catchments and the available data, regression models and the modified universal soil-loss equation (MUSLE) may be selected for arid and semi-arid regions.

MUSLE (Williams 1975) measures and estimates the data providing the rates and volumes of runoff to arrive at the runoff-energy factor. This factor is then substituted for the rainfall-energy factor in the MUSLE, and an optimization technique (DeCoursey and Synder 1969) is applied to determine the prediction equation. The prediction equation in MUSLE may be stated as

$$Y = 11.8(Q_w q_p)^{0.56} K \cdot LS \cdot C \cdot P \tag{10.2}$$

where Y is the sediment yield (metric tons), Q_w is the runoff volume (cubic meters), q_p is the peak runoff rate (cubic meters per second), K is the soil-erodibility factor, C is the crop-management factor, P is the factor governing erosion-control practice, and LS is the length-and-gradient factor of the slope. The equation implies that the grain size of the sediment in the catchments will vary from one rainfall impulse to another. The distribution of the sediment grain size exposed to any rainfall impulse will be affected by the previous rainfall impulses. Therefore, the delivered sediment volume corresponding to any rainfall impulse is very sensitive to the location of the impulse in the hydrograph of the storm. This means that this technique takes into account the effect of the temporal variation of rainfall during the storm.

10.2.1 Determination of the Soil-Erodibility Factor (K)

The sediment erodibility parameters of the subcatchments for a specific sediment grain size depend on the physical properties of the soil—its texture, aggregate size, and permeability. Knowing the K values for all soil types in the area, an area-weighted average K value can be determined for each subcatchment using Equation 10.3, according to Cambazoglu (2002):

$$K = \frac{\sum_{i}^{n} K_i DA_i}{DA_T} \tag{10.3}$$

where K is the soil-erodibility factor for the watershed, K_i is the soil-erodibility factor for an individual soil, i, DA_i is the drainage area covered by an individual soil, DA_T is the total drainage area of the watershed, and n is the number of different soils in the watershed. We found that K in arid regions ranges between .2 for fine sand and .5 for silt loam. An average value of $K = 0.35$ for alluvial material may be approximately taken.

10.2.2 DETERMINATION OF THE SLOPE LENGTH-AND-GRADIENT FACTOR

The slope length-and-gradient factor (LS) is divided into its constituent parts, which are calculated separately. The L factor, called the topographic factor, is determined using Equation 10.4 (Moore and Burch 1986):

$$L = (l/22.13)^m \tag{10.4}$$

where l is the slope length in meters and m is an exponent depending on the slope ($rn = 0.3$ for slopes < 3%, $m = 0.4$ for slopes < 4%, and $m = 0.5$ for slopes < 5%). The slope-gradient factor S is calculated initially with the corresponding slope of the boundary using Equation 10.5:

$$S = [0.04x^2 + 0.3x + 0.43]/6.613 \tag{10.5}$$

where x is the land gradient slope measured as a percentage. Then, take the averages of the results find the average S-factor values. After both L- and S-factor values are determined for each slope class, the LS factors can be determined. The only remaining parameter needed to calculate the LS factor of a catchment is the distribution of different slope classes in the area. Then, the area-weighted average LS factors can be determined using Equation 10.6, according to Cambazoglu (2002):

$$LS = \frac{\sum_{i}^{n} LS_i DA_i}{DA_T} \tag{10.6}$$

where LS is the slope length-and-gradient factor for the watershed, LS_i is the length-and-gradient factor for an individual slope class, i, DA_i is the drainage area covered by an individual slope class, DA_T is the total drainage area of the watershed, and n is the number of different slope classes in the watershed. The LS factor values, which vary between 0.05 and 0.12, can be estimated. Figure 10.2 can be used to get the LS value (the metric value in the figure has been converted).

10.2.3 DETERMINATION OF THE PARAMETERS INFLUENCING EROSION-CONTROL PRACTICES

The crop-management factor C is generally established on experimental plots or estimated using equations. The C-factor values for different end-use types in the study area were determined by comparing the previously proposed C-factor values. The C factor also depends on the percentage of ground cover. The crop-management factor for a watershed is determined by weighting the C values of each crop and management level according to the size of the area producing the crop with the same management level, using Equation 10.7:

$$C = \frac{\sum_{k=1}^{n} C_k DA_k}{DA_T} \tag{10.7}$$

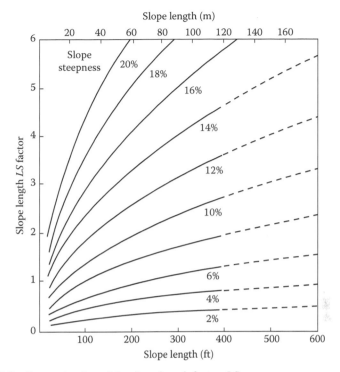

FIGURE 10.2 Determination of the slope length factor, *LS*.

where *C* is the crop-management factor for the watershed, C_k is the crop-management factor for an individual crop *k*, DA_k is the drainage area covered by an individual crop *k*, with a particular management level, DA_T is the total drainage area of the watershed, and *n* is the number of different crops and management levels in the watershed (Cambazoglu 2002). For practical reasons, the value of *CP* is considered to be unity in the context of certain problems in Sinai, Egypt, which is an arid region (WRRI 2004).

10.3 RESERVOIR SEDIMENTATION

Accumulation of sediment in reservoirs may have the following effects (Shen and Julien 1993):

- Reducing the useful storage volume for water in the reservoir
- Changing the water quality near the dam
- Increasing the flooding level upstream of the dam because of sediment aggradation
- Influencing the stability of the stream downstream of the dam
- Affecting the stream ecology in the dam region
- Causing other environmental effects by changing the water quality.

This section briefly outlines a method that can be used to estimate the potential accumulation of sediment in a reservoir. The steps are as follows:

1. Construct a flow-duration curve, which is the cumulative distribution curve of the stream runoff passing the dam (Figure 10.3).
2. Construct a sediment-rating curve, which relates sediment concentration to the stream discharge (Figure 10.3).
3. Divide the flow-duration curve into equally spaced sections of percentage, Δp (for example, for 20 sections $\Delta p = 0.05$). Read the average discharge Q_i from the flow-duration curve. Read the corresponding sediment concentration C_i from the sediment-rating curve. Repeat for each of the sections.
4. Compute the average total sediment load in weight per unit time, q_t, using the equation

$$q_t = \sum_i C_i Q_i \Delta P$$
$$= \Delta P \sum_i C_i Q_i \tag{10.8}$$

5. Determine the percentage of sediment trapped in the reservoir, which is a function of the fall velocity of sediment and the time allowed for settling. The more important factors that determine the amount of trapped sediment include the following: the relative size, shape, and operation of the reservoir and the sediment particle size. Brune (1953) developed a commonly used relationship for determining the trapping of sediments, illustrated in

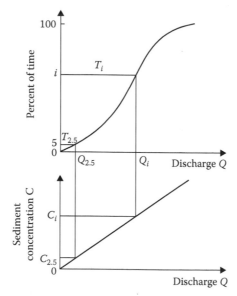

FIGURE 10.3   Flow and sediment discharge. (Adapted from Shen, H. W. and P. Y. Julien. 1993. Erosion and sediment transport. In *Handbook of Hydrology*, D. R. Maidment, ed. New York: McGraw-Hill.)

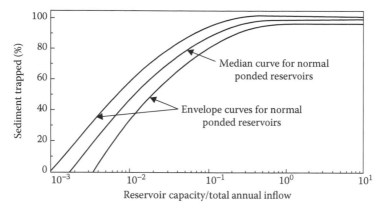

FIGURE 10.4 Efficiency of sediment traps in a reservoir. (Adapted from Shen, H. W. and P. Y. Julien. 1993. Erosion and sediment transport. In *Handbook of Hydrology*, D. R. Maidment, ed. New York: McGraw-Hill.)

Figure 10.4. Trap efficiency is the ratio between the sediment trapped in the reservoir and the total sediment entering the reservoir. Reservoir capacity is the reservoir volume at the normal operating pool level for the period considered.

The U.S. Bureau of Reclamation (1987) classifies reservoir operations according to the following criteria:

Operation	Reservoir Operation
1	Sediment always submerged or nearly submerged
2	Normally, moderate to considerable drawdown of the reservoir
3	Reservoir normally empty
4	Riverbed sediments

The proper operation number can usually be selected from the operation study prepared for the reservoir. The density of the sedimented deposits is estimated using the following equation:

$$W = W_c P_c + W_m P_m + W_s P_s \tag{10.9}$$

where W is the unit weight in pounds per cubic feet (or density in kilograms per cubic meter), P_c, P_m, and P_s are the percentages of clay, silt, and sand, respectively, of the inflowing sediment, and W_c, W_m, and W_s are the coefficients of unit weight for clay, silt, and sand, respectively, in pounds per cubic feet (or kilograms per cubic meter), as obtained from Table 10.2.

Miller (1963) developed an approximation for the average density, W_T, of the sediment deposited in T years, as follows:

$$W_T = W_0 + 0.4343K \left\{ \frac{T}{T-1} (\ln T) - 1 \right\} \tag{10.10}$$

TABLE 10.2
Reservoir Operation Number

	Initial Weight [Initial Mass in lb/ft³ (kg/m³)]		
Operation	w_c	w_m	w
1	26 (416)	70 (1120)	97 (1550)
2	35 (561)	71 (1140)	97 (1550)
3	40 (641)	72 (1150)	97 (1550)
4	60 (961)	73 (1170)	97 (1550)

Source: Adapted from Yang, C. T. 1996. Sediment Transport: Theory and Practice. New York: McGraw-Hill.

TABLE 10.3
K Values of Sand, Silt, and Clay

	K in Inch-Pound Units (Metric Units)		
Reservoir Operation	Sand	Silt	Clay
1	0	5.7 (91)	16 (256)
2	0	1.8 (29)	8.4 (135)
3	0	0 (0)	0 (0)

Source: Adapted from Yang, C. T. 1996. Sediment Transport: Theory and Practice. New York: McGraw-Hill.

where W_0 is the initial unit weight (density) derived from Equation 10.9 and K is a constant based on an analysis of the type of reservoir operation and sediment size. K is expressed as

$$K = K_c P_c + K_m P_m + K_s P_s \qquad (10.11)$$

where K_s, K_m, and K_c are found in Table 10.3.

EXAMPLE 10.1

Determine the density of sediment deposits for reservoir operation 2 with 23% clay, 40% silt, and 37% sand. Determine also the average density of sediments deposited after 100 years.

SOLUTION

Using Equation 10.9:

$$W = 561(0.23) + 1140(0.40) + 1550(0.37) = 1158 \text{ kg/m}^3$$

To determine W_{100}, get K from Equation 10.11

$$K = 0(0.23) + 29(0.40) + 135(0.37) = 62$$

$$W_{100} = 1158 + 0.4343 \times 62 \left\{ \frac{100}{100-1}(\ln 100) - 1 \right\}$$

$$= 1256.3 \ \text{kg/m}^3$$

REFERENCES

Brune, G. M. 1953. Trap efficiency of reservoirs. *Trans Am Geophys Union* 34(3):407–18.

Cambazoglu, M. K. 2002. Sediment yields of basins in the western Black Sea region, Master's thesis, Civil Engineering Department, Middle East Technical University, Turkey.

DeCoursey, D., and W. M. Snyder. 1969. Computer-oriented method of optimizing hydrologic model parameters. *J Hydrology* 9:34–56.

Fleming, G. 1969. Design curves for suspended load estimation. *Proc Inst Civ Eng* 43:1–9.

Langbein, W. B., and S. A. Schumm. 1958. Yield of sediment in relation to mean annual precipitation. *Trans Am Geophys Union* 39:1076–84.

Miller, J. F. 1963. Probable maximum precipitation and rainfall frequency data for Alaska. Technical report, U.S. Department of Commerce, Weather Bureau, Washington, DC.

Moore, I. D. and G. J. Burch. 1986. Physical basis of the length-slope factor in the universal soil loss equation. *Soil Sci Society Am J* 50:1294–8.

Shen, H. W. and P. Y. Julien. 1993. Erosion and sediment transport. In *Handbook of Hydrology*, D. R. Maidment, ed. New York: McGraw-Hill.

U.S. Bureau of Reclamation. 1987. *Design of Small Dams*, 3rd ed. Washington, DC: U.S. Government Printing Office.

Williams, J. R. 1975. Sediment-yield prediction with universal equation using runoff energy factors. In *Present and Prospective Technology for Predicting Sediment Yield and Sources*, pp. 244–52. Washington, DC: U.S. Department of Agriculture.

Water Resources Research Institute. 2004. Evaluation, development, and execution of some flash flood protection works. A project funded by the Italian government, first progress report, volume II.

Yang, C. T. 1996. *Sediment Transport: Theory and Practice*. New York: McGraw-Hill.

BIBLIOGRAPHY

Johnson, C. W., N. D. Gordon, and C. L. Hanson. 1985. Northwest rangeland sediment yield analysis by the MUSLE. *Trans ASAE* 28:1889–95.

Manar, S. B. 1958. Factors affecting sediment delivery rates in the red hills physiographic area. *Trans Am Geophys* 39:669–75.

Williams, J. R. 1981. Testing the modified universal soil loss equation. In *Estimating Erosion and Sediment Yield on Rangelands*, pp. 157–64. Washington, DC: U.S. Department of Agriculture.

Williams, X. R., and H. D. Berndt. 1972. Sediment yield computed with universal equation. *J Hydraulics Div Proc Am Soc Civil Eng* 98:2087–98.

Williams, J. R., and H. D. Berndt. 1977. Sediment yield prediction based on watershed hydrology. *Trans ASAE* 23:1100–4.

Zaki, A. 2000. Estimation of runoff and sediment discharges for the design of hydraulic structures in arid ungauged catchments, Master's thesis, Cairo University, Egypt.

11 Hydraulic Structures

11.1 INTRODUCTION

Engineering hydrology includes the segments of hydrology pertinent to the design and operation of engineering projects for the control and use of water. A brief review of some of the practical applications of engineering hydrology may provide a helpful background before undertaking a more detailed study of the subject (Linsley et al. 1975). We shall consider here a few of engineering hydrology's uses in connection with structural design, water supply, irrigation, flood control, erosion control, and environmental impacts on water resources and, finally, its applications to water-resource management in arid and semi-arid regions (Mays 2005; Chin 2006). Figure 11.1 shows many types of hydraulic structures in a watershed. All of these are needed for controlling, storing, and protecting water resources. The figure also shows crossing works such as bridges, culverts, and aqueducts.

For highways that cross a stream of a catchment in arid or semi-arid areas, the cost of constructing crossing works such as bridges, culverts, aqueducts, or Irish crossings adds up to a significant part of the total cost of hydrologic projects. Sometimes, the culverts or bridges are located so that they will not be destroyed by a flood that exceeds, even by a considerable amount, their designed capacity; in such cases, there is simply an inconvenience caused by a temporary interruption of service due to flooding of the road or rails. Here again, an economic balance must be found between the losses due to occasional interruption of service on the one hand, and the cost of larger drainage structures on the other. Hydrology provides the maximum design flow needed in order to obtain the minimum vent sizes for such structures, while the principles of hydraulics provide the basis for analyzing the effects of these structures on the flow, in addition to analyzing the structure's hydraulic design, size, and requisite protection.

The problems of highway and railroad culvert designs are similar to those of storm sewer design, although the hydrology of urban areas is a study in itself and is much more complex than the apparent similarity of such areas would suggest.

For storage works such as dams, provisions must be made for passing flood flows over or around the dam. The spillway structure is often the most expensive portion of the dam, hence, economy demands that it be as small as possible. However, overtopping the nonoverflow section is a serious matter and, in the case of an earth dam, is almost certain to cause failure, along with destruction and, frequently, deaths in the areas downstream. Clearly, the best in hydrologic design is required for a safe and economical dam design. The hydrologist must evaluate not only the probability of floods of various magnitudes, but also the effect of the reservoir on the distribution of the flood volume. This latter point is worthy of emphasis because it is sometimes

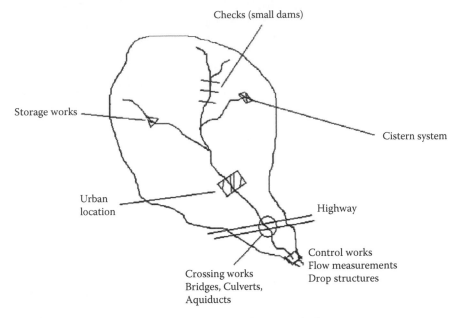

FIGURE 11.1 Locations of hydraulic structures in a watershed.

overlooked, resulting in unnecessary expenditures being made on "overdesigned" structures.

Note the distinction between hydrologic design and hydraulic design (Chow et al. 1988). The former is concerned with determining the quantities of water that must be handled; the latter proceeds from there to determine the form of the structure best suited for the job. The engineer is no more warranted in undertaking the hydraulic design of a structure without attention to hydrologic design than he would be in undertaking to determine the sizes of the structural parts without first ascertaining the loads that may come upon them. In this chapter, the hydraulic designs of different types of hydraulic structures used in hydrologic projects are given.

11.2 CROSSING WORKS

Bridges, culverts, siphons, flumes, and aqueducts are all considered *crossing works*. Bridges and culverts are used for the crossings of roads with waterways. Aqueducts or siphons are used when two waterways intersect each other (Soliman 1985). When the stream discharge is small, with a flow depth not exceeding 15 centimeters, an Irish crossing can be used where a roadway crosses a small stream. The best location for the Irish crossing is similar to that for a conventional bridge, with the exception that a wide stretch of the stream provides an easier road approach and slower, shallower water. The stream should be straight, with well-defined banks and a uniform gradient, and the bed material should be strong enough to support

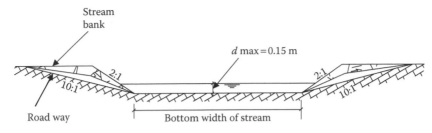

FIGURE 11.2   Irish crossing.

traffic. The roadway slope on both banks should be within 10–15%, as shown in Figure 11.2.

11.2.1   Hydraulic Design of a Bridge

Erection of the supports of a bridge (i.e., the piers and abutments) reduces the waterway area of the stream at the bridge site, which will cause an increase in the velocity at the bridge site because the discharge will be the same along the channel. Accordingly, an increase in the potential energy upstream of the bridge structure will occur. The rise in the water level is called the heading up, h. This increase in the upstream water level, h, is the difference between the original water level of the stream and the raised water level upstream of the bridge structure.

If the original waterway area of the stream is greatly reduced by the obstruction, the heading up will be relatively large. This may damage the neighboring properties, land, and buildings upstream of the structure. Moreover, the high velocity may cause scouring of the bed material just downstream of the structure. Eddies, which depend on the original velocity, the shape of piers and abutments, and the ratio of the vent velocity V_v through the bridge to the channel velocity V_c, will also be generated. Therefore, the Egyptian Ministry's Code of Practice generally recommends that h be limited to a value of 5–10 centimeters. This limitation may be included or expressed by limiting the relative increment of the velocity through the structure with respect to that of the channel. For example, if the bed material is an alluvial deposit, the velocity through the bridge vent is not allowed to exceed three times the velocity in the channel, provided that the maximum velocity in the vent does not exceed 2 meters per second.

11.2.1.1   Calculating the Heading Up

Several empirical rules have been obtained from practice. The formula adopted by the Egyptian code, however, can be written as

$$h = \frac{1}{C^2}\left(\frac{A^2}{a^2} - 1\right)\frac{V^2}{2g} \tag{11.1}$$

where h is the heading up, V is the mean velocity through the channel, a is the area of the channel waterway, A is the area of the waterway through the vents, and C is a coefficient that depends on the vent width of the bridge as follows (Figure 11.3a):

Vent Width	C
Less than 2 m	0.72
2–4 m	0.82
More than 4 m	0.92

Source: Egyptian Ministry of Irrigation Code (2003).

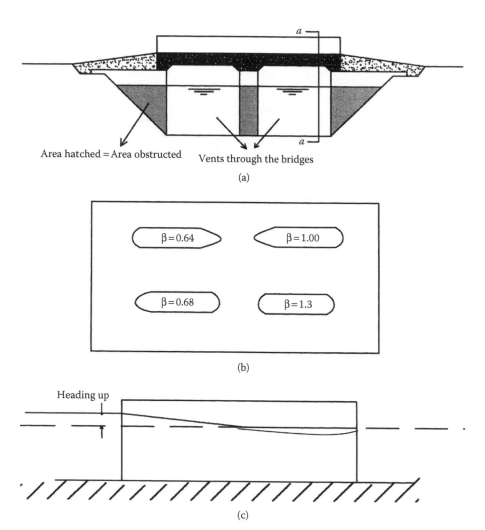

FIGURE 11.3 (a) Cross section through a bridge. (b) Plan for different pier shapes and the corresponding β values. (c) Heading up due to a bridge obstruction.

The recommended average velocity is between 0.4 and 0.8 meters per second in a channel with bed material composed of an alluvial deposit. Field observations have shown that Equation 11.1 yields slightly higher values than those observed in the field. Another empirical relation that is more frequently used is

$$h = \alpha \times \beta \times \left(\frac{V^2}{2g} \right) \tag{11.2}$$

where $\alpha = (1 - a/A)$ and β is a coefficient depending on the shape of the pier (Figure 11.3b).

The heading up (Figure 11.3c) equation is also used to derive the safe area and the number of vents for the bridge as given in Example 11.1.

EXAMPLE 11.1

A stream with bed width of 30 meters, water depth of 1.5 meters, and side slope with a ratio of 2:1 is carrying a flow of 50 cubic meters per second. A bridge must be constructed across the stream. Find the number of vents and the width of each vent so that the heading up does not exceed 10 centimeters.

SOLUTION

Stream area A	$= (30 + 1.5 \times (2/1)) \times 1.5 = 49.5 \ m^2$
Stream velocity v	$= 50/49.5 = 1.01 \ m/s$
Assume vent velocity V	$= 2 \ m/s \ \left(\text{allowable 1 to 2 m/s}\right)$ and
	take $V = 1.5 \ m/s$
Area of vents a	$= 33.33 \ m^2$
ΣB(Total vent width)	$= 33.33/1.5 = 22.2 \ m$
Select 3 vents with the vent width $= 7$ m, which gives $\Sigma B = 21.0$ m	
Therefore, a	$= (21) \times 1.5 = 31.5 \ m^2$

Use Equation 11.1 to calculate h (select $C = 0.9$)

$$h = \frac{1}{0.9^2} \left(\frac{49.5^2}{31.5^2} - 1 \right) \frac{1.01^2}{2g} = 0.0941 \ m$$

$$= 9.4 \ cm < 10 \ cm$$

Therefore, the number of vents and their widths satisfy the heading-up condition.

11.2.2 HYDRAULIC DESIGN OF CULVERTS

A culvert is a closed conduit used to pass water underneath a roadway or an earthen embankment. The culvert section can be any shape, but usually it is circular, square, or rectangular. The size of the culvert section is regularly designed when the culvert is running full and the culvert centerline is horizontal. If the culvert section is partially full it can be treated as a bridge and all the formulas from Section 11.2.1.1 are applied. If the culvert is inclined at a slope and is not running full, it can be treated as an open channel and therefore any formula related to open-channel hydraulics can be applied.

This section focuses on the design of culverts running full with submerged inlets and outlets. The water level in the canal upstream of the culvert will be raised by the heading-up amount. If the bed level of the waterway upstream of the structure is the same as the level downstream, the loss in head will be equal to the heading up, provided the velocity is kept constant, as shown in Figure 11.4. Considering the case of uniform flow through a horizontal straight culvert with a mean velocity v, the loss in the head can be expressed by the relation

$$h = \frac{v^2}{2g}\left[\eta_e + \eta_f + \eta_o\right] \tag{11.3}$$

where η_e, η_f, and η_o are the coefficients of head loss due to entrance, skin friction, and exit, respectively. For square-edge box culverts, $\eta_e = 0.5$ and can be reduced to 0.06 if a rounded entrance can be provided. Also

$$\eta_f = a\left(1 + \frac{b}{m}\right)\frac{L}{m} = f(L/m) \tag{11.4}$$

where L is the length of the culvert, m is the hydraulic radius, and a and b are coefficients depending on the material of construction of the culvert as given below:

Material	a	b
Steel pipe	0.00437	0.0256
Rusted pipe	0.00996	0.0225
Cemented barrel	0.00316	0.0306
Brick work	0.00401	0.070

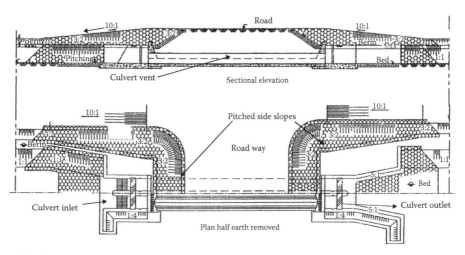

FIGURE 11.4 Sketch showing different views of a culvert.

$\eta_o = 1$ for exits with sharp corners and equals 0.2 and 0.3 for rounded and gradual exits, respectively.

The culvert can assume any shape in cross section, and various materials can be used in its construction. The most common types are the pipe and box culverts. Pipe culverts are made of steel, plain concrete, and reinforced concrete. Box culverts are usually constructed of reinforced concrete and are either single- or multivent culverts.

However, when applying Equation 11.3, v should neither exceed 2 meters per second nor be less than 1 meters per second, in order to prevent sediment deposition inside the culvert. In addition, according to the Egyptian Ministry of Irrigation code, h should not exceed 10 centimeters. The code also recommends that the upstream water level should be higher than that in the culvert invert by 25 centimeters to prevent air bubbles from entering the culvert, which will be also submerged in this case.

EXAMPLE 11.2

Find the section of a box culvert that passes a discharge of 8.8 cubic meters per second if you have the following data:
The width of the stream bed = 6 m
The water depth = 2 m
The side slope = 3:2 = 1.5:1
The roadway width = 8 m, with the culvert length = 14.5 m

SOLUTION

Step 1: Designing the cross-sectional area of the culvert.
Area of the canal waterway = $(6 + 1.5 \times 2) 2 = 18$ m²
Velocity (v) of flow in canal = 8.8/18 = 0.49 m/s
Assume barrel velocity (V) = 1.2 m/s

Velocity ratio = $\dfrac{1.2}{0.49} = 2.45$, that is, $V < 3v$ and $> 2v$.; or $2v < V > 3v$.

Area of the barrel sections = 8.8/1.2 = 7.33 m²
Choose two barrels with square sections; thus area of one barrel = 7.33/2 = 3.67 m²
= (1.91 × 1.91) m²
For simplicity, change the section to (2 × 2) m²
Actual barrel velocity = 8.8/4 × 2 = 1.1 m/s
This is still within the allowable limits.
$m = 8/16 = 4/2 \times 4 = 0.5$ m
For the culvert to run at full capacity, the level of the bottom slab should be below the water level by (2 + 0.25) = 2.25 meters, as required.

Step 2: Checking for the Heading Up.
Choose a rounded box wing wall at the inlet and a sloping splayed wing wall at the exit (Figure 11.4), with

$$\eta_e = 0.06$$

η_o is a little less than 0.5 for such outlets.

For safety, η_o will equal 0.5. Because the material of construction is reinforced concrete

$$\eta_f = 0.0032\left(1+\frac{0.03}{0.5}\right)\frac{14.5}{0.5} = 0.0099$$

and

$$h = [(1.1)^2/2 \times 9.8]\,(0.06 + 0.099 + 0.5)$$
$$= 4 \text{ cm} < 10 \text{ cm}$$

The section of the box culvert therefore satisfies all the requirements. Figure 11.4 shows a plan, the sectional elevation, and an end view of the culvert.

11.2.3 Hydraulic Design of Siphons

The objective of an inverted siphon is to carry the discharge of one stream beneath the bed of the other. If there is a lot of discharge and the canal has to cross another big waterway, for example the Suez Canal, tunnels can be used to convey the stream water from one side of the navigable waterway to the other.

An inverted siphon is similar to a pipe working under pressure. Because the cross-sectional area of the conduit is considerably less than that of the stream discharging through it, the velocity in the conduit is greater than that in the upstream and downstream channels. The result of this contraction of the section is a loss of head, which causes an increase in the water level upstream of the siphon as compared to the downstream level. The difference in the water levels is termed the heading up. Additional head-loss terms will also appear in the heading-up equation for the siphon. In the relevant formulas, the total heading-up H is represented as the sum of the head losses at different points of the siphon. These losses are represented graphically in Figure 11.5a. The inlet and friction losses for a culvert, discussed in Section 11.2.2, have the same configurations as those of the siphon, that is, the relevant formulas also apply to the siphon inlet and friction head losses, which are represented as follows:

$$H = h_e + h_f + h_{el} + h_o \tag{11.5}$$

where h_e, h_f, and h_o have the same meanings as those of the culvert;
h_{el} = head loss due to an elbow

$$h_{el} = C_{el}\left(\frac{V^2}{2g}\right) \tag{11.6}$$

The values of C_{el} for various angles α, are given below (Figure 11.5b):

α	20	40	60	80	90
C_{el}	0.03	0.14	0.37	0.75	1.0

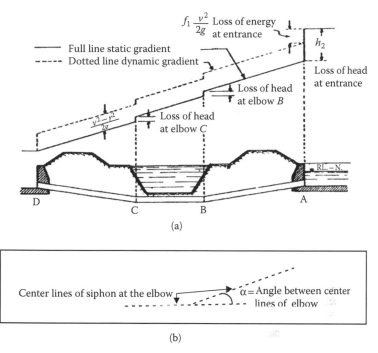

FIGURE 11.5 (a) Sectional elevation through a siphon. (b) Centerlines at the siphon elbow.

Another head loss that should be taken into consideration is that of a screen at its inlet when large amounts of debris are expected to enter the siphon. The head loss h_s of the screen can be derived on average as follows:

$$h_s = 0.15\left(\frac{V_c^2}{2g}\right)$$

(11.7)

where V_c is the velocity of approach at the siphon inlet.

The coefficient 0.15 is taken for general cases where the screen is inclined at a 60° angle with the horizontal direction. However, for the exact coefficient, the sizes of the bars and the spacing between them should be known. Note that the total head loss of the whole siphon should not be greater than 15 centimeters, as recommended by the Egyptian code. For more details, refer to any hydraulic structure code.

11.2.4 HYDRAULIC DESIGN FOR AQUEDUCTS

Flumes or aqueducts are constructed to carry stream flows across another waterway section. The sizes of the flumes depend on the flow discharges and the feasibility studies of the conveyance structures. In the past, timber and masonry were used to form aqueduct sections. Steel and reinforced concrete are now used for aqueduct construction.

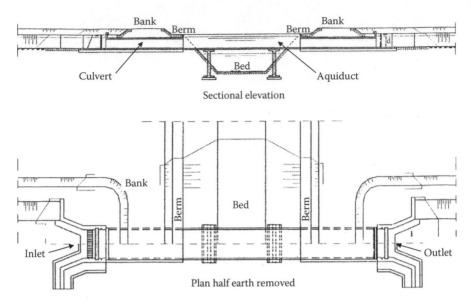

FIGURE 11.6   Cross section through an aqueduct.

Aqueducts may be divided into the following two types:

1. Pipe aqueducts
2. Rectangular section aqueducts

The first type usually yields the cheapest solution for small aqueducts, whereas the second type is more suitable for large discharges. Aqueducts sometimes serve as bridges to convey both vehicles and passersby, as shown in Figure 11.6.

The cross section of a pipe aqueduct is mostly circular. If the section is running at full capacity, it is treated as a culvert. If the aqueduct is not running full, Chezy or Manning's formula may be used to calculate the heading up of the structure. For the Chezy application

$$V = C\sqrt{mS_f}, \quad \text{and} \quad S_f = \frac{V^2}{C^2} m$$

where S_f is the frictional slope, m is the hydraulic radius, and C is the Chezy coefficient. The velocity may be assumed to be within 1–2 meters per second. The value of S_f should be selected to be always lower than the critical slope.

11.3   CONTROL AND STORAGE WORKS

11.3.1   Introduction

Control works are used to store or divert flood flow in catchments, such as weirs, checks, and barrages. Low dams are used mostly used in arid and semi-arid region catchments. Weirs can be used for measuring discharge in streams, as described in

Chapter 6. They can also be used to control the channel slopes of very steep streams to protect the channel section against erosion.

Storage works such as dams are used to store floodwater for flood control, irrigation, power development, and for urban and industrial water supply. Dam reservoirs can be used for both recreation and groundwater recharge. Dams constructed to serve several purposes are called multipurpose dams. These types of dams are large and are constructed on big rivers; the details of this category will not be analyzed in this section. For more details about the design and types of these dams, see the list of references at the end of this chapter.

Another storage work used in the catchments of arid and semi-arid regions is the cistern system, which are relatively old systems that are still in use and will be discussed in this section.

The land surface of most of the catchments in arid and semi-arid regions is composed of rocks, and only parts of these surfaces contain alluvial material. Construction of any control or storage work should take the type of soil into consideration. Any heading-up structure, such as a weir or dam, built on a rock foundation does not need a large floor (also called an apron) but does need a sufficient base width to distribute the load of the structure onto the soil within the limits of the safe allowable stresses. However, if these structures are built on alluvial or permeable soil, they must have a sufficient length of floor to ensure the safety of the structures against percolation, uplift, and scouring, as will be discussed in Section 11.3.2. Weir types and their hydraulic designs, with respect to their length and crest levels, were presented in Chapter 6. The various types of small or check dams and their hydraulic designs are mentioned here in brief to assist in the management of water resources in arid and semi-arid catchments.

11.3.2 WEIRS

Weirs constructed on permeable or porous soil must have a sufficient floor length (apron) to ensure the safety of the structure against the following:

1. Seepage or percolation that may cause undermining or piping of the pervious foundation and is followed by the collapse of the whole structure.
2. Uplift that builds up below the floor; hence, the thickness of the floor along its length should be sufficient to counteract the uplift force.
3. Scouring downstream of the weir floor that results from the weir operation.

11.3.2.1 Percolation or Seepage

Percolation is the movement of water through the voids of pervious material present below the weir floor, as shown in Figures 11.7a and 11.7b. The head lost due to the friction from this movement is proportional to the distance traveled by the percolated water and the properties of the soil. The percolation length is called the creep length.

According to Darcy's law, the velocity of percolating water increases with an increase of the hydraulic gradient (HG) or decrease of the length of percolation. A safe velocity at the floor end ensures that the water can exit without carrying

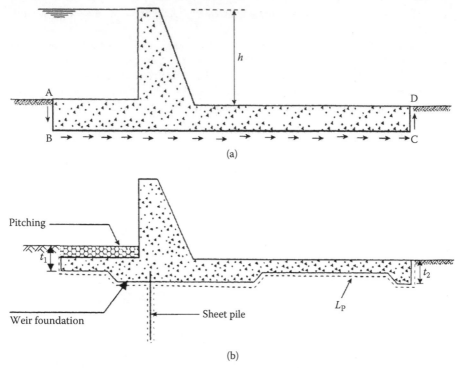

FIGURE 11.7   (a) Weir without sheet pile. (b) Weir with sheet pile.

away the soil particles. If the velocity increases beyond a certain critical value, piping will occur. The HG at this stage before piping occurs is called the critical HG, i_c. The movement of soil particles beyond this stage of critical HG will cause undermining. The following table shows the critical HG, or i_c, for different soils:

Type of Soil	i_c
Silt or clay	0.12–0.14
Fine sand	0.14–0.17
Coarse sand	0.17–0.2
Fine gravel	0.2–0.25

Two empirical formulas used in the past to calculate the safe percolation length are as follows:

1. Bligh's formula (1910): This formula states that the length of percolation L (the creep length in contact with the floor of the weir) that will prevent any undermining is the sum of the horizontal and vertical lengths of the creep, as shown below:

$$L_B = L_H + L_V$$

where L_B is the total length of the creep, L_H is the horizontal length of the floor, and L_v is the vertical length of creep, equal to twice the floor thickness. According to Bligh, L_B should have the following relation to prevent any undermining:

$$L_B = C_B H \tag{11.8}$$

where H is the seepage head and C_B is Bligh's plain creep coefficient, which depends on the nature of the soil.

The maximum value of seepage head H usually occurs when the upstream water is retained up to the highest possible level with no discharge to the downstream side (the downstream water level is taken to be the downstream bed level). The values of C_B for different types of soil are given in Table 11.1. Sheet piles can be used to reduce the floor length (Figure 11.7b). Bligh recommended that the distance between the piles should be equal to or greater than twice the effective length of the deepest sheet pile; if not, the downstream sheet pile will lose its effectiveness.

2. Lane formula (1932): Lane assumed that the vertical distance L_v is more effective than the horizontal distance L_h and suggested the following formula:

$$L_1 = \frac{L_h}{3} + L_v$$
$$= C_1 H$$

where C_1 is Lane's weighted creep coefficient, which depends on the soil type, as shown in Table 11.2.

11.3.2.2 Uplift

Uplift is the upward water pressure on the base of a structure. At any point X, the HG line measures the extent to which there is a residual head h, which causes uplift pressures, as given in Figure 11.8a. The condition is critical when the upstream water

---

TABLE 11.1

Bligh's Coefficients for Different Types of Soil

Types of Soil	C_B
Very fine sand or silt	18
Fine sand	15
Medium sand	13
Coarse sand	12
Gravel and sand mixture	9
Fine gravel	6
Medium gravel	5
Coarse gravel	4
Soft clay	9
Stiff clay	5
Puddle clay	0

TABLE 11.2

Lane's Coefficients for Different Types of Soil

Types of Soil	C_L
Very fine sand or silt	8.5
Fine sand	7
Medium sand	6
Coarse sand	5
Fine gravel	4
Medium gravel	3.5
Coarse gravel	3.0
Soft clay	3.0
Stiff clay	1.8
Puddle clay	0

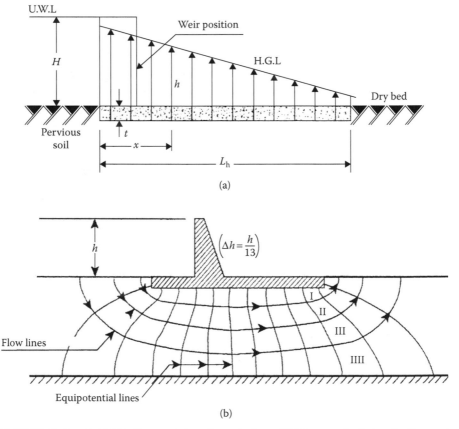

(a)

(b)

FIGURE 11.8 (a) Uplift diagram below the weir floor. (b) Diagram depicting the flow-net process below the weir floor.

level approaches the weir crest and the downstream is dry. At point X, the head h can be calculated from Bligh's formula as follows:

$$\text{Travel distance at } X = t + X$$
$$\text{Head lost at } X = C_B \times (t + X),$$
$$h = H - [C_B \times (t + X)]$$

At point X, the upward pressure should be resisted by the weight of the floor. Therefore, a unit area of the floor should have weight $= \gamma_c t \geq \gamma_w h$, where γ_c is the specific gravity of the floor material and γ_w is the specific gravity of water. For $\gamma_w = 1$, therefore, $t = h/\gamma_c$.

If the head h_1 is measured above the floor, then

$$h_1 + t = h$$

and considering the previous equation

$$t = \frac{h_1}{(\gamma_c - 1)}$$

To obtain values that are more accurate for the uplift and the exit velocity at the end of the weir floor, the flow-net theory (Figure 11.8b) can be used and can be obtained by an approximate method or by many other methods, such as numerical modeling or electric-analogy models. For more details about modeling, refer to LaMoreaux et al. (2008) and the references.

11.3.2.3   Precautions against Scouring of Downstream Weir Structures

Due to the water falling from the weir crest, the nature of the flow will change to supercritical flow above the weir floor. A hydraulic jump is created above the floor downstream of the weir. This will cause high velocities above the floor, in addition to bed eddies. If the eddies and high-velocity flows exit the floor, they will attack the bed material, causing scour holes that may endanger the weir structure itself.

Several methods can be adopted to safeguard the weir structure and the stream-bed from this effect. These methods include constructing a stilling basin, sills at the floor end, floor blocks, and many other methods, See the references for details. It is also essential to protect the bed and side slope of the stream downstream of the weir floor with concrete blocks or riprap, depending on the flow characteristics. However, the Egyptian code recommends that the length of the armoring apron upstream of the structure be twice the flow depth and the apron length downstream of the structure be three times the flow depth. This is in addition to the inclusion of the hydraulic jump length above the floor. Figure 11.9 shows different views of a weir, including the stilling basin with the up- and downstream aprons.

11.3.3   Storage Works

The first storage system discussed here is the check-dam system, and the second is the rainfall-harvesting cistern system. The first system will be stated in brief because

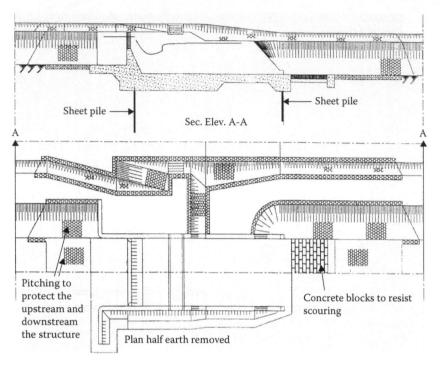

FIGURE 11.9 Drawing of a weir showing the protective works.

a lot of reading material and specifications are available in many references about dam engineering. The second system will be discussed in detail in this chapter due to the scarcity of publications about it.

11.3.3.1 Rainfall-Harvesting Storage System

Water-harvesting systems differ in their hydrologic approaches to the issue of capturing and storing rainwater. Many attempts have been made to store rainwater in storage or collection areas, which are collectively called the cistern system or Roman wells. Other attempts included spreading rainwater to recharge the groundwater in arid and semi-arid regions where there is seasonal rainfall. People were thus able to store rainwater during the rainy period. To improve the efficiency of collection of runoff, constructing water-harvesting structures such as cisterns and check dams is necessary. None of these systems captures all the precipitation, but more water can certainly be collected with an efficient system. An adequate technical design of rainwater-harvesting schemes must achieve a reduction of water losses through the collection and concentration of runoff on arable land, storage of water in the soil for use when rainfall has ceased, and reduction of the accumulation of sediments upstream of the dams. Two criteria have been established recently for the hydraulic design of rainwater-harvesting structures. The first is a protection criterion, which is based on the maximum daily rainfall, and the second is a development criterion, which is based on the average seasonal rainfall. The most limiting factor for the establishment and productivity of vegetation at watersheds is the amount of water available. At present, significant amounts of precipitation are lost as runoff from crust plateau and range areas, although some of this runoff is captured and used in cropped soils in the watershed.

Small constrictions with little intervention in the flowing pattern of the drainage system are generally the safest. These usually consist of small rock fill dams or checks that just slow down the floods. The constrained conditions of wadis with small water-harvesting reservoirs prove to be highly flexible when selecting criteria for reservoir siting. The relevant construction methodology was developed according to indigenous knowledge and local expertise.

The watershed is first studied in detail to determine its subbasin curve numbers, outlets, design storm, and water balance conditions. This is accomplished with the assistance of a hydrologic model program. Based on these data, scenario decisions are made regarding the construction of cisterns, check dams (temporary or permanent), infiltration basins, and wells. The community in the watershed should be involved in planning and selecting the type of structures, initial and maintenance costs, and locations of the structures.

A major challenge that must be taken into consideration while designing, is the critical issue of land tenure on a watershed scale when introducing water-harvesting systems. Rainfall runoff generated high up in the plateau by catchment areas that are trapped by water-harvesting structures can be used for localized agricultural production, but runoff should not be blocked from running to other downstream localities.

11.3.3.2 Check Dams

Check dams are small structures, either temporary or permanent, that are built across the direction of flow of water on shallow drainage rivulets and streams with the intention of harvesting water; these dams trap sediment, slow runoff, and reduce the longitudinal slope. Moreover, they retain the excess water flowing during rainy storms in a small catchment area to saturate the soil upstream of the structure.

Check dams are gaining popularity as a technique for rainwater harvesting. This technique reduces water losses through the collection and concentration of runoff on arable land via a diversion of an ephemeral stream into adjacent flat lands or an interception of floodwater flowing in valleys into nearby basins for crop production. The impounded water has a chance to infiltrate into the ground and to control gully erosion.

In several catchments in arid and semi-arid areas, there are main watercourses that move from higher to lower lands and large branches originating on both sides of the wadis. On the northwestern coast of Egypt, for example, the average length of the main courses of the wadis is about 19 kilometers with a 0.6% slope. Each wadi is divided into two or more sub-branches at the plateau. There are other minor branches, but with little effect on and contributions to conservation and management of land and water. There is an acute need to construct a number of flood-retention check dams along the main courses, gullies, and tributaries. Check dams are constructed from locally available materials such as earth or boulders to reduce the initial implementation and maintenance costs. Due to the plentiful existence of rocks in the area, larger loose boulders will be used to form the framework and smaller stones to pack the middle of the dams as a core. Because of the gentle slopes of the plateau in the locality and the lower speed and force of the flows, check dams can be constructed dry without mortar. Performance of dry dams increases with time. As transported sediment fills the holes and gaps, water loss declines, storage capacity rises, and water guidance improves.

The crests of check dams are about 1.5–3 meters higher than the ground level, which will in most cases be sufficient to saturate the soil profile within a single run-off event. Check dams should be extended laterally using two wings that cross most of the valley. The center of the check dam must be at least 0.10 meters lower than the outer edges. Excess water then filters through the dams or overtops during peak flows to successive terraces below. They may be permanent if designed properly and can be used in places where it is not possible to divert flow and stabilize the channel through other means.

Based on the rainfall and storm history, in addition to data from geographic information systems or any hydrologic program, the suggested check dams and their drainage basins can be obtained, as given in Chapter 12.1. By creating a reservoir at the outlet storage, routing for the outlet can be accomplished, followed by routing from the outlet of the reservoir to the next downstream outlet point. This is the point where a subbasin ends and a routing reach begins. The location of check dams is governed by land ownership, fragmentation, and the main topographic features.

As an example, the earthen check dams suggested for the Wadi Nagamish (Chapter 12.1) were installed based on site investigation, water-bearing formations, the output hydrograph at the outlet, and the storage-capacity curves. The design of these structures and the associated spillways were performed according to the volume of water that can be stored in the stream channel upstream, the locations and numbers of the structures, and the surplus flood discharge that must be evacuated safely.

The stability of the structure against various forces and the likely groundwater recharge are the main functions of dam design. Earthen dams must be designed and constructed adequately to establish and maintain satisfactory stability against sliding, overturning, foundation conditions, overtopping, and seepage through the recommended cross section, under the worst prospective conditions during rainy years. Rainwater-harvesting structures should be constructed using locally available material. The suggested dams should be constructed with clean compacted sand layers without any boulders and organic materials. The earthen dam must be protected with an envelope of cemented blocks of stone along the upstream and downstream faces at least 0.5-meters thick to help resist flash flood flow, wind-blown sand, and to conserve the stability of the earth slopes. A set of stone piles along the downstream toe should be established to help resist the evacuated flood by dissipating the falling energy of the floodwater. The earthen dam should have a side slope ratio of 3:1 on the upstream side and 2:1 on the downstream face, as given in Figure 11.10.

11.3.3.3 Cistern Systems

This type of rainwater-harvesting system is very old and is sometimes called the Roman well system. Many arid and semi-arid regions all over the world still use this method efficiently. The cistern is a big room dug into a rocky area of the catchment.

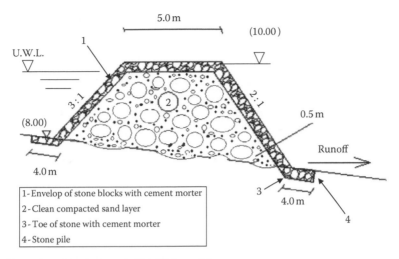

FIGURE 11.10 Check dams in Wadi Nagamish.

The capacity of the room can reach several hundred cubic meters. The size, shape, and upper slab of the room depend on the rock formation below the land surface of the catchment. The upper slab can be constructed from the rock formation if it is rigid; in other cases, a reinforced concrete slab is built to act as the upper-room boundary, where the upper slab surface will be the same as the catchment land level. An opening is made in the upper slab of the room that has a flexible cover on top and one side vent pointing toward the upstream part of the catchment to act as an inlet for the rainwater. The vent is fitted with a screen to prevent any rocks or small animals from entering the cistern. An area upstream of the vent is selected to act as a microcatchment. It can be selected with the help of a land survey. The amount of runoff that can be stored in the cistern can be calculated by a simple runoff equation as given in Chapter 5, or by a hydrologic model if the cistern will be constructed to store a big amount of water.

The cistern system in Wadi Nagamish (Figure 12.1.1) can be taken as an example (see Chapter 12.1). A major rainfall event lasting for a period of about 3 hours can be used as a design storm. The design storm can be determined according to information obtained from four meteorological stations distributed along the wadi area (Abd-Alla 1997; FAO 1970). The design storm is about 38 millimeters at the shoreline. The maximum expected depth of rainfall in the Nagamish subbasin can be calculated based on the estimated decrease in rainfall with distance toward the coast. Based on the topographic map and stream order, the area in the study was divided into 22 subbasins.

The Hydrologic Engineering Center (HEC-1) model, one of the eight modules of the watershed modeling system, has been used to simulate the surface-runoff response to the design storm by representing the basin as an interconnected system of hydrologic and hydraulic components. The result of the modeling process is the computation of stream-flow hydrographs at the desired locations, namely, the subbasin outlets. The subbasin land's surface-runoff component is used to represent the runoff movement over the land surface. The inputs of this component are the precipitation hyetograph, curve number, mode of losses, and type of basin time of concentration.

In the HEC-1 model, the rainfall and infiltration are assumed to be uniform over the entire subbasin. When editing the HEC-1 parameters, the river foothill method is used for computation of the subbasin lag time to obtain a comparison of the observed and calculated lag times. The new sites suggested for the cistern are preferred in the upper parts of the watershed, the southern plateau area, which is considered an abandoned area, and the major runoff amount is lost via evaporation. It has a high runoff coefficient due to soil crust formation from sandy loam to loam only and the presence of many rock outcrops, while the existing rocks are generally impervious.

The Nagamish plateau's runoff coefficient can be strengthened by reducing the pond surface storage and evaporation losses through mechanical treatments such as rock clearing and filling the depressions with compacted soil. This can also be achieved by lining the surface area with smooth plain concrete or treatment with any healthy, economical sealing material with an appropriate slope toward the covered cisterns. The catchment area of each cistern should be large enough to provide it with the needed amount of runoff. Some small earth dikes are sometimes necessary to lead the runoff flow into the cisterns.

The potential of subbasin rainwater harvesting for an average annual rainfall of 150 millimeters has been used to determine cistern volumes, taking into consideration that the rainfall intensity decreases rapidly from north to south in conformity with local records of scattered rainfall. Assuming that only 60% of this potential could be stored, the quantities of water available for storage at each subbasin outlet are listed in Table 11.3.

Prospective useful cisterns must be constructed in a circular shape to reduce the surface area of the inside wall, benefiting from the soil arching effect; furthermore, this shape is much easier to construct. Geological studies of the northwestern coastal zone of Egypt indicate that the top layer just below the ground surface is a horizontal layer of hard rock, followed by a thick layer of soft rock, which is characterized by

TABLE 11.3
Volumes of Subbasin Cisterns

N	Basin Name	A (km²)	Cistern Volume (m³)
1	12B	3.59	27,900
2	13B	3.42	22,656
3	14B	5.57	44,038
4	15B	8.27	74,320
5	16B	1.87	16,770
6	17B	5.43	24,766
7	18B	21.73	61,414
8	19B	0.95	3438
9	20B	8.27	23,362
10	21B	2.78	7866
11	22B	7.12	20,116

the presence of significant rock fragments mixed with sand. The availability of a horizontal sequence of soft and hard sandstone layers facilitated the establishment of underground cisterns. The hard rock layer can be used as the ceiling of the cistern, and the soft rock layer can be excavated and used as the water storage tank. In addition, silt traps must be constructed at the cistern entrances.

Cisterns should be partially watertight, as shown in Figure 11.11. Their surfaces must be covered with polyethylene, concrete plastering, or another impervious material for about two-thirds of the total height of the wall to avoid seepage loss in drought years. The upper third of the wall works as a percolation tank to recharge the soft rock aquifer that allows water to infiltrate through the infiltration zone in the rainy years.

One major challenge is the critical issue of land tenure on a watershed scale when introducing cisterns as rainwater-harvesting structures on a plateau. The nearest suggested plateau cistern is located about 17 kilometers from the Wadi Nagamish outlet, whereas the farthest one is almost 23 kilometers away.

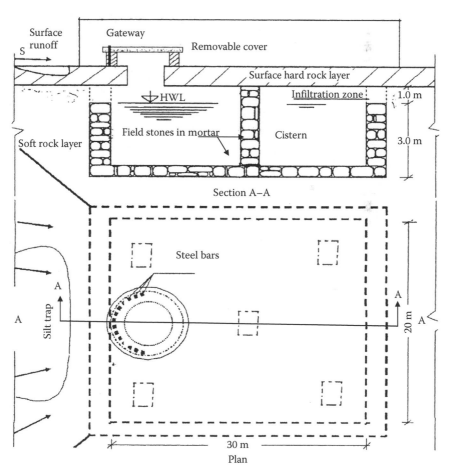

FIGURE 11.11 Drawing showing a cistern system in Wadi Nagamish.

REFERENCES

Abd-Alla, G. 1997. Assessment of rainfall runoff dynamics at East Matruh, Egypt. Ph.D. thesis, Alexandria University, Egypt.

Bligh's line creep theory. 1910. Dams, Barrages, and Weirs on Porous Foundations. Eng. News, December 29, 1910.

Chin, D. A. 2006. *Water Resources Engineering*. Englewood Cliffs, NJ: Prentice Hall.

Chow, V. T., D. R. Maidment, and L. W. Mayset. 1988. *Applied Hydrology*. New York: McGraw Hill.

Egyptian Ministry of Water Resources and Irrigation. 2003. Egyptian Code for Water Resources and Irrigation Works.

FAO. 1970. Physical conditions and water resources survey—North Coastal Region of Egypt. Survey by the Food and Agriculture Organization of the United Nations.

Lamoreaux, P. L., M. M. Soliman, B. A. Memon, J. W. LaMoreaux, and F. A. Assaad. 2008. *Environmental Hydrogeology*. Boca Raton, FL: CRC Press.

Lane, E. W. 1935. Security from under-seepage masonry dams on earth foundations. *Trance Am. Soc. Civil Engrs* 1:1235.

Linsley, R. K., M. A. Kohler, and J. L. H. Paulhus. 1975. *Applied Hydrology*. New York: McGraw-Hill.

Mays, L. W. 2005. *Water Resources Engineering*, 2nd ed. New York: John Wiley & Sons.

Soliman, M. M. 1985. *Irrigation Engineering Design*. Cairo: Ain Shams University Publishers.

BIBLIOGRAPHY

Alpsti, I. 1967. Investigation of water losses at May reservoir. In *Proceedings of the 9th International Congress on Large Dams*. Paris: International Commission on Large Dams.

Attewell, P. B., and I. W. Farmer. 1976. *Principles of Engineering Geology*. London: Chapman and Hall.

Bass, K. T., and C. W. Isherwood. 1978. The foundations of Wimbleball dam. *J Institution Water Eng Sci* 32:187–97.

Bell, F. G. 1993. *Engineering Geology*. Oxford, UK: Blackwell Science.

Binnie, G. M. 1981. *Early Victorian Water Engineers*. London: Thomas Telford.

Binnie, G. M. 1987a. *Early Dam Builders in Britain*. London: Thomas Telford.

Binnie, G. M. 1987b. Masonry and concrete gravity dams. *Ind Archaeol Rev* 10(1):41–58.

Bridle, R. C., P. R. Vaughan, and H. N. Jones. 1985. Empingham dam: Design, construction and performance. *Proc Institution Civil Eng* 78:247–89.

Clayton, C. R. I., N. E. Simons, and M. C. Matthews. 1995. *Site Investigation: A Handbook for Engineers*, 2nd ed. Oxford, UK: Blackwell Science.

Clifton, S. 2000. Environmental assessment of reservoirs as a means of reducing the disbenefit/benefit ratio. In *Proceedings of Conference Dams 2000*, pp. 190–8. London: Thomas Telford.

Coats, D. J., and G. Rocke. 1983. The Kielder headworks. *Proc Institution Civil Eng* 72:149–76.

Collins, P. C. M., and J. D. Humphreys. 1974. Winscar reservoir. *J Inst Water Eng Sci* 28:17–46.

Crager, W. P., J. Justin, and J. Hinds. 1945. *Engineering for Dams*, vols. I, II, and III. New York: John Wiley & Sons.

Fell, R., P. MacGregor, D. Stapledon 1992. *Geotechnical Engineering of Embankment Dams*. Rotterdam: Balkema.

Gosschalk, E. M., and K. V. Rao. 2000. Environmental implications: Benefits and disbenefits of new reservoir projects. In *Proceedings of Conference Dams 2000*, pp. 199–211. London: Thomas Telford.

ICOLD. 1970. Recent developments in the design and construction of dams and reservoirs on deep alluvial, karstic or other unfavourable foundations. In *Proceedings of the International Commission on Large Dams*, Paris.

ICOLD. 1988a. World Register of Dams: 1st Update. Paris: International Commission on Large Dams.

ICOLD. 1988b. Reservoirs and the environment: Experiences in managing and monitoring. In *Transactions of the 16th International Congress on Large Dams, San Francisco, International Commission on Large Dams*, Paris: International Commission on Large Dams.

Jansen, R. B., ed. 1990. *Advanced Dam Engineering for Design, Construction, and Rehabilitation*. London: Chapman and Hall.

Jansen, R. B., ed. 1992. Dams and Environment: Socio-Economic Impacts, Bulletin 86. Paris: International Commission on Large Dams.

Jansen, R. B., ed. 1994. Dams and Environment: Water Quality and Climate, Bulletin 96 I. Paris: International Commission on Large Dams.

Jansen, R. B., ed. 1998. World Register of Dams. Paris: International Commission on Large Dams.

Kennard, M. F., and J. L. Knill. 1969. Reservoirs on limestone, with particular reference to the Cow Green scheme. *J Institution Water Eng* 23:87–136.

Kennard, M. F., and R. A. Reader. 1975. Cow Green dam and reservoir. *Proc Institution Civil Eng* 58:147–75.

Kennard, M. F., C. L. Owens, and R. A. Reader. 1996. Engineering Guide to the Safety of Concrete and Masonry Dam Structures in the UK, Report 148. London: CIRIA.

Mermel, T. W. 1996. *The World's Major Dams and Hydro Plants, in Water Power and Dam Construction Handbook*. London: Reed Business.

Novak, P., A. Moffat, C. Nallurie, and R. Narazanan. 2001. *Hydraulic Structures*, 3rd ed. London: E&FN Press.

Ponce, V. M. 1999. *Engineering Hydrology*, 2nd ed. New York: Prentice Hall.

Roberson, J. A., J. J. Cassidy, and M. H. Chaudhry. 1972. *Hydraulic Engineering*, 2nd ed. New York: John Wiley & Sons.

Schnitter, N. J. 1994. *A History of Dams: The Useful Pyramids*. Rotterdam: Balkema.

Singh, B., and R. S. Varshney. 1995. *Engineering for Embankment Dams*. New Delhi: Ashgate Publishing Company.

Smith, N. A. 1971. *A History of Dams*. London: Peter Davies.

Thomas, H. H. 1976. *The Engineering of Large Dams*, vol. 2. Chichester, UK: Wiley.

Thomas, C., H. Kemm, and M. McMullan. 2000. Environmental evaluation of reservoir sites. In *Proceedings of Conference Dams 2000*, pp. 240–50. London: Thomas Telford.

United States Bureau of Reclamation. 1977. *Design of Small Dams*, 2nd ed. Washington, DC: U.S. Government Printing Office.

United States Bureau of Reclamation. 1987. *Design of Small Dams*, 3rd ed. Washington, DC: US Government Printing Office.

Vischer, D. L., and W. H. Hager. 1998. *Dam Hydraulics*. New York: John Wiley & Sons.

Wakeling, T. R. M., and C. N. D. Manby. 1989. Site investigations, field trials and laboratory testing. In *Proceedings of the Conference on Clay Barriers for Embankment Dams*. London: Thomas Telford.

Walters, R. C. S. 1974. *Dam Geology*, 2nd ed. London: Butterworth.

Weltman, A. J., and J. M. Head. 1983. Site Investigation Manual, CIRIA Special Publication 25 (PSA Technical Guide 35). London: CIRIA.

Wilson, E. M. 1990. *Engineering Hydrology*, 4th ed. New York: Macmillan/McGraw-Hill.

Zeidlev, R. B., ed. 1992. *Design of Earth Dams*. Rotterdam: Balkema.

Zipparro, V. J., and H. Hasen, ed. 1992. *Davis Handbook of Applied Hydraulics*. New York: McGraw-Hill.

12 Case Studies

12.1 CASE STUDY 1—WATER RESOURCES MANAGEMENT IN WADI NAGHAMISH AT THE NORTH COASTAL ZONE OF EGYPT*

12.1.1 INTRODUCTION

Northwestern coastal areas of Egypt are subjected to the highest rainfall intensity in the country (about 180 millimeters per year). Water harvesting has been a traditional practice in these areas. Comprehensive research and studies are required to help retain all rainfall and runoff available over the area. Development of North Western Coastal Zone (NWCZ) of Egypt is a one of the main ways to strengthen this zone's role in the national economy and strategic development. The Wadi Naghamish system is a typical representation of the wadi system along NWCZ, which was delineated using topographic map and the watershed modeling system (WMS). Wadi Naghamish receives a considerable amount of rainfall, which can be used to support some activities to enhance the sustainable development in the area. Development of Wadi Naghamish was the topic of several research studies, such as construction of storage or detention dams and drilling of some shallow wells.

The WMS has been applied to study the hydrologic conditions and flow simulation from Wadi Naghamish, and it can be used as a prognostic tool in the hydrologic forecasting services and as an analytical tool. WMS is an integrated system of software designed for interactive use in a multitasking environment. The focus of WMS is to provide a single application that integrates digital terrain models (DTMs) using a single event with a minimal dependence on rainfall-runoff historical records. Once boundaries have been created, geometric attributes such as area, slope, and runoff distances can be computed automatically.

The main objective of this research is to investigate the potential for development of water resources as well as to assess the flash flood hazards of Wadi Naghamish, which has a sustainable yield of water resources. The results obtained will then be used to determine possible harvesting techniques, potential storage, and storage locations.

---

* This case study is based on a published paper by Moghazi, H. M., H. M. Awad, and A. Fayad. 2007. Land surface process analysis in a Northwestern coastal zone: Egypt using watershed modeling systems. In *Proceedings of the Fifth International Symposium on Environmental Hydrology*, p. 255, M. M. Soliman, ed. Cairo: ASCE.

12.1.2 Model Description

Over the last 20 years, digital representations of topographic information have become increasingly available in the form of DTMs. Extracting watershed data from DTMs using computers is faster and provides more reproducible measurements than traditional manual techniques using topographic maps (Garbrecht et al. 1999).

WMS is an integrated system of software designed for interactive use in a multitasking environment. It was developed by the Environmental Modeling Research Laboratory of Brigham Young University in cooperation with the U.S. Army Corps of Engineers Waterways Experiment Station. The WMS interface is separated into several modules containing tools that allow manipulation and model creation from different data types. It includes map, geographic information system (GIS), terrain data, 2D grid, drainage, hydrologic modeling, river modeling, and scatter point modules.

Analysis of runoff per subbasin outlet is completed by representing the basin as an interconnected system of components. The output is a quantification of runoff at each outlet. The system is comprised of a GUI, separate hydraulic analysis components, data storage and management capabilities, graphics, and reporting facilities.

WMS software is capable of simulating the runoff in the watershed and determining the alternative and beneficial sites for water-harvesting structures such as reservoirs, check dams, infiltration basins, and underground cisterns. It allows the user to see and understand the domain and parameters analysis using contouring and shading.

GIS software is used for visualization of data and information on land assessment and soil characteristics. The result of watershed analysis can be represented in GIS form as three digital map layers: points, lines, and polygons. A point layer represents the watershed outlet and confluence points, a line layer represents a stream network, and a polygon layer represents watershed boundaries.

The modeling approach consists of using HEC-1-generated synthetic runoff data as a substitute for the observed runoff data needed for input into both models. The models then generate direct runoff net among other output. HEC-1 is a flood hydrograph package that enables flood forecasters to develop unit hydrographs using various methods and to calculate loss rates and route hydrographs. It is designed to simulate surface runoff from single precipitation events that have been refined over subsequent years.

Most of the flood-routing methods, the process of determining progressively the timing and shape of a flood wave at successive points along watershed subbasins, that are available in HEC-1 are based on the continuity equation and some relationship between flow and storage or stage. Watershed routing is used when the watershed basin has multiple subbasins, and hydrographs from each subbasin can be added together to determine the combined hydrograph at critical points. Common critical points are at control conveyance structures where an inflow hydrograph is required to route the discharges through the structures and to determine the location and storage capacity of the reservoirs.

12.1.3 HYDROLOGIC MODELING MODULE HEC-1

HEC-1 is a lumped parameter model and is designed to simulate the surface runoff response of a watershed to precipitation by representing the subbasins as an interconnected system of hydrologic and hydraulic components. Each component models an aspect of the precipitation-runoff process within a portion of the watershed. A component may represent a surface runoff entity, a stream channel, or a reservoir. Representing a component requires a set of parameters that specify the particular characteristics of the component and mathematical relations that describe the physical processes. The modeling process produces stream-flow hydrographs at desired locations in the watershed.

HEC-1 is capable of producing runoff hydrographs and detention pond analyses for complex watershed networks at desired locations in the watershed using the unit hydrograph method and incorporating the reservoir and channel routing procedures. It includes basin data, output control, precipitation data, loss method, unit hydrograph development, and flood routing in streams and reservoirs, as well as development of design storms and flood flows under modified conditions. A general job control HEC-1 module is input for each subbasin for specific watershed and event characteristics. Due to the absence of a detailed hydrology database for NWCZ, the HEC-1 module is considered more appropriate for simulation.

12.1.4 METHODOLOGY

Various analog maps in different scales obtained from different organizations such as the Egyptian General Survey Authority and the Military Survey Authority were converted into vector format by digitizing them in Auto Cad Map software. All the digitized data were exported into the Arc GIS software to build the spatial data.

A runoff curve number (CN) has been assigned for NWCZ wadis by the groundwater sector of the Ministry of Water Resources and Irrigation (GWS 2003). This was done using GIS techniques based on surface geological maps, soil types, and land coverage. The weighted average procedure was used to find spatial average CNs for each watershed.

All of the above data were incorporated into WMS software for the preparation of various thematic maps. Further processing and analysis were done to identify the suitable sites for water-harvesting structures and recharge installation. A methodology for locating small water-harvesting structures was developed in a Wadi Naghamish characterized by low and erratic precipitation in order to improve the domestic agricultural potential. Water-harvesting structures even out the variability of the weather by reducing high flows, increasing low flows, and generally making water available when it is needed.

In this study, a water harvesting reservoir (WHR) locating methodology was used to increase the availability of water for domestic use and supplemental irrigation during the summer months. The study involved the development and application of a system for locating and ranking suitable sites for WHRs based on the overall suitability of each reservoir, a brief consideration of construction cost, and quantification of the overall suitability for such small reservoirs through a reservoir suitability index (RSI) calculated for potential candidate sites. It is very important to help the decision makers to evaluate the variables for water-harvesting systems, including site selection, storage volume, and sustainable development.

Because the WMS method uses a composite CN value for the entire intended drainage area and due to the variety of land uses and hydrologic soil groups within the drainage basin, the watershed should be divided into a number of subbasins to achieve more accurate results. Contour information at 10-meter intervals was used to derive the digital elevation model (DEM) in a grid format to study area. Based on this DEM data, the study area was divided into four subcatchments. Figure 12.1 shows Wadi Naghamish subbasins delineated from a grid-based digital elevation model.

WMS automatically calculated the subbasins' delineations, or individual basin data needed to run the various hydrologic models including basin slope, area, length, average overland flow, elevations, weighted CN and unit hydrographs parameters, lag time, concentration time, and storage coefficients, by using the information from the digitizing map and DEM. Based on the DEM, WMS can generate the runoff flow directions using the topographic parameterizations program (TOPAZ).

To use a hydrologic model, the following watershed characteristics are required: drainage area; length, shape, and slope of the watershed; the drainage pattern;

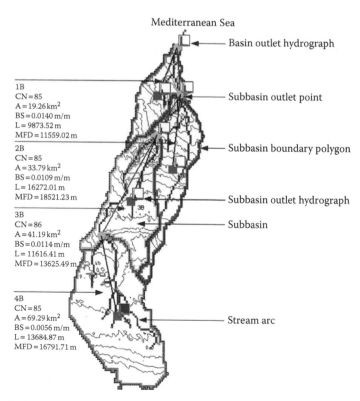

FIGURE 12.1   Subbasins of Wadi Naghamish.

TABLE 12.1
Physiographic Parameters of Studied Subbasins

Naghamish Subbasins	Area (km²)	Basin Length (km)[a]	Basin Slope (m/m)[b]	Maximum Flow Distance (km)[c]	Maximum Flow Slope (m/m)[d]
1B	19.26	9.873	0.0140	11.559	0.0070
2B	33.79	16.272	0.0109	18.521	0.0069
3B	41.19	11.616	0.0114	13.625	0.0067
4B	69.29	13.685	0.0056	16.792	0.0038
Total	163.53	29.340	0.0090	35.530	0.0047

[a] Distance to furthest point along basin perimeter.
[b] Average slope within the subbasin.
[c] Maximum flow path, including overland and stream flow.
[d] Slope along the maximum flow path as defined above.

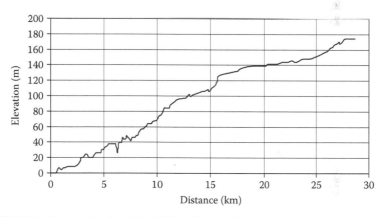

FIGURE 12.2 Longitudinal profile of Wadi Naghamish.

time and flow parameters; land use; and hydrologic soil types. The physiographic parameters of the subbasins are shown in Table 12.1. The study area is generally characterized by low relief and mild topography (see Figure 12.2).

12.1.5 HYDROLOGIC STUDIES

The socioeconomic design of rainwater-harvesting structures requires knowledge of the probability of occurrence of storms of various intensities, durations, and frequencies. Occasionally rebuilding the structure is cheaper than designing it to handle the largest expected rainfall event. Hydrologic analysis is typically done using the available rainfall data and the HEC-1 lumped parameter models through WMS.

The study area was subjected to the surface runoff, which must be stored. The cisterns that have been constructed throughout the study area are efficient for collecting

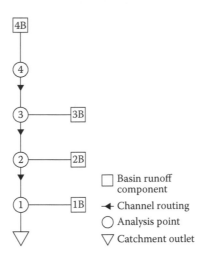

FIGURE 12.3 Hydrologic modeling tree of Naghamish watershed.

water, reaching efficiencies of up to 43%, but have insufficient capacity and leakage problems that increase the risk for people and animals (FAO 1993). The interconnectivity between the catchments and the various components of the watershed-runoff process, and how these components are combined into a basin model, are schematically shown in Figure 12.3.

A basic schematic diagram of these components is developed by first delineating the watershed boundary. This can be done with a topographic map. Segmentation of the watershed into a number of subbasins determines the number and types of stream network components to be used in the model. Each subbasin is represented by a combination of model components. The subbasins and their components are linked together to represent the connectivity of the watershed area. HEC-1 has a number of methods for combining or linking together the outflow from different components. The unit hydrograph technique produces a runoff hydrograph at the most downstream point in the subbasin.

The general arrangement of the HEC-1 model is as follows. Rainfall-runoff calculations are performed by overlaying a "precipitation model" into a "basin model." The precipitation model represents the design storm used in calculations and the basin model calculates rainfall losses and transforms the rainfall excess into an outflow hydrograph for the subbasin. Subbasin hydrographs are then routed through the channel network to the basin outlet using the hydrologic routing methods. A variety of calculation methods are available for each step of the process.

12.1.6 RAINFALL ANALYSIS

The analysis of rainfall data is an important element in hydrologic studies. It is used as an input to determine the amount of infiltration and runoff volumes for each subbasin in the study area. Evaporation during rainfall storms is generally small and can be neglected, as it does not affect the water balance of the system.

Monthly rainfall records and the number of rainy days for the study area are available from Mersa Matrouh meteorological station and Mersa Matrouh airport meteorological station for a period of about 50 years. To estimate the potential for water harvesting, the daily rainfall depth is required for at least 30 years. Because it is impossible to deduce daily rainfall from monthly records, two assumptions must be made. The first is that the conservative side of flood (extreme event) analysis will guide decision making in flood protection. Based on this assumption, 95% of the total monthly rainfall is assumed to occur in a single day. The second assumption relates to sustainable firm yield conditions for development of the area. It is assumed that the total monthly rainfall is distributed evenly over the total number of rainy days of that month. In flood risk analysis, the maximum monthly record from each year is selected and manipulated according to the first assumption. For assessment of water-harvesting potential, both assumptions are tested (GWS 2003).

According to long-term monthly recorded rainfall, the maximum total monthly rainfall, 146.8 millimeters, occurred in 1989–1990. Table 12.2 gives a summary of the results corresponding to the average 100-year daily rainfall depth for the three fitted distributions (GWS 2003). The results of this analysis were used in hydrologic simulation run for watershed behavior prediction. Yearly rainfalls of the three main meteorological stations in the NWCZ were analyzed. The first group of catchments were in the area of the Sidi Barrani meteorological station. The second group of catchments were in the area of Mersa Matrouh airport meteorological station. The third group of catchments were in the area of Alexandria meteorological station. For each catchment in all the groups, four parameters are calculated: namely, maximum specific yield, minimum specific yield, specific yield at 25% of ranked series, and specific yield at 75% of ranked series.

The results for each of the three groups were plotted in box plot format (see Figure 12.4). Figure 12.4 illustrates that the first quartile and third quartile values are based on accumulated monthly rainfall. The first assumption results skewed toward the lower values are more promising, provided one accepts the assumption. The range bounded by maximum and minimum specific yields is much bigger than that of the first and third quartiles.

TABLE 12.2

Average 100-Year Daily Rainfall Depth

Station Name	Probability Distribution		
	Gumble (mm)	Log-Pearson (mm)	Log-Normal (mm)
Sidi Barrani	173	195	178
Salloum Town	189	214	180
Ras El Dabaa	158	165	180
Mersa Matrouh	149	148	165

Source: Adapted from Ground Water Sector (GWS), Ministry of Water Resources and Irrigation (MWRI). 2003. Hydrological Analysis for the Northwestern Coastal Zone, Egypt.

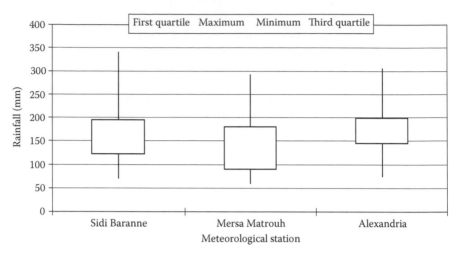

FIGURE 12.4   Box plot for the main three meteorological stations rainfall (in millimeters).

12.1.7   OVERVIEW OF RAINFALL-RUNOFF REGIME

The average precipitation for a year at Wadi Naghamish is about 141.2 millimeters, but it was as low as 34.1 millimeters in 1950–1951 and as high as 274.6 millimeters in 1989–1990. There are number of ways to determine the amount of water that runs off of a surface. In addition to collections in the field, computer models can also be used to estimate the runoff. One method of converting the rainfall volume into runoff volume is the U.S. Department of Agriculture, Soil Conservation Services (SCS) runoff CN method (SCS 1993). The GWS has been calculated CNs for a number of wadis in NWCZ. The CN is about 86 for Wadi Naghamish. The CN, longest flow path, and slopes are used to calculate the lag time of the watershed. The precipitation model can be defined using girded precipitation (actual or calculated), user-specified rain gauge weighing methods, and hypothetical storms. Precipitation losses can be calculated using initial or constant, SCS CN, girded SCS CN, and Green–Ampt algorithms.

Measuring wadi runoff is a suitable approach to obtain rainfall–runoff relationships and runoff coefficients because the relationship between rainfall and runoff differs quite markedly from catchment to catchment. These relationships provide an important mean for calculating the sizes of catchments for cultivated areas, which is particularly important in dry years.

12.1.8   APPLICATION OF WMS PROGRAM

The research conducted aimed to develop a new methodology for locating and ranking suitable sites and volumes for small water-harvesting structures by using WMS as a hydrologic model. Water-harvesting structures are necessary to conserve potential rainfall, which is mostly lost by evaporation and surface run-off towards the Mediterranean Sea, especially in wet years. To begin with, a two topographic map for Wadi Naghamish based on aerial photographs at a scale of 1:50,000 was used as the base map. In addition, a field inspection was conducted to assess the feasibility of building small water-harvesting structures versus other means of finding new water resources for sustaining inhabitants and agriculture in the area.

The locations of water-collecting structures are determined by a multitude of factors, often less than intuitive and involving a compromise of interests. The method used in this study was based on quantifying the overall suitability of a site for rainwater-harvesting structures through a suitability index calculated for potential candidate sites. This index was developed using hydrologic modeling in conjunction with WMS and hydrologic and geological data. The fulfillment of justification and ranking of structure alternatives must be determined according to construction and maintenance costs, economic feasibility, side effects, and water preservation. It provides a systemic approach in conducting multicriteria analysis and decision making and excludes sites where reservoirs can not be built due to physical constraints and/or restrictive land-use policies and regulations.

Farmers in the lower areas may be subjected to an unequal water supply due to excessive water being diverted by intended water-harvesting structures in the upstream area. To solve this problem, farmers and other key stakeholders in the watershed area suggested several water management options and assessment propositions. Therefore, research aims to find the most acceptable management option to eliminate water allocation conflicts among watershed inhabitants beside rainwater conservation without the degradation of a fragile environment. A critical step in developing and implementing a watershed management plan is to perform a comprehensive assessment of existing environmental conditions before a decision can be made on the best locations for RWH structures and management of the important aspects of the intended watershed. All selection criteria and their weights were based on indigenous knowledge and expertise in the area to be analyzed, other than knowledge of hydrologic models, to determine the best construction site and to fine-tune the threshold values used for site attribute classification. The objective was to ensure the economic feasibility of reservoirs in the context of the low-input agriculture practiced in NWCZ.

According to the field monitoring of the study area, many underground cisterns could become water reserves by enlarging their storage capacity, and new cisterns in the upper neglectful tableland, which has favorable topographical and geological conditions, could be constructed to store late rainfall events in wet years. There is some additional capacity to extend the construction of check dams to enlarge the cultivated area. The water retention should be an indigenous system without large capital investment and should be focused near the upstream areas of the wadis because there are a lot of harvesting structures in the lower portions of the wadis. In addition to serving as a tool to verify the developed methodology, building the reservoirs would be a tangible output of this project that lead to direct benefits for the local inhabitants of Wadi Naghamish. Plastic sheets must line the bottom and walls of the reservoirs, called cisterns, to prevent or diminish water loss by seepage and to conserve storage water quality.

The watershed was divided into four subwatersheds in series according to geographical properties. Outlet points were used to define the locations where hydrographs were combined and then routed downstream. The input to the hydrographic routing component was an upstream hydrograph created from individual contributions of subbasin runoff. The hydrograph was routed to a downstream point based on the characteristics of the channel, and routing data was entered in order to simulate the movement of a flood wave through the watershed reaches. Data entry for the WMS model, including rainfall hyetograph or WMS arid synthetic hyetograph, job control, runoff CN, and any other parameters not defined as attributes in the shapefiles, can be completed using WMS's hydrologic modeling interface for each subbasin. The hydrographs at the outlet of each subbasin, according to the rainfall depth at different return periods, were obtained as shown in Figure 12.5. The watershed routing method was used; subbasin outlet and confluence points were used to define the locations where hydrographs were combined and then routed downstream until reaching the Naghamish watershed outlet. Subbasin routing hydrographs were obtained using Muskingum–Cunge method and the combination of the different hydrographs was then estimated as illustrated in Figure 12.6. Predictions of the runoff amount of the subbasins were obtained to determine the water-harvesting structures. Locations of check dams were suggested upon field investigations according to the most practical and appropriate sites. Criteria included availability of construction materials and process and transportation of the equipment and materials.

Due to the absence of a detailed hydrology database, it is often necessary to have recourse to formulas by which the maximum volume discharge of the study area can be predicted on the basis of its physiographic characteristics and the maximum expected depth of precipitation. Consequently, some methods have been created to

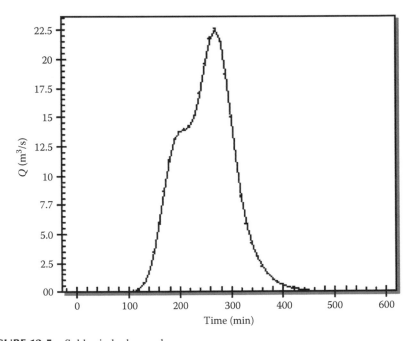

FIGURE 12.5 Subbasin hydrographs.

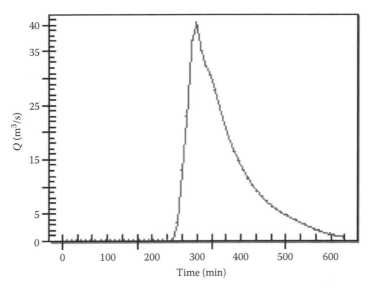

FIGURE 12.6 Subbasin routing hydrographs.

estimate the runoff coefficients and volume for the catchments of ungauged streams. The estimation of runoff volume from Matrouh watersheds has been studied by Ball (1937), Swidan (1978), and Zaki (2000). They derived the following empirical relations:

$Q = 0.75(P - 8)$	Ball 1937 ($p > 8$ mm)
$Q = 0.06(P - 1.12) - 0.013t$	Economides 1968
$Q = (P - 7.29)^2/(P + 139.64)$	Sewidan 1978
$Q = 0.3(P - 10)$	Wakil and Shaker, 1989 ($p > 10$ mm)
$Q = (P - 4.01)^2/(P + 63.22)$	Zaki 2000 (2–5 hours duration)
$Q = (P - 5.40)^2/(P + 39.55)$	Zaki 2000 (6–8 hours duration)

where Q is predicted surface runoff in millimeters, P is storm rainfall in millimeters, and t is the duration of rainfall in hours.

WMS was used to predict the surface runoff for all Naghamish watershed subbasins, 1B, 2B, 3B, and 4B, and the predicted results were compared with measured values using the empirical formula under the same amount of rainfall, P (in millimeters). Table 12.3 illustrates the runoff volume that resulted from applying the WMS model and Zaki's empirical equation to subbasins. Table 12.4 depicts the runoff quantity obtained using Wakil's and Zaki's empirical equations and their average value compared with the value obtained by applying WMS program in Wadi Naghamish subbasins. The table demonstrates that the WMS model matches with the measured model. The WMS program in NWCZ is recommended, as is reevaluating the subbasins runoff CN according to variable conditions such as land cover and antecedent moisture content.

TABLE 12.3

Comparison between Results of the Watershed Modeling System and Empirical Equations

Subbasin Name	Watershed Modeling System				Zaki Empirical Equation
	P (mm)	Q (m³)	A (km²)	Q (mm)	Q (mm)
1B	31.6	148,282	19.26	7.70	8.03
2B	25.0	142,741	33.79	4.22	4.99
3B	23.4	166,954	41.19	4.05	4.34
4B	15.0	105,731	69.29	1.53	1.54

TABLE 12.4

Comparison between Runoff Obtained by the Watershed Modeling System and Empirical Equations

Subbasin	P (mm)	Zaki: Q (mm)	Wakil: Q (mm)	Average Q (mm)	WMS: Q (mm)
1B	31.60	8.03	6.48	7.26	7.70
2B	25.00	4.99	4.50	4.75	4.22
3B	23.40	4.34	4.02	4.18	4.05
4B	15.00	1.54	1.50	1.52	1.53

12.1.9 CONCLUSIONS

In this study, a series of hydrologic and hydraulic simulations and computations for the Naghamish watershed storm that occurred in 1990 were carried out in order to suggest the most acceptable and economical solutions for rainwater-harvesting structures and to examine their effectiveness. The HEC-1 model was used to generate the runoff at each subbasin outlet. The simulated flood hydrograph seemed to relatively agree with the measured hydrograph and the empirical equations in terms of overall discharge rates and shapes of hydrographs.

Based on the potential rainfall analysis, observation during site visits, and the WMS simulation model, the following conclusions and recommendations can be stated. There must be more effort to develop the area of NWCZ and to prevent possible water losses and flood disasters. The surface of upstream tableland must be treated to improve the runoff efficiency. Cisterns and detention dams, especially in the upstream zones, should be used to store unexploited surface runoff. Shallow wells should be drilled near the coastal line to diminish groundwater seepage to the sea. Rubble circular cisterns should be constructed, substituting the traditional construction for cisterns, and low cost and effective underground plastic barrels should be used. A lack of historical data for weather and hydrology forces the use of a large time step and the use of approximation techniques for analysis. Field testing of the initial results is the next logical step.

This can be done by building one or two reservoirs and monitoring precipitation and run-off for 2 or 3 years to calibrate and refine the developed methodology.

REFERENCES

Ball, J. 1937. The water supply of Mersa Matrouh. Cairo: Survey and Mines Department.

Economides, P. 1968. Final report: S.F. project. Unpublished internal report of the Pre-Investment Survey of the Northwestern coastal region of Egypt. Presented to the Matrouh government, Egypt.

Food and Agriculture Organization of the United Nations. 1993. Feasibility and design considerations for establishing small storage dams in Wadi Gnoda and Om-Marzook: Agricultural Development in the North-West Coastal Zone and Siwa Project in Mersa Matrouh. Rome: FAO.

Garbrecht, J., L. W. Martz, K. H. Syed, and Goodrich, D. 1999. Determination of Representative Catchment Properties from Digital Elevation Models. 1999 International Conference on Water Resources Engineering, American Society of Civil Engineers, Session HY-3, Geographic Information Systems (GIS) Applications in Hydrologic Engineering, Seattle, Washington. Published on CD-ROM by ASCE.

Moghazi, H. M., H. M. Awad, and A. Fayad. 2007. Land surface process analysis in a Northwestern coastal zone: Egypt using watershed modeling systems. In *Proceedings of the Fifth International Symposium on Environmental Hydrology*, p. 255, M. M. Soliman, ed. Cairo: ASCE.

Sewidan, A. S. 1978. Water budget analysis for the Northwestern coastal zone of the Arab Republic of Egypt. Ph. D. thesis, Cairo University.

Soil Conservation Service. 1993. SCS *National Engineering Handbook*, Section 4: Hydrology. Washington, DC: U.S. Department of Agriculture.

Wakil, M. and M. R. Shaker. 1989. Rainfall–runoff relationship of the Wadi Mehleb experimental basin. Preliminary study presented at the University of Alexandria, Egypt.

Zaki, M. H. 2000. Assessment of surface water runoff in Mersa Matrouh area, northwestern coastal zone. Master's thesis, Alexandria University, Egypt.

12.2 CASE STUDY 2—URBANIZATION IMPACTS ON THE HYDROLOGICAL SYSTEM OF CATCHMENTS IN ARID AND SEMI-ARID REGIONS*

12.2.1 INTRODUCTION

The 2005 revision of the UN World Urbanization Prospects report described the twentieth century as witnessing "the rapid urbanization of the world's population" as the global proportion of urban population increased dramatically from 13% (220 million) in 1900 to 29% (732 million) in 1950 and to 49% (3.2 billion) in 2005. The same report projected that the figure is likely to rise to 60% (4.9 billion) by 2030. The simple definition of urbanization is the removal of the rural character of an area. This process is usually associated with the development of civilization. Demographically,

---

* This study is based on a paper by A. Elzawahry. 2009. Urbanization impacts on Wadi systems. In *Proceedings of the Sixth International Conference on Environmental Hydrology*, M. M. Soliman, ed. Cairo: ASCE.

the term denotes redistribution of populations from rural to urban settlements with all needed structural activities.

Urbanization within a watershed increases the area of impervious surfaces (Paul and Meyer 2001), which decreases the infiltration of precipitation and increases the runoff (Dunne and Leopold 1978; Gordon et al. 1992). Runoff increases in proportion to the cover of impervious surface in a watershed (Arnold and Gibbons 1996), and the increased storm runoff increases peak discharges and flood magnitudes. Increases in flood magnitudes are greater for floods with shorter recurrence intervals than those with long recurrence intervals (Hirsch et al. 1990). Reduced infiltration of precipitation into the groundwater aquifers may reduce groundwater recharge and stream base flow. However, importing water into an urban watershed for landscape irrigation may increase the stream base flow.

Urbanization may take place within the wadi watershed and/or within the wadi course and flood plain. The impact of watershed urbanization is measured by two main factors. The first factor is the degree of urbanization as measured by the percent of impervious surface cover, which determines wadi peak floods and flow hydrographs. The second factor is the location of urbanization within the watershed, which plays an equivalent important role. Sheeder et al. (2002) found that dual urban and rural hydrograph peaks are based on the degree and spatial location of urbanization within the watershed. The first peak is a result of drainage from the urban area followed closely by drainage from the rural area. Flood duration may increase as a result of the two flooding peaks occurring consecutively. If the lower reaches of a basin become urbanized with little development in the upper basin, water from the lower urban portion could drain before the arrival of water from the upper portion, thus decreasing the flooding magnitude but increasing the duration.

Urbanization within the flood plain has severe impacts on the urbanized area. For low floods, the wadi is obstructed, increasing both water level and velocity. For high floods, the flood plain is flooded with water, causing additional damage. In such cases, the flow is obstructed, causing backwater curve with increasing water level and causing damage to upstream zones of the urbanized areas.

This case study addresses the urbanization impacts for two different study cases. The first case investigates the level of urbanization and the location of the urbanized subcatchments within the watershed of Wadi El-Arish in Egypt. The second case reflects the impact of urbanization within the flood plain and main channel of Wadi Adai in Sultanate of Oman, during an extreme cyclone event.

12.2.2 WADI EL-ARISH STUDY CASE

12.2.2.1 Catchment Characteristics

Wadi El-Arish is situated in northerneastern Egypt in the north part of the Sinai peninsula. It stretches between the Suez gulf in the west, the Akaba gulf in the east, and the Mediterranean Sea in the north. The area of Wadi El-Arish drainage basin is about one-third of the Sinai area. It has a drainage basin of 23,633 square kilometers.

Due to the variety of climatic conditions, arid in the south and wet in the mountains and the Mediterranean seaside, a variety of soils can be found within the basin of Wadi El-Arish. Table 12.5 summarizes the areal percentages of soil groups in the study basin.

The topography of Wadi El-Arish drainage basin has an elevation difference of 1400 meters between its highest and lowest points, with a slope of about 1:25 toward the north. The elevation attains it highest altitude, 1400 meters, at Jabal Umm Mafud at the south edge of the basin and drops to 230 meters at Jabal El Thamila and 500 meters at Jabal Ash Shaiira at the eastern edge of the basin, where Wadi El Jeria joins Wadi El-Arish. At the west of the basin, the elevation ranges between 330 meters at Jabal Umm Zaranik and 520 meters at Jabal Umm Ali, where the El Buruk tributary joins Wadi El-Arish.

To obtain the basin characteristics, the watershed management system (WMS) package has been used. WMS is a comprehensive environment for hydrologic analysis. It was developed by the Environmental Modeling Research Laboratory of Brigham Young University in cooperation with the U.S. Army Corps of Engineers Waterways Experiment Station. WMS is used to delineate the catchment and wadi streams. The input data to the WMS model were obtained from Shuttle Radar Topography Mission (SRTM3). SRTM3 data give the elevations in reference to the mean sea level in the center of a grid of 90×90 meter spacing. Table 12.6 summarizes the

TABLE 12.5
Areal Percentages of Soil Groups in the Study Basin

Soil Group	Area (km²)	Area (%)
Tertiary limestone	4017.3	17
Cretaceous sandstone	10,634	45
Rock escarpment	1890.5	8
Quaternary alluvial deposits	4253.6	18
Gravels of denuded rocks and gravel	2363.1	10
Sand dunes and sand sheets	472.62	2

TABLE 12.6
Geomorphology Characteristics of Wadi El-Arish

Area	23,633 km²
Perimeter	1206.3 km
Basin length	237.82 km
Maximum flow length	347.75 km
Maximum stream length	329.1 km
Basin slope	0.0424 m/m
Maximum flow slope	0.004 m/m
Maximum stream slope	0.003 m/m
Shape factor	2.39 km²/km²
Sinuosity factor	1.38
Average overland flow	8.81 km

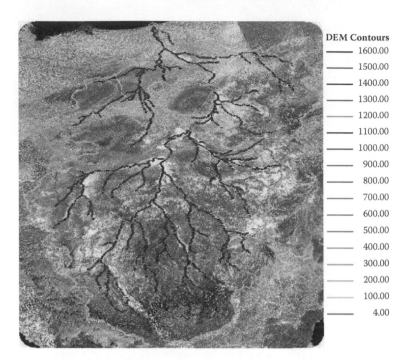

DEM Contours
—— 1600.00
—— 1500.00
—— 1400.00
—— 1300.00
—— 1200.00
—— 1100.00
—— 1000.00
—— 900.00
—— 800.00
—— 700.00
—— 600.00
—— 500.00
—— 400.00
—— 300.00
—— 200.00
—— 100.00
—— 4.00

FIGURE 12.7 Wadi El-Arish catchment.

geomorphology characteristics of Wadi El-Arish. Figure 12.7 shows the delineated wadi catchment and streams overlaid on the satellite image.

12.2.2.2 Rainfall Analysis

Rainfall data are available for eight rain gauges within and around the Wadi El-Arish catchment. Figure 12.8 and Table 12.7 summarize the rain gauge data. A frequency analysis was carried out for El-Arish rain gauges to obtain the intensity–duration–frequency (IDF) curves representing Wadi El-Arish. The results are given in Figure 12.9.

In addition to the data obtained from the ground rain gauges, the Tropical Rainfall Measuring Mission (TRMM) data were obtained and used to generate similar IDF curves. TRMM is an international project jointly sponsored by the Japan National Space Development Agency (NASDA) and the U.S. National Aeronautics and Space Administration (NASA) Office of Earth Sciences. The rainfall data can be obtained in millimeters per hour by using the algorithm 3B42-V6, which produces TRMM-merged high-quality infrared precipitation and root mean square precipitation-error estimates. These girded estimates are at a 3-hour temporal resolution and a 0.25° × 0.25° spatial resolution in a global belt extending from 50° south to 50° north latitude. The available TRMM data covers 9 years starting from 1998.

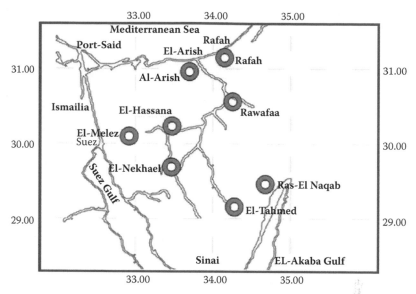

FIGURE 12.8 Wadi El-Arish ground rain gauges.

TABLE 12.7
Wadi El-Arish Rain Gauges

Station Name	Lat.	Long.	Elevation (m)	Record Period	
El-Arish	31.1	33.8	27.93	1982	2006
El-Nekhel	29.9	33.7	402	1997	2006
Rawafaa	30.8	34.4	136	1990	2006
El-Hassana	30.5	33.8	25.6	1999	2006
El-Tahmed	29.3	34.3	625	1990	2006
Ras-El-Naqab	29.6	34.7	743.6	1982	2006
Rafah	31.2	34.2	73.26	1990	2006
El-Melez	30.4	33.2	320	1982	2006

The generated IDF curves in the vicinity of El-Arish rain gauge are given in Figure 12.10. The ground rain gauges gave higher values, around 12% more than those obtained from TRMM data.

Figure 12.11 shows the maximum storm obtained from the analysis of the 9-year TRMM data in the vicinity of Wadi El-Arish. The storm started on January 14 at 18:00 UTC (Coordinated Universal Time) and ended on January 15 at 9:00 UTC.

12.2.2.3 Runoff Analysis

Based on the surface geology and land use, the Wadi El-Arish basin was divided into 36 subbasins. The delineated wadi streams and catchments were overlaid the surface geology of the area, and values of curve numbers (CN) were assigned to each subbasin (Figure 12.12).

The WMS package was used to obtain the runoff hydrograph at the outlet of the Wadi El-Arish basin. The input data to the model of the 2002 TRMM storm are given in Figure 12.10, the catchment and wadi delineations are given in Figure 12.7,

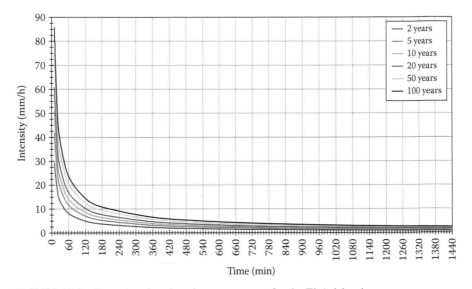

FIGURE 12.9 Intensity–duration–frequency curve for the El-Arish rain gauge.

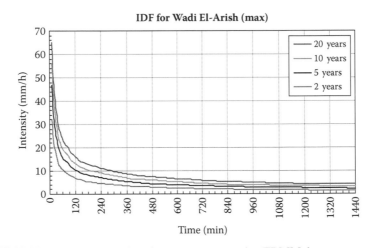

FIGURE 12.10 Intensity–duration–frequency curves using TRMM data.

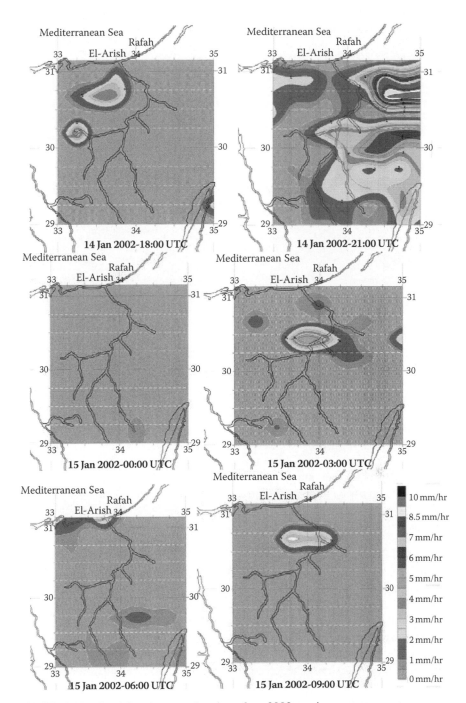

FIGURE 12.11 Spatial and temporal regimes for a 2002 maximum storm event.

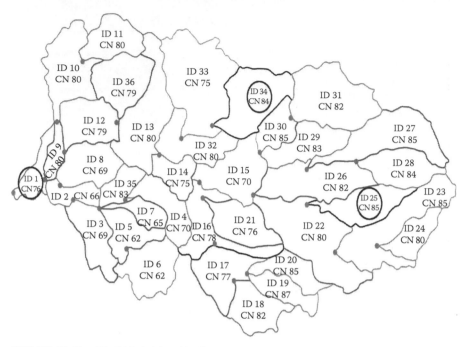

FIGURE 12.12 Wadi El-Arish subbasins and curve number values.

and subbasins characteristics and CN values are given in Figure 12.12. The obtained runoff hydrograph (basic run) is given in Figure 12.13.

To study the impact of urbanization on the runoff at the wadi outlet, three subbasins were selected to be urbanized. The selection of the subbasins was based on their remoteness from the outlet and their altitude. The three selected subbasins numbers are 1, 25, and 34 (Figure 12.12). For each subbasin, the level of urbanization is tested by changing the CN values assigned to each subbasin. Table 12.8 summarizes the main characteristics of the studied subcatchments. The results are summarized in Figures 12.13 through 12.15.

Figure 12.13 depicts the variation in the flood peak due to the change in the urbanization level (CN) for subcatchment 1. The CN values vary between 60 (low urbanization) and 100 (fully urbanized). Because subcatchment 1 is present in the vicinity of the wadi outlet, the urbanization impact is relatively high, with a 30% increase in the peak flood.

Figure 12.14 depicts the same facts for subcatchment 25, located at 760 meters above sea level, near the highest catchment edge. The variation in the peak was limited to 17% due to the transmission losses along the wadi, which are a function of the length, slope, cross section, and bed material of the wadi. The same trend is observed for subcatchment 34 (Figure 12.15), which is located at catchment edge and attains an average level of 480 meters above sea level. The variation in the peak was limited to about 4%.

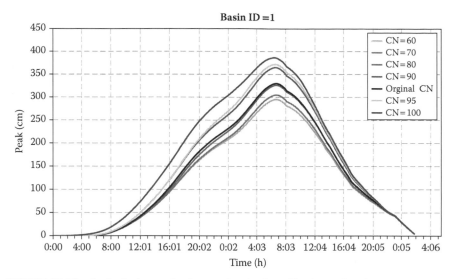

FIGURE 12.13 Outlet hydrographs due to urbanization of basin number 1.

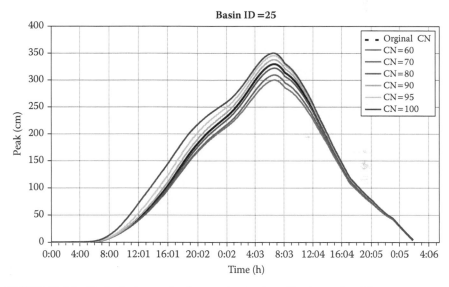

FIGURE 12.14 Outlet hydrographs due to urbanization of basin number 25.

TABLE 12.8

Subcatchments' Characteristics

Subcatchment	Catchment Area (km²)	Curve Number	Catchment Average Level (m)
A1	294.7	82	79
A25	624.7	85	760
A34	910.0	81	480

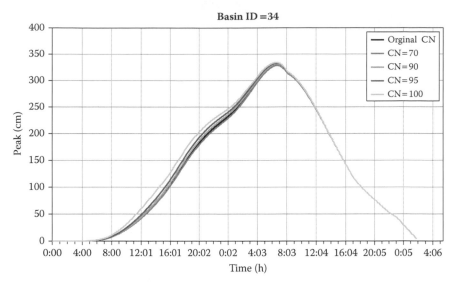

FIGURE 12.15　Outlet hydrographs due to urbanization of basin number 34.

12.2.3　Wadi Adai Study Case

12.2.3.1　General Remarks

Wadi Adai has a catchment area of 320 square kilometers measured at the wadi flow gauge and 380 square kilometers measured at its outlet to the sea. The catchment area of Wadi Adai is surrounded on all sides by mountains up to 1700 meters high. The foothills of the surrounding mountains form a mild sloping catchment toward the top of the western side of the catchment (the outlet). The outlet of the entire catchment is very narrow and twists to move parallel to the sea and turn to cross the main road (Sultan Qaboos) near the commercial area of Al Qurm. The low area of the catchment (nonmountainous) is relatively wide and has several relatively low scattered hills that can be connected to form dikes for flow diversion or even a dam body. The soil formation is the typical alluvium in the deep wadi channels: boulders, gravel, and big rocks. Figure 12.16 depicts the catchment contours and characteristics obtained from the SRTM data and WMS software.

12.2.3.2　Rainfall Analysis

Rainfall records are available for nine stations in the vicinity of the catchments contributing to the dam sites. Table 12.9 summarizes the general data of the nine stations. The time period of the record varied between 12 and 54 years of measurements. The maximum recorded daily rainfall was 187.7 at the FUWAD rain gauge on April 26, 1997. Figure 12.17 depicts the locations of the rain gauges on the Map Source program. The ratio between the maximum daily rainfall and the annual total varied between 0.224 and 1, with a global average of 0.625.

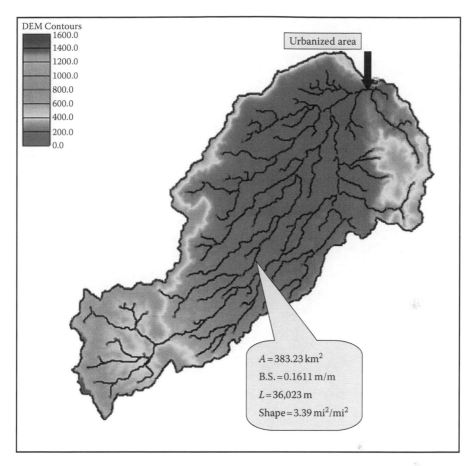

FIGURE 12.16 Wadi Adai catchment area obtained from Shuttle Radar Topography Mission and watershed management system.

TABLE 12.9
Wadi Adai Rain Gauges Data

Rain Gauge	Coordinates	Record Period		Maximum (mm/d)
Amrat	N23.42712 E58.49955	1994	2006	66.8
Fuwad	N23.34053 E58.58169	1994	2006	187.7
Hayfadh	N23.32617 E58.74189	1994	2006	126.5
Mina Al Fahl	N23.62831 E58.51947	1967	2002	87.6
Mina Qaboos	N23.62785 E58.56553	1976	2004	76
Muscat	N23.61222 E58.59181	1952	2006	110
Madinat Al Nahda	N23.42764 E58.44278	1995	2006	120
Ruwi	N23.59831 E58.53969	1982	2006	100
Wadi Al Jannah	N23.37858 E58.47453	1986	2006	95

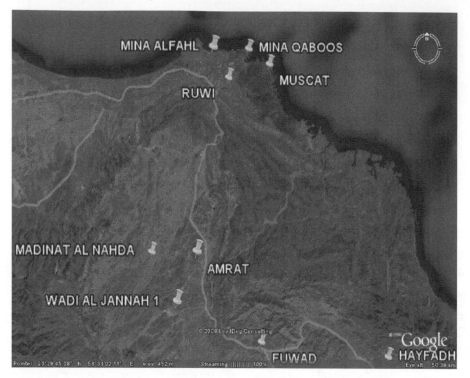

FIGURE 12.17   Rain gauges in the vicinity of Wadi Adai.

A frequency analysis is carried out for Muscat, the longest record. The applied theoretical probability distributions include generalized extreme value (GEV), three-parameter lognormal (LN3), log-Pearson type III (LP3), and the Pearson type III (P3). The location, scale, and shape parameters for each of these distributions were determined using the above fitting techniques, and the resulting equations were used to estimate the flow values associated with certain return periods. The application of the above frequency analysis approach for the Muscat rain gauge is presented in Table 12.10.

The distance between Ruwi and Muscat rain gauges is 5.5 kilometers, which is smaller than the size of the storms generated in the area. The records of Ruwi are used to complete the missing data of the Muscat station to form a longer record of 40 years. Frequency analysis of such records suggests the values listed in Table 12.11.

12.2.3.3   Runoff Analysis

The peak flood records of Wadi Adai flow gauge are available from 1979 to 2003 with missing records for 6 years. Frequency analysis was carried out for the peak flood to obtain the results given in Table 12.12. The results indicate a 100-year return flood of about 1600 square miles per second.

TABLE 12.10
Maximum Daily Rainfall Frequency Results

Return Period	GEV	LN3	LP3	P3
2	50.84	48.88	47.47	48.53
5	81.81	80.68	83.21	78.84
10	100.67	101.92	106.84	98.78
20	117.62	122.39	128.59	117.59
25	122.78	128.91	135.25	123.5
50	138.02	149.14	155	141.44
100	152.25	169.46	173.45	158.99
200	165.57	190.02	190.71	176.26
500	181.94	217.75	211.84	198.73

GEV = generalized extreme value, LP3 = log-Pearson type III, P3 = the Pearson type III.

TABLE 12.11
Recommended Maximum Daily Rainfall Frequency Results

Return Period (years)	2	5	10	20	25	50	100	200	500
Daily Rainfall (mm/d)	46	75	94	112	117	134	151	168	189

TABLE 12.12
Wadi Adai Peak Floods (m³/s)

RP	GEV	LN3	LP3	P3
2	125	93	121	123
5	311	343	411	358
10	489	670	623	554
20	718	1164	799	762
25	806	1366	847	832
50	1133	2163	970	1053
100	1562	3267	1058	1282
200	2129	4765	1119	1520
500	3163	7527	1167	1843

GEV = generalized extreme value; LP3 = log-Pearson type III; P3 = the Pearson type III.

The frequency analysis of the maximum daily volume (runoff) and the estimated runoff coefficients are summarized in Table 12.13. The runoff depth (in millimeters) was obtained by dividing the maximum daily volume recorded by the contributing catchment. The runoff coefficient is the ratio of the runoff depth and the rainfall.

12.2.3.4 Cyclone Gonu Analysis

The available data, during the cyclone event, are limited to the 12 rain gauges surrounding the area of Muscat. The rainfall records included the rainfall in millimeters against time in minutes (hyetograph) for the entire period of the cyclone, around one day. Additionally, the daily volume was recorded in two wadi gauges—Wadi Adai and Wadi Al Ansab. Table 12.14 summarizes the general data of the rain gauges. Figure 12.18 shows the contours of the recorded rainfall on June 6, 2007.

Both ground rain gauges and TRMM daily rainfall records demonstrate similar spatial distribution. However, the maximum recorded rainfall by the ground rain gauges is more than double of the TRMM data (615 millimeters against 250 millimeters). This may be due to the fact that the TRMM records are averaged over a grid of $0.25° \times 0.25°$, against point rain gauge records.

The frequency analysis for maximum daily rainfall was carried out for two sets—with and without Gonu. Figures 12.19 and 12.20 show the results for the Ruwi

TABLE 12.13
Runoff Coefficient

Return Period (year)	Runoff (mm)	Rainfall (mm) (GEV)	Runoff Coefficient
30	28.57	122	0.23
100	48.57	152	0.32
500	114.28	180	0.64

TABLE 12.14
Rain Gauges Data (June 6, 2007)

No.	Station	Coordinates		June 6, 2007
		Zone 40		
		E	N	Rainfall (mm)
1	FL609065AF Yiti	—	—	286.2
2	FL672275AF Buei near Buei	58 35 25.6	23 15 13.4	378.4
3	FL681271CF Fuwad 2 at Fuwad	58 34 54.1	23 20 25.9	373.8
4	FL771702AF Taba at Taba	58 40 19.5	23 17 43.2	514.8
5	FL788017AF Hayfadh at Hayfadh	58 44 30.8	23 19 34.2	613.2
6	FL950429CF Mazara 3 at Mazara	58 51 25.1	23 05 30.8	322.0
7	FL974085AF Qurayat near Qurayat	58 54 13.9	23 13 55.9	187.8
8	FM104840AF Alkhaud	58 07 15.9	23 34 42.3	118.8
9	FM415095AF Al Khuwair	—	—	168.8
10	FM517016AF Ruwi at Ruwi	58 32 22.9	23 35 53.9	223.8
11	FM593824AF Amrat	58 30 00.9	23 29 18.7	438.4
12	FM612285BF Muscat near Muscat	58 35 30.5	23 36 44.0	139.4

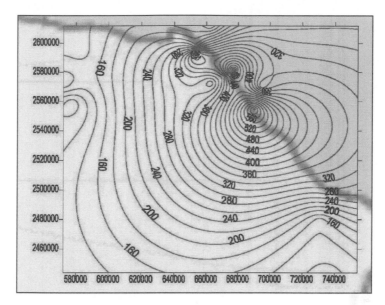

FIGURE 12.18 Daily rainfall distribution (June 6, 2007).

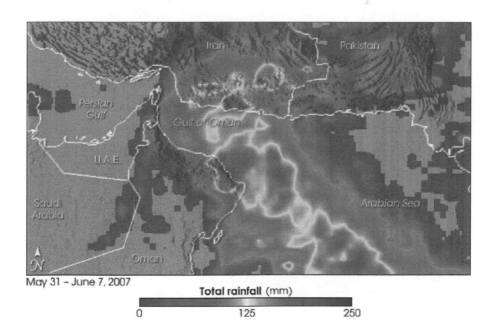

May 31 – June 7, 2007

Total rainfall (mm)

0   125   250

FIGURE 12.19 Daily rainfall distribution (June 7, 2007).

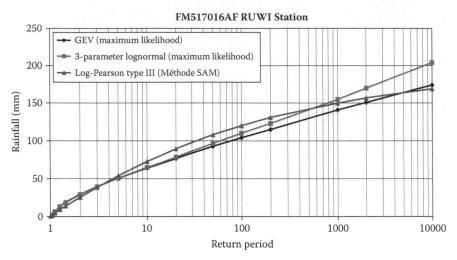

FIGURE 12.20 Rainfall frequency analysis without considering Gonu for Ruwi rain gauge.

rain gauge. The 100-year daily rainfall at Ruwi station reached 110 millimeters for all records up to 2006. The addition of the record for the year 2007 (Gonu) raised the 100-year daily rainfall up to 200 millimeters. This value corresponds to 10,000-year return period in Figure 12.19 (without Gonu). Similar results have been observed in all other stations. The daily runoff volume for Wadi Adai reached 67.824 million cubic meters (June 6, 2007). The maximum volume recorded up to 2006 was 10.8 million cubic meters. Figure 12.21 shows the daily volume for different return periods with and without Gonu cyclone for Wadi Adai.

The results indicated a tremendous increase in the daily runoff volume. The values indicated that the runoff volumes generated from Gonu had not been experienced over the available period of records. The 100-year daily runoff volume including Gonu exceeds four times the values previously adopted for the design. The volume recorded for Gonu corresponds to the year 2000. The recorded peak flood during Gonu was 2350 cubic meters per second. This value corresponds to a 500-year flood (according to a Gumbel frequency analysis without Gonu, Table 12.12).

12.2.3.5 Urbanized Study Area

Figure 12.22 presents the urbanization within the flood plain and along the Wadi Adai course of the study area, located about 20 kilometers from Muscat. The layout also shows that Wadi Adai intersects with a number of streets where culverts, bridges, and Irish bridges have been constructed to provide road flood protection. Wadi Adai is currently undergoing excessive development along its flood plain within the study area. Major commercial centers are located at the flood plain side of the

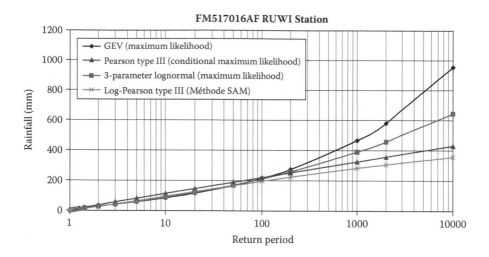

FIGURE 12.21   Rainfall frequency analysis considering Gonu for Ruwi rain gauge.

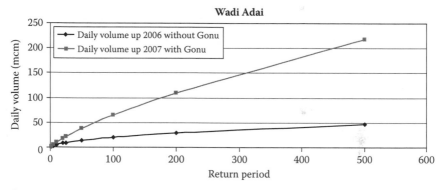

FIGURE 12.22   Daily volume frequency analysis for Wadi Adai (with and without Gonu).

wadi. The columns of the raised parking lot of one of these centers are constricting the wadi. Workers in the commercial centers say that their stores have been flooded within the last 5 years.

The entire reach of the wadi with its flood plain in the vicinity of the urbanized area has been modeled using two models. The first model is Hydraulic Engineering Center River Analysis System (HEC-RAS) developed by the Hydraulic Engineering Center of the U.S. Army. The second is the CARIMA model developed by Sogreah Consultants. A detailed actual survey of the area was carried out and used as input to the models. The simulations covered the current urbanized zones for different

FIGURE 12.23   Wadi Adai urbanization (Al Qurum commercial; refer to Figure 12.10).

return periods. Additionally, the anticipated future urbanization within the area was simulated. The results of the simulations are presented in Figures 12.23 and 12.24. Figure 12.23 shows that the urbanized areas are not flooded for only the 5-year return period flood. For higher return periods, the flood levels increase, covering the commercial areas. The entire urbanized areas are almost fully flooded for the 100-year return period flood.

A flood protection and recharge dam scheme for Wadi Adai is currently under investigation by the Ministry of Water Resources. The feasibility study recommended two dams be constructed within the Wadi Adai catchment to retain majority of its flood before it reaches the urbanized area. Additionally, a wadi training scheme with dikes and bridge construction is proposed, which would replace the Sultan Qaboos Road Irish bridge adjacent to the urbanized area.

Figure 12.24 presents the simulation results for a 100-year flood based on current and future urbanization. The simulation covered two cases: with and without the flood protection scheme. The proposed scheme indicated that the flood limit will be mainly limited to the wadi main channel. Figure 12.25 shows the measured water depth at the wadi flood plain during Gonu cyclone. An increase of about 3 meters above the simulated 100-year flood is observed. Figure 12.26 shows the actual scene during the cyclone at the location.

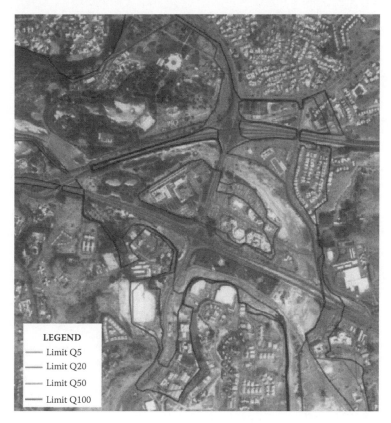

LEGEND
— Limit Q5
— Limit Q20
— Limit Q50
— Limit Q100

FIGURE 12.24 Current urbanization flood limits for 5, 20, 50, and 100-year return period.

12.2.4 CONCLUSION

This case study addressed the subject of urbanization within the wadi catchment, wadi flood plain, and wadi channel. Two cases were presented, for Wadi Al-Arish (Egypt) and Wadi Adai (Oman). The analysis revealed that urbanization within a very small wadi subcatchment may cause a relatively high increase in the peak flood. In addition to the urbanization level, the location of the subcatchment relative to the wadi outlet and its altitude played a major role in the peak flood changes.

Urbanization within the wadi flood plain and its main channel is extremely risky, especially in big wadis. The damage to the urbanized areas and its upstream zones could be disastrous. The mitigation measures in such cases are usually very expensive. The mitigation schemes may include a combination of dams, dykes, drainage structures, and wadi training, as demonstrated in Wadi Adai case. Construction and urbanization should be totally avoided within the wadi channel. Flood plain urbanization must be located outside the 100-year flood limits. Subcatchment urbanization must be studied carefully, before implementation, to ensure its safe impact on upstream and downstream zones of the wadi basin.

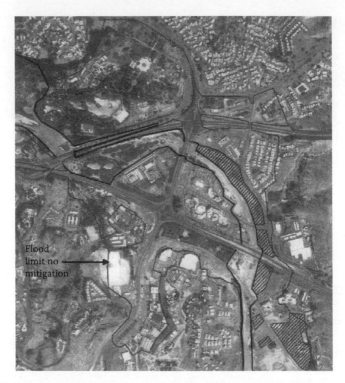

FIGURE 12.25   Current and future urbanization flood limits for 100-year return period flow (with and without mitigation).

FIGURE 12.26   Spot water depth during Gonu cyclone (about 10 meters from wadi bed). The arrow in the map indicates the location of the building seen in the picture to the right during the flood.

REFERENCES

Arnold, C. L., and C. J. Gibbons. 1996. Impervious surface coverage: The emergence of a key environmental indicator. *J Am Plann Assoc* 62:243–258.

Dunne, T., and L.B. Leopold. 1978. *Water in Environmental Planning*. New York: *W.H.* Freeman.

Elzawahry, A. 2009. Urbanization impacts on Wadi systems. In *Proceedings of the Sixth International Conference on Environmental Hydrology*, M. M. Soliman, ed. Cairo: ASCE.

Gordon, N. D., T. A. McMahon, and B. L. Finlayson. 1992. *Stream Ecology: An Introduction for Ecologists*. New York: John Wiley & Sons.

Hirsch, R. M., J. F. Walker, J. C. Day, and R. Kallio. 1990. The influence of man on hydrologic systems. In *Surface Water Hydrology* (The Geology of North America. Vol. 1), eds. M. G. Wolman and H. C. Riggs, pp. 329–59. Boulder, CO: The Geologic Society of America.

Paul, M. J., and J. L. Meyer. 2001. Streams in the urban landscape. *Ann Rev Ecol Syst* 32:333–65.

Sheeder, S. A., J. D. Ross, and T. N. Carlson. 2002. Dual urban and rural hydrograph signals in three small watersheds. *J Am Water Resour Assoc* 38:1027–40.

Watershed Management System. 2002. Hydraulic Engineering Center, U.S. Army Corps of Engineers, Version 7.1.

World Urbanization Prospects. 2005. The 2005 Revision Population Database, http//esa. un.org/unup.

12.3 CASE STUDY 3—RUNOFF SIMULATION USING DIFFERENT PRECIPITATION LOSS METHODS*

12.3.1 INTRODUCTION

Rainfall losses are significant in the rainfall–runoff simulation process. Rainfall losses that are significant in analyzing flood events include land-cover interception, depression storage, and infiltration. Interception and depression storage result from water retention on the vegetation surfaces and water retention in local depressions in the ground surface, pavement cracks, and so on. Infiltration, the most significant loss during flood events, is the movement of water into the ground surface. The rate at which water can infiltrate into the ground surface is highly dependent on surface conditions. Several methods have been used to determine the amount of rainfall converted to runoff. Selecting a method to find the loss and estimating the parameters are critical steps in developing a hydrologic model. This case study will discuss three different methods: the initial/constant method, the curve number method, and the Green–Ampt method.

The initial/constant method estimates losses based on an initial abstraction followed by a constant or uniform loss rate. The advantages of this method are that it is easy to set up and use and it includes only two variables. The disadvantages are that it is difficult to develop parameters without calibration to gauged data and the

---

* Based on a paper by E. A. El-Sayed. 2007. Runoff simulation using different infiltration models. In *Proceedings of the Fifth International Symposium on Environmental Hydrology*, ed. Soliman, M. M. Cairo: ASCE.

methodology may be too simplified to provide a reasonable estimate of losses during an event. Because of its simplicity, this loss rate function is widely used for design storms. It is best suited for large areas and studies where lumped estimates of losses are acceptable (Hoggan 1989).

The Soil Conservation Services (SCS) curve number method estimates losses based on a function of cumulative precipitation, soil cover, land use, and antecedent moisture. The advantages of this method are that it is simple, predictable, and stable. It is a well-established, widely accepted method that relies on only one parameter that is representative of all of the factors affecting precipitation losses. The disadvantages are that the estimated loss accumulation is not in accordance with classical unsaturated flow theory, and the infiltration rate will approach zero during a storm of long duration rather than achieve a constant rate as expected. This method was developed for small agricultural watersheds in the midwestern United States and may not be applicable to other areas. Also, the rainfall intensity is not considered, so 1 millimeter of rainfall over an hour produces the same amount of loss as 1 millimeter of rainfall in a day (Hoggan 1989).

The Green–Ampt method estimates losses based on a function of soil texture and the capacity of the given soil type to convey water. The advantage of this method is that the parameters can be estimated for ungauged watersheds from information about the soils. The disadvantages are that it is not a widely used method and it is not as simple as the other methods (Hoggan 1989).

The modeling approach followed herein will be global at the scale of the study plot, which will be considered one entity represented by average values of soil moisture content and soil hydrodynamic properties. The output of the model will be a simulated hydrograph, which will be compared to the original measured hydrograph to assess the model performance. However, to compare the hydrograph parameters, the excess rainfall used in the model was matched with the observed volume of runoff for a given storm. The observed total runoff volume was found by integrating the direct runoff hydrograph. Several methods are available to estimate the rainfall excess, but each method has assumptions and limitations. The application of each method depends on the availability and type of infiltration data. Most runoff simulation models transform the rainfall excess into runoff. This is done using physically based models (e.g., Green–Ampt or Philip) or using empirical relations (e.g., Horton, Holton, and SCS). The three different methods of excess rainfall estimations are described in Sections 12.3.2 through 12.3.4.

12.3.2 Initial and Constant Loss Method

In this rainfall loss method, all rainfall is lost until an initial loss is satisfied; then the remainder is lost at a constant rate. This method assumes that the rainfall losses process can be simulated as a two-step procedure (see Figure 12.27). The initial loss (STRTL) is the sum of all losses prior to the onset of runoff and is made up of surface retention loss and some amount of infiltration. The uniform loss rate (CNSTL) represents the long-term, equilibrium infiltration capacity of the soil. The values of STRTL and CNSTL are estimated according to soil texture classification and the hydraulic soil group.

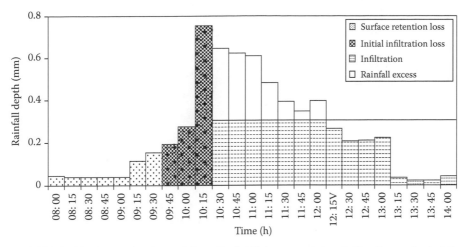

FIGURE 12.27 Example of separation of rainfall hyetograph into infiltration and rainfall excess (runoff) using the initial/constant method.

12.3.3 SCS Loss Method

The rainfall–runoff relationship in this method is derived from the water balance equation and a proportionality relationship between retention and runoff. The SCS rainfall–runoff relationship is given as shown in Figure 12.28:

$$P_e = \frac{(P - I_a)^2}{(P - I_a) + S} \tag{12.1}$$

where P is the rainfall depth, P_e is the depth of excess rainfall, I_a is the initial abstractions, and S is the volume of total storage, all in millimeters.

Storage includes both the initial abstractions and total infiltration. The initial abstraction is a function of land use, infiltration, detention storage, and antecedent soil moisture. The initial abstraction and the total storage are related in an empirical statistical equation as follows:

$$I_a = 0.2S \tag{12.2}$$

Substituting Equation 12.2 into Equation 12.1 yields

$$P_e = \frac{(P - 0.2S)^2}{(P + 0.8S)} \tag{12.3}$$

The storage S (in millimeters) is obtained using the following formula:

$$S = \frac{25,400}{CN} - 254 \tag{12.4}$$

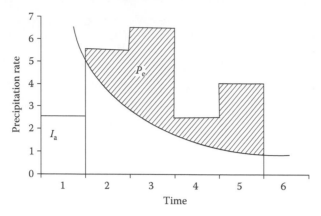

FIGURE 12.28 Example of separation using the SCS method.

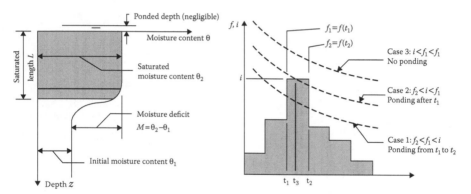

FIGURE 12.29 Example of separation parameters and losses using the Green–Ampt method.

where CN is the curve number that can be obtained from standard tables for different combinations of land use, land cover, hydrologic soil groups, and hydrologic conditions (see Chapter 4). The hydrologic soil group reflects the soil's permeability and surface runoff potential.

12.3.4 GREEN–AMPT METHOD

Another method to determine the amount of rainfall that is converted to runoff is the Green–Ampt method (Figure 12.29). In 1911, W. H. Green and G. A. Ampt developed this method to determine the amount of water that infiltrates into the soil during a rainfall event. The basic equation used in the Green–Ampt model is given by

$$f = K_s\left(1 + \frac{\psi\theta}{F}\right) \quad \text{for} \quad f < i \tag{12.5}$$

where f is the infiltration rate (cm/h), I is the rainfall intensity (cm/h), K_s is the hydraulic conductivity, wetted zone, steady-state rate (cm/h), ψ is the average capillary suction in the wetted zone (cm), θ is the dimensionless soil moisture deficit, which is equal to the effective soil porosity times the difference in final and initial volumetric soil saturations, and F is the depth of rainfall infiltrated into the soil since the beginning of rainfall (cm).

12.3.5 CHARACTERISTICS OF THE FLOOD EVENTS STUDIED

The watershed study of Wadi Sudr was equipped with rainfall and runoff measurements to be an experimental basin (pilot area). The basin area was 450 square kilometers and was located in the western part of Sinai as shown in Figure 12.30. The catchment had a Red Sea climate with a dry summer season characterized by high-intensity and short-duration storms causing flash floods. Rainfall thus had a bimodal temporal distribution with two major rainy periods, one in winter and another in spring. However, strong inter- and intra-annual variations of rainfall were observed from the data collected since 1989. Flood events were selected based on the following criteria:

- Isolated storm event (single storm)
- At least 4 millimeters of rainfall
- Peak flow rate larger than 10 cubic meters per second
- Single-peak hydrograph

The selected events, with at least 6 hours between the consecutive storms, were considered and four events were selected according to the above criteria. Table 12.15

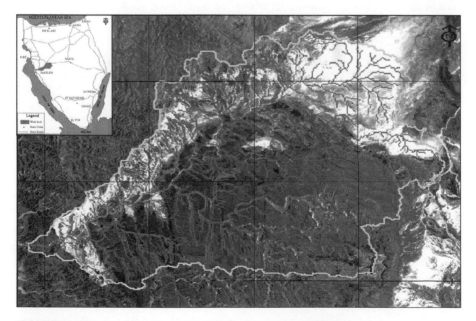

FIGURE 12.30 Location map of Wadi Sudr, the study watershed.

TABLE 12.15
Characteristics of the Selected Storm Events

	Rainfall Data		Runoff Data	
Storm No.	Depth (mm)	Duration (h)	Depth (mm)	Peak Flow (m³/s)
1	11.96	12.75	0.50	10.5
2	04.29	5.750	0.60	18.0
3	09.02	10.00	0.70	18.2
4	17.72	12.75	5.50	280

gives a summary of the characteristics of the selected events, including a significant rainfall amount associated with observable runoff.

Another type of data is field infiltration measurements and site investigations, which have been carried out. Soil infiltration was measured using a standard double-ring infiltrometer according to ASTM infiltration protocol. Infiltration rate was found from the distance the water level decreased by the time required for this decrease (the difference between the start and end times). The Excel worksheet is used to record, calculate the infiltration data, and plot the values. These data are used for assuming the initial conditions of the model's parameters for different infiltration models.

12.3.6 DEVELOPMENT OF MODELS

One of the most famous hydrologic models, the Hydrologic Modeling System (HEC-HMS), was used to carry out this study. HEC-HMS was developed by the U.S. Army Corps of Engineering. It provides a powerful optimization technique for estimating the loss rate and unit hydrograph parameters when ungauged precipitation and runoff data are available. The steps in the initial development of a simulation model are as follows:

1. *Assessment of data requirements and availability*: Types of data required for developing a model include historical precipitation and flow data, runoff parameter data from past studies, and data associated with watershed characteristics such as drainage area, soil type, and land use. Information acquired from field observations helps in understanding of the runoff-response characteristics of the watershed.
2. *Acquisition and processing of data*: All precipitation and stream flow data are stored on electronic media, which can greatly facilitate data acquisition.
3. *Development of basin configuration*: Nondistributed models use lumped (spatially averaged) values for precipitation and loss (infiltration) parameters. The basin is relatively small so that spatial averaging of this information is reasonable.

4. *Development of initial estimates for parameter values*: After defining the basin configuration, a skeleton input file is developed, which contains all the required information (such as drainage area and soil type) except values for runoff parameters. Such parameters are required for defining unit hydrograph, kinematic wave, loss rate, base flow, or routing relationships. At this point, initial estimates of the values for runoff parameters are made and entered into the input file. Estimates are derived from past studies, application of previously developed regional relationships, and the physical characteristics of the basin.

12.3.7 Optimization of Models

The optimization process begins with initial parameters estimates and adjusts them so that the simulated results match the observed flow as closely as possible. The HEC-HMS program has two different search algorithms (univariate gradient and the Nelder–Mead method) that move from the initial estimates to the final best estimates. These search methods are available for minimizing the objective function and finding the optimal parameter values. The univariate gradient method evaluates and adjusts one parameter at a time while holding the other parameters constant. The Neder–Mead method uses a downhill simplex to evaluate all parameters simultaneously and determine which parameter to adjust.

The objective function measures the goodness of fit between the computed outflow and observed flow. Six different functions that measure the goodness of fit in different ways are provided. The peak-weighted RMS error function is a modification of the standard root mean square error that gives greatly increased weight to flows above average and less weight to flows below average. The sum of squared residuals function gives increased weight to large errors and less weight to small errors. The sum of absolute residuals function gives equal weight to both large and small errors. The percent error in peak flow function ignores the entire hydrograph except for the single peak value. The percent error in volume function ignores peak flow or timing considerations but favors volume. The time-weighted function gives greater weight to errors near the end of the optimization time window and less weight to errors in the window. However, a minimum objective function is obtained when the parameter values that are best able to reproduce the observed hydrograph are found. It is equal to zero if the hydrographs are exactly identical.

12.3.8 Calibration Phase

The objective of model calibration is to select parameter values so that the model simulates the measured hydrograph as closely as possible. The quality of the data and the simplifications and errors inherent in the model structure also put limitations on how closely the model is actually able to simulate the hydrograph. A calibration scheme can include optimization of multiple objectives that measure different aspects of the hydrologic response of a unit (Madsen 2000). In this respect, it is important to note that, in general, trade-offs exist between different objectives. For instance, one may find a set of parameters that provides a very good simulation of volume but a poor simulation of the hydrograph shape or peak flow and vice versa. To obtain a successful calibration using automatic optimization routines, numerical performance measures or objective criteria must be formulated that reflect the calibration objectives. This can be done by considering the calibration problem in a multiobjective framework (Yapo et al. 1998; Madsen 2000).

In this phase, the selected data have been used to develop values for infiltration parameters. Parameter values are adjusted in automatic calibration to minimize the magnitude of the objective function. Each parameter, to be estimated, is decreased by 1% and then by 2%, the system response is evaluated, and the objective function is calculated for each change, respectively. The best value of the parameter is then estimated using Newton's method. This continues until no single change in any parameter yields a reduction of the objective function of more than 1%. Figure 12.31 shows the HMS interface to enter optimization parameters for the model calibration. The figure also shows the objective function and the optimized parameters results.

The values of the calibrated parameters of the three models for each objective function are presented in Table 12.16. The calibrated values of the parameters for all the three models change with the objective criterion, but in general they give similar calibrated parameter values, especially for Green–Ampt method. This means that parameter values obtained by the Green–Ampt method may not be sensitive to whether equal weight is given to errors computed in calibration.

To compare the performance of the calibration procedures made with different objective criteria, the relative error on both peak flow and runoff depth for a combination of models and objective criteria was computed: for a given event i, the error on peak flow (RE_{Qi}) and runoff depth (RE_{Vi}) are defined, respectively, by the following equations:

$$RE_{Qi} = \frac{Q_{si} - Q_{oi}}{Q_{oi}} \tag{12.6}$$

where Q_{si} is the simulated peak flow rate and Q_{oi} is the observed peak flow rate, and

$$RE_{Vi} = \frac{V_{si} - V_{oi}}{V_{oi}} \tag{12.7}$$

where V_{si} is the simulated runoff depth and V_{oi} is the observed runoff depth.

Figure 12.32 illustrates the findings for all the three calibrated models. These results show that error values range between 0 and 1.33. The minimum error in peak flow and runoff depth is obtained when using percent error in peak flow and volume objective functions, respectively. The peak-weighted RMS error and time-weighted RMS error objective functions give the minimum error values for all the infiltration models.

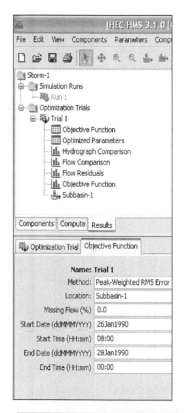

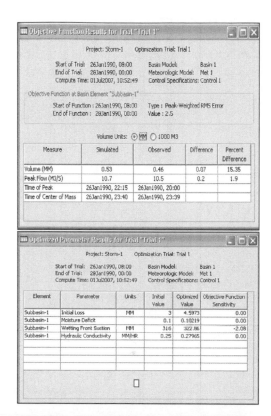

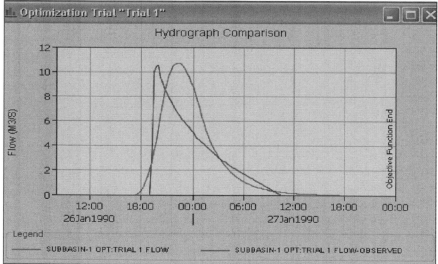

FIGURE 12.31 Hydrologic modeling system optimization parameters.

TABLE 12.16

Calibrated Parameter Values Corresponding for All Models and Objective Functions

Objective Function	Initial and Constant		SCS		Green and Ampt			
	STRTL (mm)	CNSTL (mm)	I_a (mm)	CN	I_a (mm)	q	Y (mm)	K_s (mm/h)
Peak-weighted RMS error	8.90	1.7	1.9	91.1	2.2	0.2	328.6	0.3
Sum of squared residuals	10.8	0.7	2.2	86.6	5.5	0.2	294.0	0.2
Sum of absolute residuals	8.60	0.4	0.0	94.2	0.0	0.1	300.0	0.3
Percent error in peak flow	7.20	0.3	0.0	85.7	6.0	0.1	300.0	0.3
Percent error in volume	5.30	2.6	1.1	89.7	4.5	0.1	304.5	0.3
Time-weighted RMS error	7.00	0.9	2.5	86.8	2.9	0.1	279.4	0.2
Maximum	10.8	2.6	2.5	94.2	6.0	0.2	328.6	0.3
Minimum	5.30	0.3	0.0	85.7	0.0	0.1	279.4	0.2

12.3.9 VALIDATION PHASE

In this phase, the other historical precipitation and flow data are used to test the calibrated values for infiltration parameters. Model validation was carried out on one event. The calibrated parameters for the three models were used. The simulated hydrographs for the same event is plotted in Figure 12.33. When overlaying the simulated hydrographs for the same event by the three models, note that in some instances the models' outputs are quite similar due to the fact that during split-sample tests, models may exhibit similar performance (Donnelly 1997).

Table 12.17 summarizes the simulated values of runoff depth and peak flows for all the models. The results, in terms of consistency of the simulated runoff depth and peak flows for all the models, show that Green–Ampt model performs slightly better than the other two models. Consequently, in comparison with the simulated runoff depth and peak flows for all models, the Green–Ampt formulation is more suitable than the other formulations, initial/constant and the SCS curve number.

A comparison of how each of the three infiltration models accumulates loss for a 9 millimeter, 10-hour duration rainfall is shown in Figure 12.34, which adjusts loss parameters to the calibrated parameters for the three models. From these graphs, it is apparent that at the beginning the losses have the same values, after which they have slight differences in values.

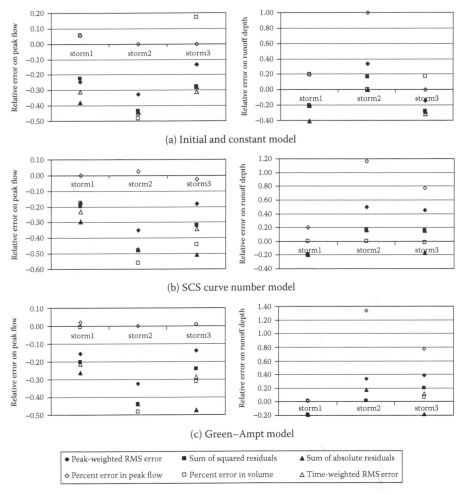

FIGURE 12.32   Comparison of the relative error on peak flow and runoff depth obtained for different infiltration models.

12.3.10 Discussion and Conclusion

The main purpose of this research is to construct a hydrologic model that can be used to simulate the hydrologic phenomenon. The performance of three infiltration models was compared with rainfall–runoff data obtained on a 450-square kilometer watershed located in the western part of the Sinai Peninsula. The models were the SCS curve number method, the initial/constant method, and the Green–Ampt method. They were used to obtain the unit hydrographs for four flood events. Alternative optimization techniques and alternative objective functions are currently being evaluated using the HEC-HMS model. The model was been calibrated and the computed and observed hydrographs were compared. The calibration included six different objective functions corresponding to runoff volume, peak flow conservation, and

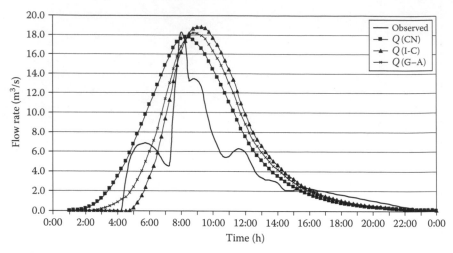

FIGURE 12.33 Comparison between the simulated and measured hydrographs for all three models using the calibrated parameters.

TABLE 12.17
Simulated Runoff Depth and Peak Flows for All the Models

Simulated models		Runoff Depth (mm)	Peak Flows (m³/s)
Simulated models	Data observed	0.7	18.2
	Green and Ampt	0.8	18.2
	Initial and constant	0.9	18.6
	SCS	1.0	17.7

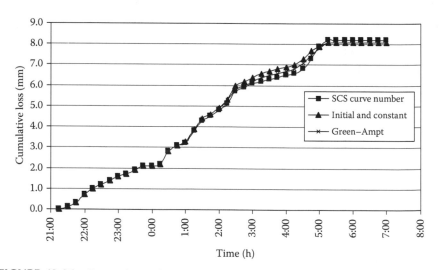

FIGURE 12.34 Comparison of the simulated losses for the three models using the calibrated parameters.

sum of residuals and weighted RMS. A comparison of the calibration and validation results indicates the following:

- When analyzing the performance of the six objective criteria, namely, peak-weighted RMS error and time-weighted RMS error give minimum error values for all infiltration models.
- According to the validation process, no degradation was observed between the calibration and validation results.
- The calibrated parameter values of a given model vary slightly according to the selected objective criterion but lie within similar bounds.
- When comparing the overall performance of the three tested models, the Green–Ampt formulation ranked the best on average in terms of accuracy of simulated runoff volume and peak flow.

REFERENCES

American Society for Testing and Materials. 1994. *Standard Test Method for Infiltration Rate of Soils in Field Using Double-Ring Infiltrometer.* Philadelphia: ASTM.

Donnelly, M. L. 1997. Comparison of three rainfall-runoff modeling techniques in small forested catchments. Master's thesis, Simon Fraser University, Burnaby, British Columbia, Canada.

El-Sayed, E. A. 2007. Runoff simulation using different infiltration models. In *Proceedings of the Fifth International Symposium on Environmental Hydrology*, ed. Soliman, M. M. Cairo: ASCE.

Hoggan, D. H. 1989. *Computer-Assisted Floodplain Hydrology Hydraulics, Featuring the U.S. Army Corps of Engineers, HEC-1 and HEC-2 Software System.* New York: McGraw-Hill.

Madsen, H. 2000. Automatic calibration of a conceptual rainfall-runoff model using multiple objectives. *J Hydrol* 235:276–88.

U.S. Army Corps of Engineers Hydrologic Engineering Center. 1998. HEC-HMS Hydrologic Modeling System User's Manual, Davis, California.

Yapo, P., H. Gupta, S. Sorooshian. 1998. Multi-objective global optimization for hydrologic models. *J Hydrol* 204:83–97.

12.4 CASE STUDY 4—DESIGN OF SALBOUKH FLOOD CONTROL SYSTEM, IN RIYADH CITY, KINGDOM OF SAUDI ARABIA*

12.4.1 INTRODUCTION

A new 3-square kilometer community in the Salboukh region, 20 kilometers northwest of Riyadh in the Kingdom of Saudi Arabia (KSA), is being developed. The project plot exists on the confluence of medium- or small-size wadis that run in a diverse topography and geology, making the area an exceptionally natural and wild landscape. Accordingly, the development is going to be one of the most stirring to

---

* Based on a paper by M. Gad and M. Abdel-Hamid. 2009. Design of Salboukh flood control system in Riyadh. In *Proceedings of Sixth International Conference on Environmental Hydrology*, M. M. Soliman, ed. Cairo: ASCE.

residential communities in the neighborhoods of Riyadh. The project area is located west of the intersection of the Salboukh highway (100 meters wide) and Al-Ammaria road (40 meters wide), so it is bound by the Salboukh highway in the east and by Al-Ammaria road in the south. The project area is bounded from the north by the Dar Al-Arkan residential development, which was undergoing construction at the time of this study. Undeveloped wadi pond areas followed by farmlands bound the project area from the west. Figure 12.35 presents the project location.

Because flood waters that periodically run through the project area may pose a threat to the activities inside the community, flood control design and the corresponding earthworks must be finished concurrently with the master plan design before any other design works start. In addition, flood protection works are a sensitive part of the project because they highly interact with the landscape, which is the most attractive feature in the project.

This case study discusses all the hydrologic and hydraulic design issues, including the data, methodologies adopted, analysis, and design recommendations.

12.4.2 Data Processing and Site Reconnaissance

12.4.2.1 LandSat Image

This section presents the data and information collected for designing the flood protection scheme. The satellite photography reported here are LandSat images (28.5 meter resolution) from the year 2000 (Figure 12.36).

These images were used to understand the project area, the characteristics of the catchment areas, and the boundary conditions upstream and downstream from the project location. They were used as the base for the study and also to help conduct the hydrologic and hydraulic analysis. The images were georeferenced and rectified to overlay properly on the common coordinate system. The common horizontal coordinate system chosen for the project was Ain EL-Abd-Z36 north. Specifying a common coordinate system is essential in order to properly overlay data from multiple sources. The vertical coordinate system used was the mean sea datum.

12.4.2.2 Hydrometeorological Data

Winter rainfall events over the Arabian peninsula are associated with southeastward propagating weather systems from the Mediterranean region (i.e., the southeastward propagating frontal system). The rainfall over most of the Arabian region is generally restricted to a few wet spells during the winter season, brought on by weather disturbances originating over the Mediterranean Sea. The mountainous regions in Asir and Hejaz provinces along the Red Sea coast and the mountains of Yemen and Oman, rising to the height of 2.5–3.5 kilometers, receive periodic monsoon-like downpours and an annual rainfall of about 400 millimeters. Across the mountains, rainfall is low and unreliable; average rainfall below 200 millimeters and often less than 100 millimeters over a basin-shaped interior with alternating steppe and desert landscape make the climate arid.

The project area is classified as semi-arid, where average annual rainfall is about 100–150 millimeters. Figure 12.37a presents the average monthly rainfall. Although the monthly averages reported are low, historical records in nearby regions in Kuwait

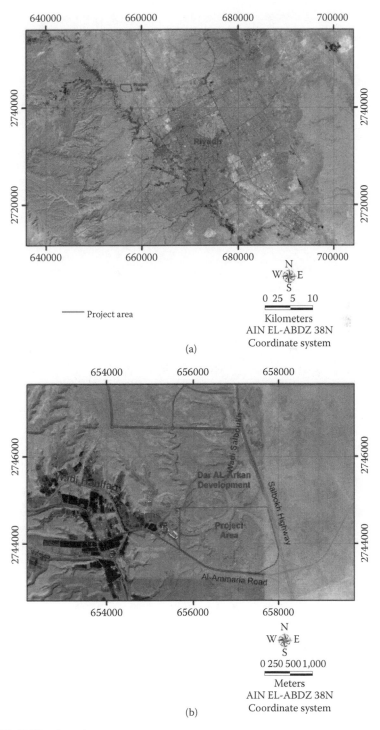

FIGURE 12.35 Location maps.

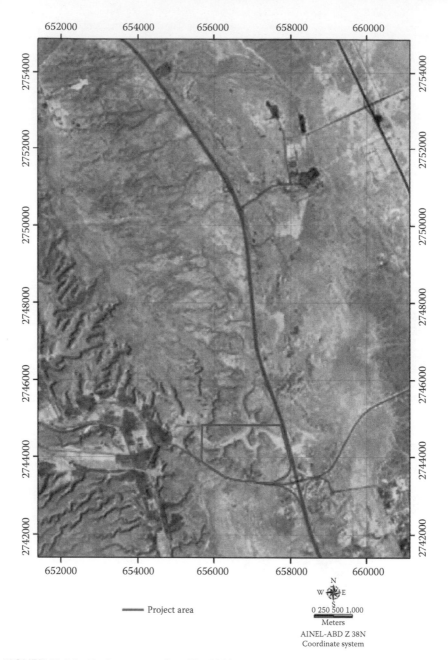

FIGURE 12.36   Project area on LandSat 2000.

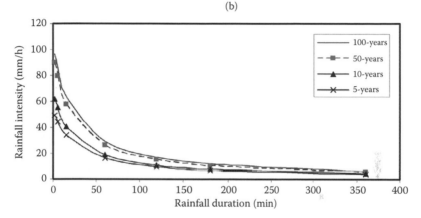

	Data Period	Jan	Feb	Mar	Apr	May	Jun	Jul	Aug	Sep	Oct	Nov	Dec
Average Rainfall Depth (mm)	1981–2000	12.3	5.8	30.2	23.3	6.2	0	0	0.3	0	2.3	7.4	11.2
Average No. of Rainy Days	1981–2000	5.8	4.1	10.8	10.6	3.9	0	0.1	0.3	0.1	1.7	3.4	6.5

(a)

Constant	100-years	50-years	10-years	5-years
a	2399	2101	1556	1397
b	22.82	21.47	23.16	26.45

(b)

FIGURE 12.37 Curve fitting the IDF data: (a) average monthly rainfall and (b) best-fitting parameters.

(which has similar monthly averages) indicate that the maximum monthly rainfall can reach up to 100–120 millimeters in an extreme month containing a very wet day of about 50–60 millimeters rainfall daily.

The region experiences frequent thunderstorms especially from March 10 to April 8, which are sometimes accompanied by severe dust storms during which the visibility drops to zero. Thunderstorms are characterized by severe rainfall intensity and small spatial and temporal extents.

The intensity–duration–frequency (IDF) curves for the Riyadh meteorological station were obtained from the Presidency of Meteorology and Environment.

12.4.2.2.1 Curve Fitting of Intensity–Duration–Frequency Data

The obtained IDF data were tabulated, and nonlinear regression was used to fit an IDF equation in the following form:

$$i = \frac{a}{(D+b)}$$

where i is the rainfall intensity in millimeters per hour, D is the rainfall duration in minutes, and a and b are constants. The best-fitting constants and the corresponding fitted IDF curves are shown in Figure 12.37b.

12.4.2.2.2 Upstream and Downstream Boundary Conditions and Wadi Crossings

Six wadis with routes through the project area threaten it. In addition to the flow from these wadis, wadi A receives its main flow from its northern upper catchments. The flood of the northern upper catchments crosses Salboukh highway via bigger culverts just north of the project area. The flood then travels south in the wadi until it reaches the project area, where it joins the flow from wadis B, C, D, E, F, and G. Figure 12.38 shows the existing upstream crossings and the directions of flood flow, beginning from upstream and flowing downstream through the project area.

Figure 12.38 also presents the capacity of the worst case combinations of the existing culverts on Salboukh highway, which, assuming that these culverts had been designed properly in the past, give a first prediction of the flows of the two wadis.

Downstream and outside the project area, Wadi Salboukh ($A = 62$ square kilometers, $L = 14$ kilometers, $T_c = 7$ hours, and $i = 5.41$ millimeters per hour) proceeds in between farmland until it discharges into Wadi Abou-Hanifa, which is a major flood route in the region. Wadi Salboukh can be considered a minor tributary to Wadi Abou-Hanifa. The joint flow of Wadi Salboukh (i.e., wadi A) and Wadi Abou-Hanifa is also called Wadi Abou-Hanifa ($A = 858$ square kilometers, $L = 65$ kilometers, $T_c = 18$ hours, and $i = 2.18$ millimeters per hour), which runs south, crossing Al-Ammaria road via a reinforced concrete bridge (Figure 12.39). The bridge is about 4 meters high above the upstream invert level and has active 11 vents; each is approximately 4.5 meters wide by 3 meters high. By entering this information into the hydraulic design equations of a bridge, and assuming 0.5 meters free board, the capacity of the bridge can be estimated to be about 400 cubic meters per second. From the comparison of two wadis by the rational method, assuming the same runoff coefficient, the flow of Wadi Salboukh is found to be 71.3 cubic meters per second. This agrees with the flow calculated upstream (66.6 cubic meters per second; see Figure 12.38). The assumption of a similar runoff coefficient is actually a conservative assumption, because the Wadi Abou-Hanifa upper subcatchments, although bigger than the Salboukh area, are the severe flood-producing type, as opposed to Wadi Salboukh, which should have a lower runoff coefficient. However, the hydrologic analysis presented in Section 12.4.3 will determine the exact design discharges for all involved wadis.

12.4.2.2.3 Hydrologic Characteristics

Wadi Salboukh is on average a mildly sloped wadi with many widespread alluvial pond areas that reduce its flood severity due to percolation, in addition to the flood peak attenuation caused by the storage effect. However, on the other side, significant areas of its subcatchments have sandstone or limestone and rock cover that have low infiltration losses. This diversity of land cover requires special care when selecting the infiltration parameters of its subcatchments. Land use patterns in the upper catchment show an urbanization trend beginning, with urban growth expected to rapidly increase in the near future in about 30–50% of its upper catchment areas. The wadi routes inside the project area have wide and very deep sections that are surrounded on two sides by very steep ridges that rise up to a much higher land.

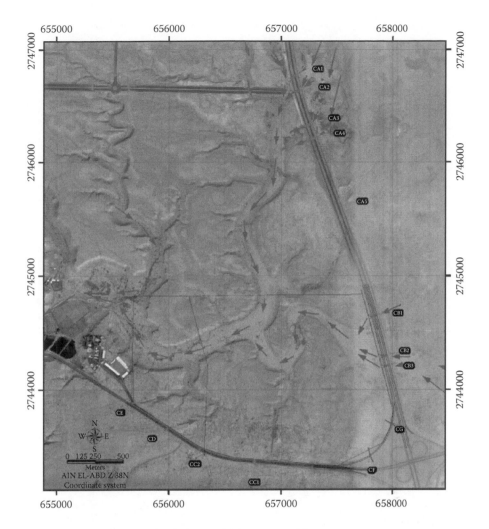

Maximum capacity of the existing culverts leading to Wadi A (worst case scenario)

Crossing	n	W (m)	D (m)	Road Level (m)	U.S. Invert L (m)	Estimated Capacity (m³/s)
CA1	2	3	1.5	679.1	677	15
CA2	2	3	1.5	678.8	677	10.2
CA3	2	2	1.5	679.5	676	19.8
CA4	2	2	1.5	679.8	676	21.6
					Total Flow =	66.6

Note:
All other crossing are minimum crossings with flows that range from 1 to 3 m³/s.

Culvert capacity is not only function in its sectional area but also function in the water head upstream (i.e., Road L. -U.S. Inv. L.)

Maximum capacity of the existing culverts leading to Wadi B (worst case scenario)

Crossing	n	W (m)	D (m)	Road Level (m)	U.S. Invert L (m)	Estimated Capacity (m³/s)
CB3	1	3	1.5	677.2	675	7.2
CB2	3	3	1.5	676.8	674.8	20.4
					Total Flow =	27.6

FIGURE 12.38 Existing crossings on Al-Ammaria and Salboukh roads (master plan is superimposed to clarify conflict areas).

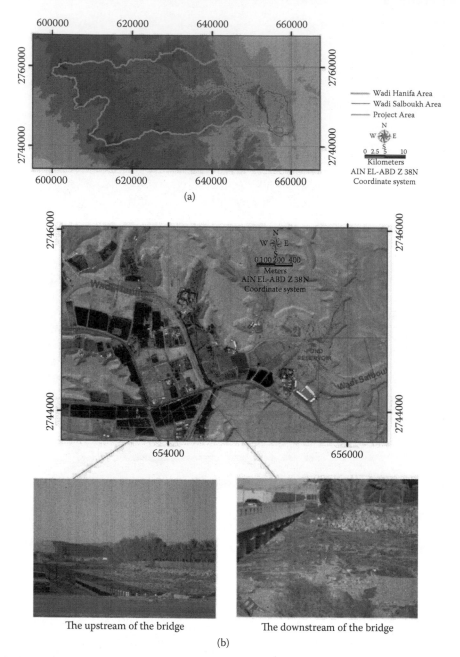

FIGURE 12.39 Project downstream conditions. (a) The catchment areas of Wadi Abou-Hanifa bridge versus Wadi Salboukh and (b) the location of Wadi Abou-Hanifa bridge crossing with Al-Ammaria road.

The wadies are at about 10–20 meters below the adjacent land level. The adjacent high land (allocated for residential activities) is composed of rock or stone formations. The steep ridges have evidence of cavities, and there is some possibility of existence of some faults or fractures, which should be brought to the attention of the geotechnical investigators.

Traces and watermarks inside the wadis indicate that the average water depth across the section during extreme floods is very shallow and rarely exceeds 0.3–0.5 meters. Only fine sediments are brought with the flood, and bigger boulders ($d50 > 5$ centimeters) are rarely brought with the flood because most of them are dumped upstream near the Salboukh highway.

12.4.3   HYDROLOGIC ANALYSIS

In this section, a quasidistributed rainfall–runoff model with spatially uniform rainfall input is constructed for Wadi Salboukh using the Hydrologic Modeling System (HEC-HMS) package (originally known as HEC-1 software). The HEC-1 flood hydrograph hydrological package (developed by the Center of the U.S. Army Corps of Engineers) has frequently been revised and updated since its inception in 1967. Different versions have been released and incorporated into other surface water hydrologic packages (e.g., graphical HEC-1, Haestad methods, and others). HEC-HMS is part of the new-generation software recently released by HEC that uses the Windows interface. The HEC-1 series implements the unit hydrograph concept (i.e., the watershed response concept), in which the watershed hydrologic system is broken down into components of determinable responses that can be estimated using hydrologic parameters. The hydrologic parameters are calculated from the topographical and land cover characteristics.

The objective in this section is to simulate the real rainfall–runoff transforming behavior of Wadi Salboukh (i.e., wadi A) to precisely determine the extreme design discharges at the key points that affect the project area.

12.4.3.1   Construction of the Hydrologic Model

When using the unit hydrograph principle, a large basin should be subdivided because of the size and complexity of the physical system, including the heterogeneity of the hydrologic parameters and the input rainfall, which may vary in larger systems. Basins with major tributaries and a diversity of topography and land cover must be broken into smaller components to fit the assumptions of spatial homogeneity of the hydrologic parameters, the standard watershed shape assumed in the unit hydrograph methods, and the spatial uniformity of rainfall over a single watershed.

In addition, a very important aspect is to simulate reality as much as possible in terms of the arrival time of each single unit of the system. The smaller the subdivision, the closer to a distributed system (reality) in which excess rain falls at different locations in the watershed and does not arrive at the watershed outlet at the same time. This difference in arrival time affects the peak discharge calculated via the distributed system as opposed to a lumped system in which all areas contribute at the same time at the watershed outlet, leading to a much higher calculated peak discharge. This again emphasizes the importance of subdividing the hydrologic system in order to simulate reality as much as possible.

12.4.3.2   System Hydrologic Parameters

A GIS interface was developed to extract the hydrologic parameters of the system components (i.e., the subwatersheds and the routing reaches). Figure 12.40b shows the hydrologic system chart.

The loss rate method used for estimating excess rainfall hyetographs was the SCS loss rate method and the transfer from excess rainfall to runoff hydrograph was done using the SCS unit hydrograph method. Accordingly, the hydrologic parameters involved for each subcatchment were area, curve number (CN), and lag time. For flood routing, the Muskingum method was implemented. Muskingum routing requires the specification of two parameters for each reach (time shift of the hydrograph and the storage coefficient). Land-use polygons were extracted from LandSat satellite images, and the corresponding CNs were assigned based on the standard tables developed by the U. S. Soil Conservation Service. The developed background CN grid of the study area is shown in Figure 12.40a. The subcatchments intersected on the CN grid, and an average CN for each subcatchment was summarized.

The unit hydrograph lag parameter was calculated using the Kirpich formula (which is recommended by the Saudi Code of Practice, the MOC) as 0.6 of the time of concentration calculated. The routing lags of the reaches were also estimated using the Kirpich formula. Figure 12.41 presents the extracted hydrologic parameters.

12.4.3.3   Rainfall Input

Most meteorological stations worldwide archive long-term daily records (records for shorter rainfall durations are usually not available). In such cases, the extreme daily depths (i.e., the 100-year daily depth) are determined using frequency analysis, and the calculated 24-hour extreme depth across the design storm duration is distributed using an internationally or regionally critical design storm distribution (e.g., SCS 24-hour type-I, I-A, II, and III design storm distributions). Note that type-II storms are the most critical type, generating higher flood peaks when compared to other distributions for the same rainfall depth and simulating severe thunderstorm-convective rainfall activities.

When shorter duration records are available, a design storm can be constructed by determining the extreme rainfall intensity for each duration and using this information to develop a design storm that implements the maximum intensity corresponding to each duration, such as in the alternating block method design storm (ABM). In addition, different SCS distributions are available for the 6 hour storm, 2 hour storm, and so on.

Because the available rainfall data for the area is in the form of IDF curves for the Riyadh station, which is based on only 37 years, two critical design storms (i.e., the SCS-II storm and the 6-hour ABM storm) will be considered and the highest corresponding hydrograph will be taken for design purposes. In addition, because the available data cover only 37 years, the 100-years values will be considered when designing the flood-accommodating structures. Figure 12.42 shows two design storms.

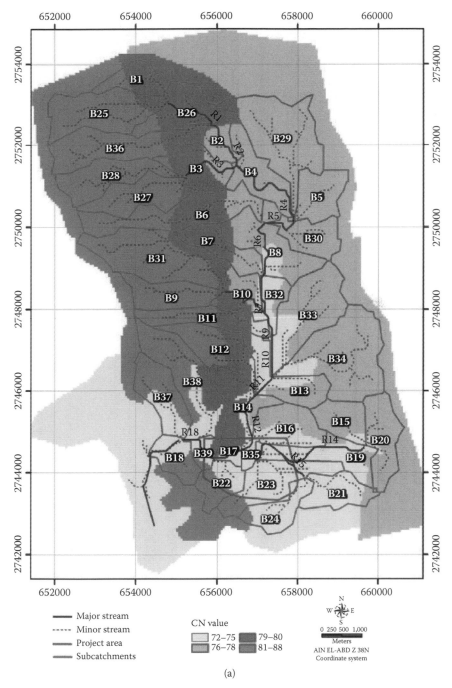

(a)

FIGURE 12.40 (a) CN values based on land cover and (b) Wadi Salboukh modeled hydrologic system.

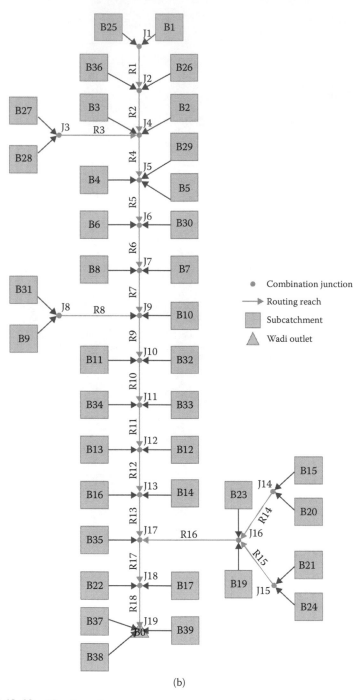

(b)

FIGURE 12.40   (*Continued*)

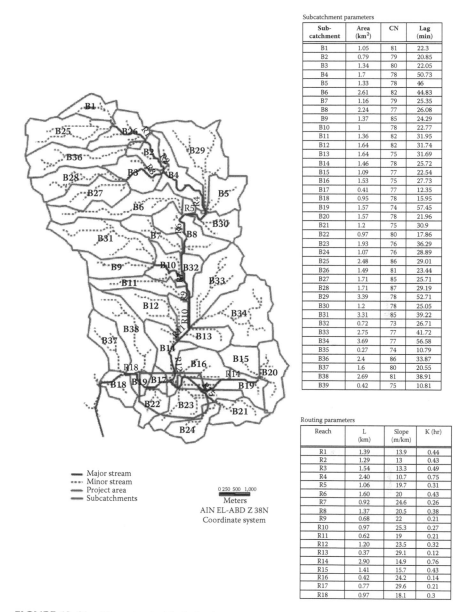

Subcatchment parameters

Sub-catchment	Area (km²)	CN	Lag (min)
B1	1.05	81	22.3
B2	0.79	79	20.85
B3	1.34	80	22.05
B4	1.7	78	50.73
B5	1.33	78	46
B6	2.61	82	44.83
B7	1.16	79	25.35
B8	2.24	77	26.08
B9	1.37	85	24.29
B10	1	78	22.77
B11	1.36	82	31.95
B12	1.64	82	31.74
B13	1.64	75	31.69
B14	1.46	78	25.72
B15	1.09	77	22.54
B16	1.53	75	27.73
B17	0.41	77	12.35
B18	0.95	78	15.95
B19	1.57	74	57.45
B20	1.57	78	21.96
B21	1.2	75	30.9
B22	0.97	80	17.86
B23	1.93	76	36.29
B24	1.07	76	28.89
B25	2.48	86	29.01
B26	1.49	81	23.44
B27	1.71	85	25.71
B28	1.71	87	29.19
B29	3.39	78	52.71
B30	1.2	78	25.05
B31	3.31	85	39.22
B32	0.72	73	26.71
B33	2.75	77	41.72
B34	3.69	77	56.58
B35	0.27	74	10.79
B36	2.4	86	33.87
B37	1.6	80	20.55
B38	2.69	81	38.91
B39	0.42	75	10.81

Major stream
Minor stream
Project area
Subcatchments

0 250 500 1,000
Meters
AIN EL-ABD Z 38N
Coordinate system

Routing parameters

Reach	L (km)	Slope (m/km)	K (hr)
R1	1.39	13.9	0.44
R2	1.29	13	0.43
R3	1.54	13.3	0.49
R4	2.40	10.7	0.75
R5	1.06	19.7	0.31
R6	1.60	20	0.43
R7	0.92	24.6	0.26
R8	1.37	20.5	0.38
R9	0.68	22	0.21
R10	0.97	25.3	0.27
R11	0.62	19	0.21
R12	1.20	23.5	0.32
R13	0.37	29.1	0.12
R14	2.90	14.9	0.76
R15	1.41	15.7	0.43
R16	0.42	24.2	0.14
R17	0.77	29.6	0.21
R18	0.97	18.1	0.3

FIGURE 12.41   The extracted hydrologic parameters.

12.4.3.4   Design Flood Hydrographs

The hydrologic model was run on the two rainfall inputs separately. The calculated hydrographs corresponding to the two storm scenarios are presented in Figures 12.43 and 12.44. The hydrograph of Wadi Salboukh (i.e., Wadi A) is characterized by a double-peak signature as the upper catchments and lower catchments do not contribute simultaneously at the wadi outlet. The lower catchments' peak arrives

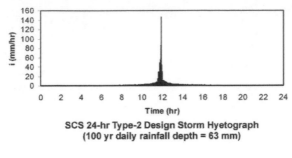

SCS 24-hr Type-2 Design Storm Hyetograph
(100 yr daily rainfall depth = 63 mm)

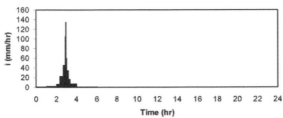

SCS 6-hr Design Storm Hyetograph
(100 yr 6 hr rainfall depth = 50 mm)

FIGURE 12.42   Two 100-year design storms.

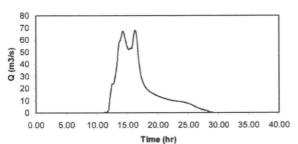

WADI-A Calculated Hydrograph Corresponding to
the 24 hr SCS-II Design Storm (Peak Discharge = 67.98 m3/s)

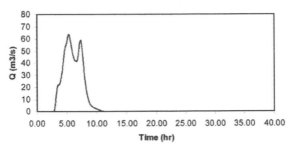

WADI-A Calculated Hydrograph Corresponding to
the 6 hr SCS Design Storm (Peak Discharge = 63 m3/s)

FIGURE 12.43   Estimated 100-year hydrographs for wadi A (Salboukh) using two design storms.

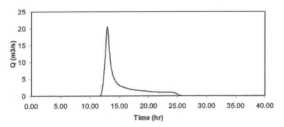

**WADI-B Calculated Hydrograph Corresponding to
the 24 hr SCS-II Design Storm (Peak Discharge = 20.47 m3/s)**

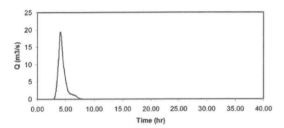

**WADI-B Calculated Hydrograph Corresponding to
the 6 hr SCS Design Storm (Peak Discharge = 19.33 m3/s)**

FIGURE 12.44 Estimated 100-year hydrographs for wadi B using two design storms.

at the outlet about 2 hours earlier than the arrival time of the upper catchments. As shown in the hydrographs, the calculated peak discharges amount to 68 cubic meters per second and 21 cubic meters per second for wadis A and B, respectively.

12.4.4 FLOOD CONTROL WORKS

12.4.4.1 Design Criteria

- All flood protection works will accommodate the 100-years flood at any possible existing or future flood route and safely discharge all water flowing downstream in the project area.
- Flood protection works will comply with the master plan and will not affect the designated land use and/or activities.
- Buried or closed culverts are the preferred option to accommodate flood waters within the project area.
- Landscaping will be considered a main constraint and the flood protection works must try to achieve beautiful landscape scenes.
- The works will link properly with the flood protection works currently being constructed upstream, where Dar Al-Arkan is enabling its site.
- Cost and earthworks will be minimized as much as possible by providing an optimum, functional, and safe design.
- Water velocity will not exceed the scouring or depositing limits, depending on the type of bedding.

12.4.4.2 Design Concept

Seven wadis run through the project area (A, B, C, D, E, F and G), two of which are main wadis (A and B). The remaining wadis are small with fewer catchment areas and peak discharges. In general, the width and depth of the wadis complies with what is hydraulically required to accommodate a 100-year flood or even a 1000-year flood. This large formation of wadis is attributed to the ancient historical rainy era (million of years ago) in which terrestrial rain occurred, in addition to other geological reasons. For example, the width of the existing Wadi Salboukh (i.e., wadi A) inside the project area is about 200 meters on average. The wadi is bounded from the sides by high and very steep rocky ridges that rise to about 10–20 meters above the wadi bed level. Land inside the wadi is almost flat across the cross section with a natural downstream longitudinal slope. The hydraulic analysis in Section 12.4.4.3 reveals that only a 20–30-meter-wide channel is required for flood accommodation.

Accordingly, the main concept design of the flood protection works is as follows:

- The main wadis A and B must be kept and their existing bed levels raised by relatively small amount (2.5–3 meters). The flood protection works will be placed below the raised wadi level. This will ensure recreational activities within the wadi area are protected against flooding while maintaining the same wadi formation and its environmental existence. This will also greatly facilitate storm drainage inside the project area. The cross sections shown in Figure 12.45 present this concept design.
- The routes of the other small wadis inside the project area will be filled, and the waters will be connected or diverted to the nearest connection with A and/or B. The diversion may be done along the southeastern border road (i.e., in the strip between Al-Ammaria road and road (RD) B/S in the south and between Salboukh highway and RD B/E in the east) by the proper profiling of the bounding roads B/S and B/E (see Figure 12.46).

12.4.4.3 Minimum Hydraulic Dimensions

The proposed flood accommodating works may include either open channels or closed culverts. Accordingly, this section presents the minimum "hydraulic" dimensions required to accommodate the 100-year flow rates. Both alternatives are considered here (i.e., open channels and closed culverts) regardless of the recommended alternative that will be proposed later in this section. Note that the minimum dimensions reported here are considered the lowest limit for selecting the optimum dimensions. The finally selected optimum dimensions (presented in the Section 12.4.4.4) may be bigger than the minimum dimensions reported here, depending on other constraints such as landscaping and master plan requirements.

The FlowMaster software is used to design the open channels, whereas the CulvertMaster/Stormcad software (culvert equations including secondary losses equations) is used to design the closed culverts. For more details on the software, please refer to the technical or user manuals.

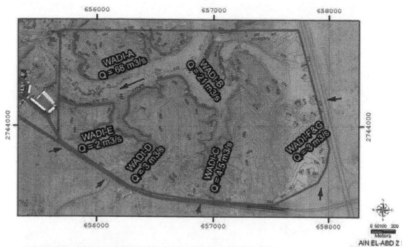

MINIMUM OPEN CHANNEL HYDRAULIC SECTIONS
(Manning's Roughness = 0.05)

ID	b (m)	H (m)	Z	Long. Slope (m/m)	Water Depth (m)	Velocity (m/s)	Froud No. -
Salboukh	25	2.5	2	0.001786	1.97	1.19	0.29
Trib. A	8	2	2	0.001893	1.55	1.19	0.35
Trib. B3	3	1.5	2	0.003150	1.02	0.87	0.33
Trib. B2	1	1	2	0.020000	0.75	1.61	0.88
Trib. B1	1	1	2	0.020000	0.75	1.61	0.88
Joint B1&B2	2	1.5	2	0.006900	1.03	1.22	0.47
Trib. C	2	1.5	2	0.006000	0.82	1.01	0.43

FILLED EMBANKMENT LEVEL

FILL   H   EXSISTING WADI LEVEL   z 1   FILL

b

MINIMUM CULVERT SECTIONS (Manning's Roughness = 0.013)

WADI	n × W × D (m)	H (m)	Total Width (m)	Min Long. Slope (m/m)	U.S/D.S.Depths (m)	D.S Exit V. (m/s)	Flow Regime
A	6 × 3 × 1.75	3	20	0.001786	1.96 / 1.13	3.33	Subcritical
B	3 × 3 × 1.5	2.5	10	0.001893	1.43 / 0.81	2.81	Subcritical
C	1 × 1.5 × 1.5	2.5	2	0.003150	1.68 / 0.97	3.08	Supercritical
D	Min 1 × 1 × 1	2.25	1.5				
E	Min 1 × 1 × 1	2.25	1.5				
D&E	1 × 1.5 × 1.5	2.5	2	0.008000	1.82 / 0.88	3.77	Supercritical
F&G	1 × 1.5 × 1	1.5	2	0.006000	1.30 / 0.67	2.98	Supercritical

FILLED EMBANKMENT LEVEL

FILL   D   H   EXSISTING WADI LEVEL   FILL

W

FIGURE 12.45   Minimum hydraulic requirements.

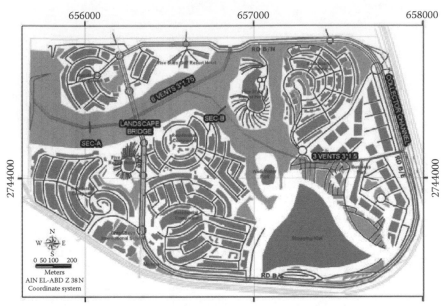

Alternative-2 for main wadies A and B (master plan background)

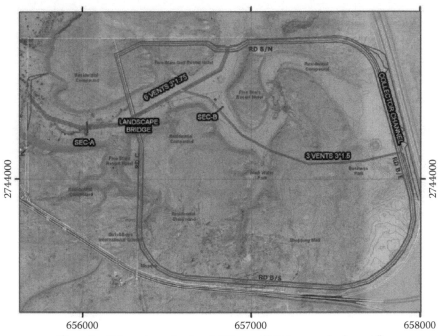

Alternative-2 for main wadies A and B (Ikonos 2006 & survey background)

(a)

FIGURE 12.46 (a) Closed culverts of main wadis A and B (layout) and (b) closed culverts of main wadis A and B (sections).

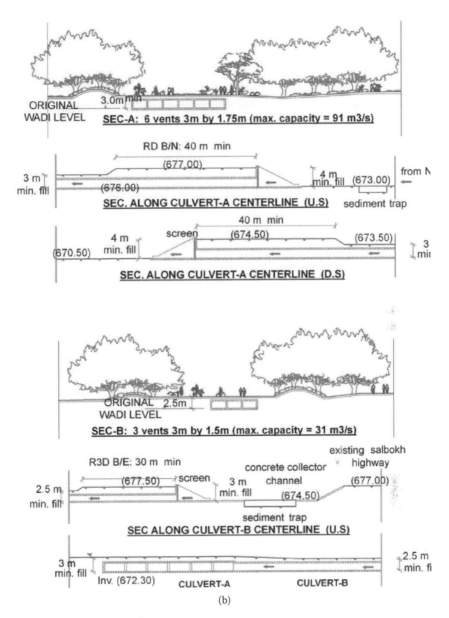

FIGURE 12.46 *(Continued)*

12.4.4.4 Proposed Alternatives for Main Wadis A and B

Wadis A and B highly affect the project area because they run through its heart. Wadi A flows from a level of 673 upstream just at the northern boundary to a level of 670.5 downstream at the western boundary via a 1400 meter reach (average longitudinal slope = 0.00178 meters per meter). Wadi B runs from a level of 674.5 at the eastern boundary on Al-Ammaria road to a level of 672.35 at its confluence with wadi A (about a 1400 meter shortcut distance).

Two alternatives are available for flood accommodation: (1) open channels or (2) buried culverts. Buried culverts are the preferred alternative because the quality of flood waters may suffer from a level of pollution due to the urbanization upstream. The polluted water may then have negative impacts on activities within the project area. The following points summarize the concept design of the buried culverts:

- Fully closed (buried) culverts are used to collect flood waters from the upstream entrances at RD B/N (wadi A) and at RD B/E (wadi B) and discharge the water downstream from the project area. The culvert will be placed almost on the existing low wadi levels and fill will be required beside and above the culverts. Wadi A's culvert has 6 vents (each 3 meters wide by 1.75 meters high; length is 1390 meters). Wadi B's culvert has 3 vents (each 3 meters wide by 1.5 meters high; length is 1440 meters).
- All areas inside the wadis will be raised (i.e., filled) by about 2.5–3 meters in order to protect activities within the wadis from flooding. The depth of the fill can be slightly decreased in the project area as long as sufficient fill is used above the upstream and downstream ends (i.e., wing walls) to prevent water from entering the raised wadis. Figure 12.46a and b show the closed culverts. Note the upstream and downstream longitudinal sections along the wadi centerline in Figure 12.46b, which present the idea of a variable filling depth along the centerline.
- The strip between RD B/E and the Salboukh highway will be used as a collector channel (concrete lined) to direct flows coming from the existing culverts beneath the Salboukh highway to the entrance of the wadi B culvert.
- The alignment of the centerline of wadi B culvert can be modified by the planner per any future master plan modifications.
- The depth of water in the culverts in the 100-year flood is less than the culverts' heights (i.e., the culverts are "running free"). This guards against any spills from within the access holes that will be provided along the culverts (150–200 meter spacings).
- The maximum capacities of the culverts (i.e., when the upstream embankment roads are about to be overtopped by water) are 91 and 31 cubic meters per second for culverts A and B, respectively, while the 100-year design discharges are 68 and 21 cubic meters per second. This provides additional safety against any weather changes or extreme longer-cycle events in the future.

12.4.4.5 Alternatives for Small Wadis

All small wadis that run through the project area are characterized by small discharges (i.e., less than 5 cubic meters per second in the 100-year design) although their existing wadis show distinctive wide or deep formations. This is supported by the small sizes of the existing culverts beneath Al-Ammaria road (wadis C, D, and E). These large wadis' formations are attributed to the rainy era millions of years ago, in addition to the geological compositions. Accordingly, parts of these low wadi areas should be filled and used for residential development because they occupy a considerable part of the land. The current master plan already allocates residential activities in these areas, which agrees well with this study's recommendation. Hence, two alternatives are available here for discharging the small wadies.

12.4.4.5.1 Alternative 1 for Small Wadis: Culverts under the Wadis

- The first alternative is to use underground culverts that receive water from the existing culverts beneath Al-Ammaria road (wadis E, D, and C) and to continue within the residential areas following the interior streets, in other words, to extend the existing culverts under Al-Ammaria road via extensions below the interior streets (Figure 12.47).

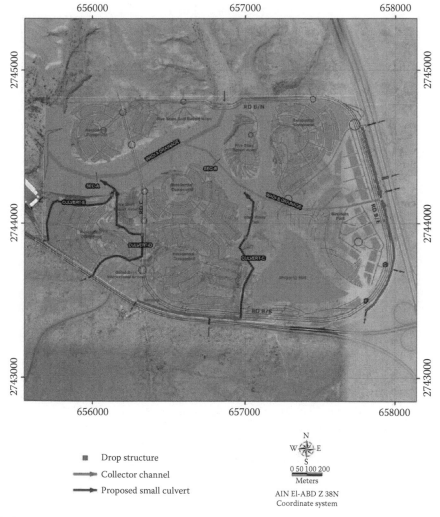

■   Drop structure

⟶   Collector channel

⟶   Proposed small culvert

N
W ⊕ E
S

0 50 100 200
Meters

AIN El-ABD Z 38N
Coordinate system

FIGURE 12.47 Alternative 1 for small wadis: small culverts through the project area.

- The downstream wing walls of the existing culverts (D and E) beneath Al-Ammaria road must be demolished in order to connect the old and new culverts, because there is no space to use a collecting channel adjacent to Al-Ammaria road. For culvert C, a space is available to use a collecting channel adjacent to Al-Ammaria, leading to the culvert C entrance.
- Drop structures and/or stilling basins are essential for when the culverts reach the steep ridges, so they can drop flood waters about 7–12 meters to safely reach raised levels of wadis A and B. These drop structures will be difficult to construct in the steep rocky configurations at the wadi ridges, but it is possible.
- The advantage of this alternative is that the culverts can be used for storm drainage inside the residential compounds.
- The existing culverts F and G beneath the Salboukh highway ramp will be directed to the collecting channel leading to the culvert B entrance.

12.4.4.5.2 Alternative 2 for Small Wadis: Diversion along a Bounding Collector Channel

This alternative diverts all water coming from beneath Al-Ammaria road along the bounding roads (RD B/S and RD B/E). This can be achieved by constructing a collector channel in the strip between these bounding roads and the existing highways (i.e., Al-Ammaria and Salboukh highway). The first reach of the channel is in the form of an underground culvert, as well as in the crossings with entrance roads (Figure 12.48).

- The natural land slopes along the direction of the channel, which facilitates construction. The channel ultimately delivers water to main culvert B.
- No drop structures are needed.
- The disadvantage of this solution is that the channel will have to be replaced by culverts in the future when more development occurs in the neighborhood.
- The existing culverts F and G beneath the Salboukh highway ramp will be directed to the collecting channel leading to the culvert B entrance.

12.4.4.6 Recommended Flood Protection Scheme

Based on the analysis performed in this study, the recommended flood protection scheme for this project is as follows (Figure 12.49):

- For main wadis A and B, the buried culverts should be chosen as an alternative to avoid the negative impacts of unclean flood waters.
- For the small wadis, alternative 1 is recommended although it is more costly than alternative 2. Alternative 1 will facilitate storm drainage inside the development and avoid separating the whole project area from the neighborhood, which will create a problem in the future when urban growth occurs.

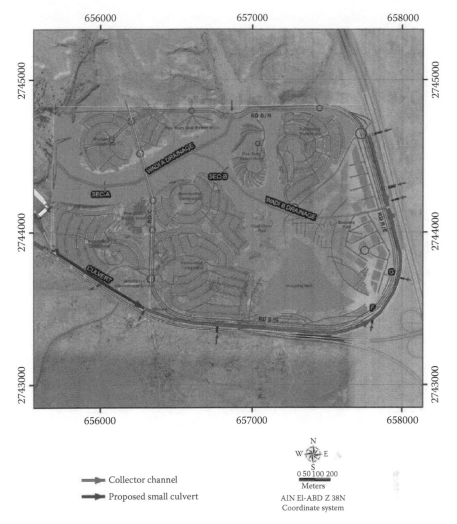

FIGURE 12.48 Alternative 2 for small wadis: diversions along a bounding collecting channel adjacent to Al-Ammaria.

12.4.4.7 Fill Methodology

To fill the wadis to the proposed raised levels (2.5–3 meters), approximately 2 million cubic meters of fill material will be needed to fill all low areas. The surface fill material must be cultivatable since landscaping will take place in this soil. The cultivatable soil recommended is sand, which is not available on site and must be borrowed from outside the project area. To reduce the cost, the fill depth can be broken into two layers. The fill material for the lower layer can be taken from the project area, whereas for the upper layer, it must be borrowed from outside.

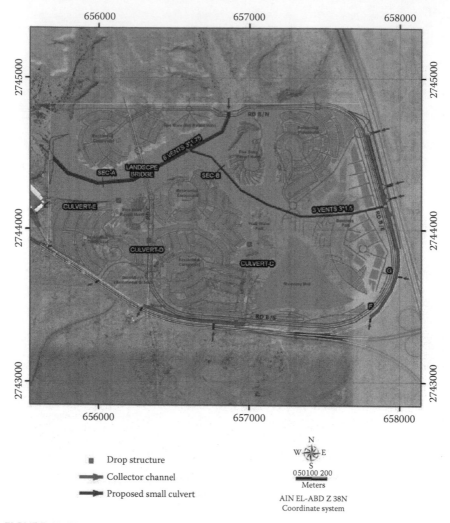

FIGURE 12.49 The recommended flood protection scheme.

12.4.4.8 Comments on the Preliminary Proposed Master Plan

- The alignment of the centerline of the wadi B drainage culvert can be modified by the planner according to any modifications in the master plan in the future. The current version of the master plan needs a slight adjustment to accommodate the wadi B culvert.
- The master plan needs to provide more access roads to the low wadi areas. These roads will not only benefit visitors, but they will be of great importance for maintenance activities for the wadis in addition to the benefits during construction.
- A bridge is needed on wadi A to pass traffic on RD C across the wadi. The design of this bridge is mainly governed by landscape and that a long span

is needed because it interferes with the wadi visual continuity as well as the landscape scenes. Accordingly, they recommend implementing a "piece-of-art" architecture for this bridge (cable-suspended or at least arch design) to provide the required artistic touch for the whole wadi continuity and its image.

12.4.4.9 Effect on the Downstream

Wadi A is currently blocked by some small-dam earthworks just at the downstream western boundaries. The earthworks have been completed by the downstream farms to prevent floods from flashing the farms. The blocking confines flood waters into the wide pond area just at the western border of the project area. The pond area acts as a reservoir. Available information for the previous decades indicates that flood waters periodically cut through these small dams almost every 10 years and flash flooded the downstream farms.

Because the wadis will be raised inside the project area to 2.5–3 meter levels that will be higher than the small earth dams downstream, the possibility of cutting through the small earth dams increases. In addition to the small reduction in the pond areas, an effect of the flood protection works of the project is to remove attenuation by eliminating the storage reservoirs, leading to a more concentrated, higher peak flow downstream. However, this impact is relieved by the existence of the remaining pond area, created by the small earth dams just downstream the project area, which is capable of absorbing a considerable amount of flood water. A permanent solution to this problem is to ensure a clear route for flood waters downstream until the outlet on Wadi Abou-Hanifa. Hence, the downstream neighbors should be alerted of the flash flood problem (although they may be already aware of it) to urge them to take proper precautions instead of using the temporary solution of small dams.

REFERENCES

Gad, M., and M. Abdel-Hamid. 2009. Design of Salboukh flood control system in Riyadh. In *Proceedings of the Sixth International Conference on Environmental Hydrology*, M. M. Soliman, ed. Cairo: ASCE.

BIBLIOGRAPHY

Hydrologic Engineering Center. 2001. HEC-HMS Hydrologic Modeling System User's Manual. Technical report. U.S. Army Corps of Engineers, Davis, CA.
Soil Conservation Service. 1993. *SCS National Engineering Handbook,* Section 4: Hydrology. Washington, DC: U.S. Department of Agriculture.
U.S. Army Corps of Engineers. 1982. Simulation of Flood Control and Conservation Systems, HEC-5. Hydrologic Engineering Center, Davis, CA.

Appendix A: Conversion Tables

Basic Units	
SI	**Imperial**

Distance

    1 meter (1 m) = 10 decimeters (10 dm)   12 inch (in.) = 1 feet (1 ft)

                   = 100 centimeters (100 cm)   3 ft = 1 yard (1 yd)

                   = 1000 millimeters (1000 mm)   5280 ft = 1 mile

                                                  1760 yd = 1 mile

    1 decameter (1 dam) = 10 m

    1 hectometer (1 hm) = 100 m

    1 kilometer (1 km) = 1000 m

Conversions:

    1 in. = 25.4 mm

    1 ft = 30.48 cm

    1 mile = 1.61 km

    1 yd = 0.914 m

    1 m = 3.28 ft

Area

    $1 \text{ m}^2 = 10,000 \text{ cm}^2$ $1 \text{ ft}^2 = 144 \text{ in}^2$

            $= 1,000,000 \text{ mm}^2$ $1 \text{ yd}^2 = 9 \text{ ft}^2$

                                          1 square mile = 640 acre = 1 section

    $1 \text{ hm}^2 = 10,000 \text{ m}^2$

              = 1 hectare (1 ha)

    $1 \text{ km}^2 = 1,000,000 \text{ m}^2$

Conversions:

    $1 \text{ in.}^2 = 6.45 \text{ cm}^2 = 645 \text{ mm}^2$

    $1 \text{ m}^2 = 10.8 \text{ ft}^2$

    1 acre = 0.405 ha

    $1 \text{ square mile} = 2.59 \text{ km}^2$

Volume

    $1 \text{ m}^3 = 1,000,000 \text{ cm}^3$ $1 \text{ ft}^3 = 1728 \text{ in}^3$

            $= 1 \times 10^9 \text{ mm}^3$ $1 \text{ yd}^3 = 27 \text{ ft}^3$

    $1 \text{ dm}^3 = 1$ liter 1(liquid) U.S. gallon (gal) = 231 in^3

    $1 \text{ liter} = 1000 \text{ cm}^3 = 4$ (liquid) quarts

    $1 \text{ mL} = 1 \text{ cm}^3$ 1 U.S. barrel (bbl) = 42 U.S. gal

    $1 \text{ m}^3 = 1000$ liters   1 imperial gallon = 1.2 U.S. gal

Conversions:

    $1 \text{ in.}^3 = 16.4 \text{ cm}^3$

    $1 \text{ m}^3 = 35.3 \text{ ft}^3$

(Continued)

(Continued)

Basic Units	
SI	**Imperial**

1 liter = 61 in³
1 U.S. gal = 3.78 liters
1 U.S. bbl = 159 liters
1 liter/s = 15.9 U.S. gal/min

Mass and weight

   1 kilogram (1 kg) = 1000 grams   2000 lb = 1 ton (short)

   1000 kg = 1 ton   1 long ton = 2240 lbs

Conversions:

   1 kg (on Earth) results in a weight of 2.2 lb

Density

   Mass density = mass/volume   Weight density = weight/volume

$$\rho = \frac{m}{V}\left(\frac{\text{kg}}{\text{m}^3}\right) \qquad\qquad \rho = \frac{W}{V}\left(\frac{\text{lb}}{\text{ft}^3}\right) = \text{kg wt/m}^3$$

Conversions:

   On Earth, a mass density of 1 kg/m³ results in a
      weight density of 0.0623 lb/ft³

Relative density

In SI, relative density is a comparison of mass density to a standard. For solids and liquids, the standard is fresh water.	In Imperial, the corresponding quantity is specific gravity; for solids and liquids, it is a comparison of weight density to that of water.

Conversions:

   In both systems, the same numbers hold for
      relative density as for specific gravity because
      these are equivalent ratios.

Relative Density (Specific Gravity) of Various Substances

Water (fresh)	1.00
Water (sea average)	1.03
Aluminum	2.56
Antimony	6.70
Bismuth	9.80
Brass	8.40
Brick	2.1
Calcium	1.58
Carbon (diamond)	3.4
Carbon (graphite)	2.3

Relative Density (Specific Gravity) of Various Substances (*Continued*)

Carbon (charcoal)	1.8
Chromium	6.5
Clay	1.9
Coal	1.36–1.4
Cobalt	8.6
Copper	8.77
Cork	0.24
Glass (crown)	2.5
Glass (flint)	3.5
Gold	19.3
Iron (cast)	7.21
Iron (wrought)	7.78
Lead	11.4
Magnesium	1.74
Manganese	8.0
Mercury	13.6
Mica	2.9
Nickel	8.6
Oil (linseed)	0.94
Oil (olive)	0.92
Oil (petroleum)	0.76–0.86
Oil (turpentine)	0.87
Paraffin	0.86
Platinum	21.5
Sand (dry)	1.42
Silicon	2.6
Silver	10.57
Slate	2.1–2.8
Sodium	0.97
Steel (mild)	7.87
Sulfur	2.07
Tin	7.3
Tungsten	19.1
Wood (ash)	0.75
Wood (beech)	0.7–0.8
Wood (ebony)	1.1–1.2
Wood (elm)	0.66
Wood (lignum-vitae)	1.3
Wood (oak)	0.7–1.0
Wood (pine)	0.56
Wood (teak)	0.8
Zinc	7.0

FORCE

Force is a vector quantity, a push or pull that changes the shape and/or motion of an object.
In SI, the unit of force is newton, N, defined as a kg·m/s².
In Imperial, the unit of force is pound, lb.
9.81 N = 2.2 lb = kg wt (kilogram weight)

WEIGHT

Weight is the gravitational force of attraction between a mass, m, and the mass of Earth.
In SI, weight can be calculated by
Weight = F = mg, where g = 9.81 m/s²
In Imperial, the mass of an object (rarely used), in slugs, can be calculated from the known weight in pounds:
Mass = Weight/g, where g = 32.2 ft/s²

PRESSURE

Pressure, a vector quantity, is the force per unit area.
In SI, the basic units of pressure are pascals, Pa, and kPa.
1 Pa = 1 N/m²
In Imperial, the basic unit is the pound per square inch, psi.

ATMOSPHERIC PRESSURE

At sea level, atmospheric pressure equals 101.3 kPa or 14.7 psi = 10.13 meters of water.

PRESSURE CONVERSIONS

1 psi = 6.895 kPa
Pressure may be expressed in standard units or in units of static fluid head, in both the SI and Imperial systems.
Common equivalencies are:
1 kPa = 0.294 in. mercury = 7.5 mm mercury
1 kPa = 4.02 in. water = 102 mm water
1 psi = 2.03 in. mercury = 51.7 mm mercury
1 psi = 27.7 in. water = 703 mm water
1 m H_2O = 9.81 kPa
Other pressure unit conversions:
1 bar = 14.5 psi = 100 kPa
1 kg·wt/cm² = 98.1 kPa = 14.2 psi = 0.981 bar
1 atmosphere (atm) = 101.3 kPa = 14.7 psi = 10.13 m of water column

Appendix B: Glossary

Acid precipitation: Any atmospheric precipitation that has an acid reaction due to the absorption of acid-producing substances such as sulfur dioxide.

Aggradation: The general building up of the land by deposition processes.

Air instability: A state existing in a body of air, marked by a strong vertical temperature decrease and high moisture content. Unstable air tends to rise.

Albedo: The amount of light reflected by a given surface compared to the amount received by that surface.

Alluvium: Material deposited by running water (gravel, sand, silt, or clay).

Anaerobic condition: Absence of air or free oxygen.

Andesitic basalt: A fine-grained extrusive igneous rock composed of plagioclase feldspars and ferromagnesian silicates.

Aquiclude: A formation that, although porous and capable of absorbing water, does not permit its movement at rates sufficient to furnish an appreciable supply for a well or spring.

Aquifer: A porous, permeable, water-bearing geologic body of rock. Generally, restricted to materials capable of yielding an appreciable amount of water.

Aquifuge: A formation that has no interconnected openings and hence cannot absorb or transmit water.

Aquitard: A confining bed that retards but does not prevent the flow of water to or from an adjacent aquifer; a leaky confining bed. It does not readily yield water to wells or springs, but may serve as a storage unit for groundwater.

Artesian: Groundwater confined under sufficient hydrostatic pressure to rise above the upper surface of the aquifer.

Artesian aquifer: An aquifer in which the water is confined by an overlying, relatively impermeable bed and is, therefore, under sufficient pressure to raise the water in wells above the top of the confined aquifer.

Artesian head: The level to which water from a well will rise when confined in a standing pipe.

Artesian well: A well in which water from a confined aquifer rises above the top of the aquifer. Some wells may flow without the aid of pumping.

Avalanche: A large mass of snow, rock debris, soil, or ice that detaches and slides down a mountain slope.

Barometer: An instrument that measures atmospheric pressure. The first liquid barometer was designed by Torricelli in 1644.

BOD (biochemical oxygen demand): The amount of oxygen used in meeting the metabolic needs of aquatic aerobic microorganisms. A high BOD correlates with accelerated eutrophication.

Brackish water: Water with a salinity between that of freshwater and seawater.

Brine: The concentrated salt solution remaining after removal of distilled product; also, concentrated brackish saline or seawater containing more than 100,000 mg/L of total dissolved solids.

Canopy fire: A forest fire that involves the crowns of trees, also called a crown fire.

Carbon dioxide (CO_2): A gaseous product of a combustion about 1.5 times as heavy as air. A rise in CO_2 in the atmosphere increases the greenhouse effect.

Carbon monoxide (CO): A product of incomplete combustion. CO is colorless and has no odor. It combines with hemoglobin in the blood, leading to suffocation caused by oxygen deficiency.

Centipoise: A unit of viscosity based on the standard of water at 20°C (which has a viscosity of 1.005 centipoises).

Chemical treatment: Any process involving the addition of chemicals to obtain a desired result.

Climate: The statistical sum of meteorological conditions (averages and extremes) for a given point or area over a long period of time.

Cloud seeding: The artificial introduction of condensation nuclei (dry ice or silver iodide) into clouds to cause precipitation.

Cold front: The boundary on the earth's surface, or aloft, along which warm air is displaced by cold air.

Colloidal dispersion: The process of extremely small particles (colloids) being dispersed and suspended in a medium of liquids or gases.

Compressibility (β): The reciprocal of bulk modules of elasticity.

Concentration: The amount of a given substance dissolved in a unit volume of solution; or, the process of increasing the dissolved solids per unit volume of solution, usually by evaporating the liquid.

Concentration tank: A settling tank of a relatively short detention period in which sludge is concentrated by sedimentation of floatation before treatment, dewatering, or disposal.

Confined aquifer: An aquifer bounded above and below by impermeable beds or lower-permeability beds than that of the aquifer itself; an aquifer containing confined groundwater.

Confined groundwater: A body of groundwater overlain by a material that is sufficiently impervious to sever a free hydraulic connection with overlying groundwater except at the intake. Confined water moves in conduits under the pressure caused by the difference in head between intake and discharge areas of the confined water body.

Confining bed: A body of impermeable or distinctly less permeable material stratigraphically adjacent to one or more aquifers. Comparable terms include aquitard, aquifuge, aquiclude.

Convection: Mass motion within gases and liquids caused by differences in density brought about by cooling or heating.

Corrasion: Wearing away of the earth's surface, forming sinkholes and caves, and their widening due to running water.

Corrosion: The gradual deterioration or destruction of a substance or material by chemical action, frequently induced by electrochemical processes. The action proceeds inward from the surface.

Creep: A slow movement of unconsolidated surface materials (soil, rock fragments) under the influence of water, strong wind, or gravity.

Darcy: A standard unit of permeability equivalent to the passage of one cubic centimeter of fluid of one centipoise viscosity flowing in one second under a pressure differential of one atmosphere through a porous medium having an area of cross section of one square centimeter and a length of one centimeter. A millidarcy is one-thousandth of a darcy.

Darcy's law: A formula for the flow of fluids derived on the assumption that the flow is laminar and inertia can be neglected. The numerical formulation of this law is generally used in studies of gas, oil, and water production from underground formations.

Debris slide: A sudden downslope movement of unconsolidated earth materials or mine waste, particularly once it becomes water saturated.

Deep-well injection: A technique for disposal of liquid waste materials by pressurized infusion into porous bedrock formations or cavities.

Degradation: The general lowering of the land by erosional processes.

Desalination: Any process capable of converting saline water to potable water.

Desertification: The creation of desert-like conditions, or the expansion of deserts as a result of human actions, including overgrazing, excessive extraction of water, and deforestation.

Desertization: A relatively new term that denotes the natural growth of deserts in response to climatic change.

Deserts: Permanently arid regions of the world where annual evaporation by far exceeds annual precipitation. Deserts cover about 16% of the earth.

Detrital: A term that relates to deposits formed of minerals and rock fragments and transported to the place of deposition.

Discharge: The volume of water passing a given point within a given period of time.

Doppler radar: An instrument that emits a radar frequency that changes as the wave is bounced back from a moving object. The frequency lengthens when the distance between transmitter and object increases, and it shortens as the distance decreases.

Downdrafts: Downward and sometimes violent cold air currents frequently associated with cumulonimbus clouds and thunderstorms.

Downwind effect: Severe turbulence that develops on the downwind (leeward) side of large buildings and mountains. This turbulence is also called the "snow-fence" effect; it can be dangerous to aircraft.

Drainage basin: The area that is drained by a river and its tributaries.

Drilling fluid: A heavy suspension, usually in water but sometimes in oil, used in rotary drilling, consisting of various substances in a finely divided state (commonly bentonitic clays and chemical additives such as barite), introduced continuously down the drill pipe under hydrostatic pressure, out through openings in the drill bit, and back up in the annular space between the pipe and the borehole walls and to a surface pit where cuttings are removed. The fluid is then reintroduced into the pipe. Drilling fluid is used to lubricate and cool the bit, to carry up the cuttings from the bottom, and to prevent sloughing and cave-ins by plastering and consolidating the walls with a clay lining, thereby making casing unnecessary during drilling and also offsetting pressures of fluid and gas that may exist in the subsurface. A synonym is "drilling mud."

Drought: An extended period of below-normal precipitation especially in regions of sparse precipitation. Prolonged droughts can lead to crop failures, famines, and sharply declining water resources.

Dust storm: A severe weather system, usually in dry area, characterized by high winds and dust-laden air. Major dust storms were observed during the 1930s in the Dust Bowl region of the United States.

Earthquake: Sudden movements and tremors within the earth's crust caused by fault slippage or subsurface volcanic activity.

Ecosystem: A functional system based on the interaction between all living organisms and the physical components of a given area.

Effective porosity: The measure of the total volume of the interconnected void space of a rock, soil, or other substance. Effective porosity is usually expressed as a percentage of the bulk volume of material occupied by the interconnected void space.

Effective stress: The average normal force per unit area transmitted directly from particle to particle of a soil or rock mass. It is the stress that is effective in mobilizing internal friction. In a saturated soil in equilibrium, the effective stress is the difference between the total stress and the neutral stress of the water in the voids; it attains a maximum value at complete consolidation of the soil.

Elastic rebound: A concept that implies that rocks, after breaking in response to prolonged strain, rebound back to their previous position or one similar to it. This sudden breaking and rebound may cause earthquakes.

Environmentalism: A concept, also called environmental determinism, that proposes that the total environment is the most influential control factor in the development of individuals or cultures.

Epicenter: A point on the earth's surface located directly above the focus on an earthquake.

Evapotranspiration: A term that describes the sum of evaporation from wetted surfaces and transpiration by vegetation.

Extension fault: A branch rupture extending from a major fault line.

Eye (of a hurricane): The mostly cloudless and calm center area of a hurricane, surrounded by near-vertical cloud walls.

Facies: A term used to refer to a distinguished part or parts of a single geologic entity, differing from other parts in some general aspect, for example, any two or more significantly different parts of a recognized body of rock or stratigraphic composition. The term implies physical closeness and genetic relation or connection between the parts.

Facies change: A lateral or vertical variation in the lithologic or paleontologic characteristics of contemporaneous sedimentary deposits. It is caused by, or reflects, a change in the depositional environment. "Facies evolution" is a comparable term.

Facies map: A broad term for a stratigraphic map showing the gross areal variation or distribution (in total or relative content) of observable attributes or aspects of different rock types occurring within a designated stratigraphic unit, without regard to the position or thickness of individual beds in the vertical succession, specifically, a HT/HP facies map. Conventional facies maps are prepared by drawing lines of equal magnitude through a field of numbers representing the observed values of the measured rock attributes. A comparable term is "vertical-variability map."

Fault: A surface or zone of rock fracture along which there has been displacement from a few centimeters to a few kilometers.

Filtrate: The liquid that has passed through a filter.

Filtration: The process of passing a liquid through a filtering medium (which may consist of granular material such as sand, magnetite, or diatomaceous earth, or may be finely woven cloth, unglazed porcelain, or specially prepared paper) for the removal of suspended or colloidal matter.

Firestorm: A violent and nearly stationary mass fire that develops its own inblowing wind system. It develops mostly in the absence of preexisting ground wind.

Fishery: The commercial extraction of fish in a given region.

Fission: The splitting of an atom into nuclei of lighter atoms through bombardment with neutrons. Enormous amounts of energy are released in this process, which is used in the development of nuclear power and weapons.

Fissure eruption: A type of volcanic eruption that takes place along a ground fracture instead of through a crater.

Flank eruption: A type of volcanic eruption that takes place on the side of a volcano instead of from the crater. This typically occurs when the crater is blocked by previous lava eruptions.

Flash flood: A local, very sudden flood that typically occurs in dry river beds and narrow canyons as a result of heavy precipitation generated by mountain thunderstorms.

Floe: Small masses, commonly gelatinous, formed in a liquid by the reaction of a coagulant through biochemical processes or by agglomeration.

Flood crest: The peak of a flood event, also called a flood wave, which moves downstream and shows as a curve crest on a hydrograph.

Flood plain: A stretch of relatively level land bordering a stream. This plain is composed of river sediments and is subject to flooding.

Flood stage: The stage at which overflow of the natural banks of a stream begins to cause damage in the reach in which the elevation is measured.

Flow rate: The volume per time given to the flow of water or other liquid substance, which emerges from an orifice, pump, or turbine, or passes along a conduit or channel, usually expressed as cubic feet per second (cfs), gallons per minute (gpm), or million gallons per day (mgd).

Focus: A point of earthquake origin in the earth's crust from where earthquake waves travel in all directions.

Foliation: A textural term referring to the planar arrangement of mineral grains in metamorphic rock.

Formation: A body of rock characterized by a degree of lithologic homogeneity; it is usually, but not necessarily, tabular and is mappable on the earth's surface or traceable in the subsurface.

Formation water: Water present in a water-bearing formation under natural conditions, as opposed to introduced fluids such as drilling mud.

Fossil fuel: Fuels such as natural gas, petroleum, and coal that developed from ancient deposits of organic deposition and subsequent decomposition.

Fuel load: The total mass of combustible materials available to a fire.

Geophysical logs: The records of a variety of logging tools that measure the geophysical properties of geologic formations penetrated and their contained fluids. These properties include electrical conductivity, resistivity, the ability to transmit and reflect sonic energy, natural radioactivity, hydrogen ion content, temperature, and gravity. These geophysical properties are then interpreted in terms of lithology, porosity, fluid content, and chemistry.

Geothermal gradient: The rate of increase of temperature in the earth with depth. The gradient near the surface of the earth varies from place to place depending upon the heat flow in the region and the thermal conductivity of the rocks. The approximate geothermal gradient in the earth's crust is about 25°C/km.

Glacial drift: A general term applied to sedimentary material transported and deposited by glacial ice.

Glacis: A protective earthen bank that slopes away from the outer walls of a fortification.

Graben: A down-faulted block, which may be bounded by upthrown blocks (horsts).

Gradation: The leveling of the land through erosion, transportation, and deposition.

Granite: A light-colored or reddish coarse-grained intrusive igneous rock that forms the typical base rock of continental shields.

Greenhouse effect: The trapping and reradiation of the earth's infrared radiation by atmospheric water vapor, carbon dioxide, and ozone. The atmosphere acts like the glass cover of a greenhouse.

Ground avalanche: An avalanche type that involves the entire thickness of the snowpack and usually includes soil and rock fragments.

Ground fire: A type of fire that occurs beneath the surface and burns rootwork and peaty materials.

Group (general): An association of any kind based upon some similar feature or relationship (stratigraphy). Lithostratigraphic units consist of two or more formations, a more or less informally recognized succession of strata too thick or inclusive to be considered a formation, and subdivisions of a series.

Grout: A cementitious component of high water content, fluid enough to be poured or injected into spaces such as fissures surrounding a well bore and thereby filling or sealing them. Specifically, a pumpable slurry of portland cement, sand, and water forced under pressure into a borehole during well drilling to seal crevices and prevent the mixing of groundwater from different aquifers.

Horst: An up-faulted block, which may be bounded by downthrown blocks (grabens).

Hot spot (in geology): Excessively hot magma centers in the asthenosphere that usually lead to the formation of volcanoes.

Humus: Partially or fully decomposed organic matter in soils, dark in color and partly of colloidal size.

Hurricane: A tropical low-pressure storm (also called Baguio, tropical cyclone, typhoon, willy). Hurricanes may have a diameter of up to 400 mi (640 km) and a calm center (the eye), and must have wind velocities higher than 75 mph (120 km/h). Some storms have wind velocities of 200 mph (320 km/h).

Hydrate: A term that refers to compounds containing chemically combined water.

Hydraulic (in general): Pertaining to a fluid in motion, or to movement or action caused by water.

Hydraulic action: The mechanical loosening and removal of weakly resistant material solely by the pressure and hydraulic force of flowing water, as by a stream surging into rock cracks or impinging against the bank on the outside of a bend, or by ocean waves and currents pounding the base of a cliff.

Hydraulic conductivity: Ratio of flow velocity to driving force for viscous flow under saturated conditions of a specified liquid in a porous medium.

Hydraulic gradient: In an aquifer, the rate of change of total head per unit of distance of flow at a given point and in a given direction.

Hydraulic head: The height of the free surface of a body of water above a given subsurface point. Also, the water level at a point upstream from a given point downstream or the elevation of the hydraulic grade line at a given point above another given point of a pressure pipe.

Hydraulics: An aspect of engineering that deals with the flow of water or other liquids; the practical application of hydromechanics.

Hydrocarbon: Organic compounds containing only carbon and hydrogen, commonly found in petroleum, natural gas, and coal.

Hydrodynamics: The aspect of hydromechanics that deals with forces that produce motion.

Hydrogeology: A science dealing with subsurface waters and related geologic aspects of surface waters; also used in the more restricted sense of groundwater geology only. The term was defined by Mead (1919) as the study of the laws of the occurrence and movement of subterranean waters. More recently, it has been used interchangeably with geohydrology.

Hydrograph: A graph that shows the rate of river discharge over a given time period.

Hydrography: A science dealing with the physical aspects of all waters on the earth's surface, especially the compilation of navigational charts of bodies of water; or, the body of facts encompassed by hydrography.

Hydrologic cycle: A constant circulation of water from the sea, through the atmosphere, to the land, and eventually back to the atmosphere by way of transpiration and evaporation from sea and land surfaces.

Hydrologic system: A complex of related parts—physical, conceptual, or both—forming an orderly working body of hydrologic units and their human-related aspects such as the use, treatment, reuse, and disposal of water and the costs and benefits thereof, and the interaction of hydrologic factors with those of sociology, economics, and ecology.

Hydrology: A science dealing with global water (both liquid and solid), its properties, circulation, and distribution, both on and under the earth's surface and in the atmosphere, from the moment of its precipitation until it is returned to the atmosphere through evapotranspiration or is discharged into the ocean.

In recent years, the scope of hydrology has been expanded to include environmental and economic aspects. At one time, there was a tendency in the United States (as well as in Germany) to restrict the term "hydrology" to the study of subsurface waters (DeWeist 1965). Also, the sum of the factors studied in hydrology; the hydrology of an area or district.

Hydrosphere: The waters of the earth, as distinguished from the rocks (lithosphere), living things (biosphere), and the air (atmosphere). Includes the waters of the oceans, rivers, lakes, and other bodies of surface water in liquid form on the continents; snow, ice, and glaciers; and liquid water, ice, and water vapor in both the unsaturated and saturated zones below the land surface. Water in the atmosphere, which includes water vapor, clouds, and all forms of precipitation while still in the atmosphere, is included by some but excluded by others.

Hydrothermal: Of or pertaining to hot water, to the actions of hot water, or to the products of such actions, such as a mineral deposit precipitated from a hot aqueous solution, with or without demonstrable association with igneous processes, also said of the solution itself. This term is generally used for any hot water but has been restricted by some to water of magmatic origin.

Hydrothermal processes: Processes associated with igneous activities that involve heated or superheated water, especially alteration, space filling, and replacement.

Hygroscopic particles: Condensation nuclei in the atmosphere that attract water molecules (carbon, sulfur, salt, dust, ice particles).

Impermeable: Impervious to the natural movement of fluids.

Induction: The creation of an electric charge in a body by a neighboring body without having physical contact.

Injection well: A recharge well; or, a well into which water or a gas is pumped to increase the yield of other wells in the area; or, a well used to dispose of fluids in the subsurface environment by allowing them to enter the well by gravity flow or injection under pressure.

Intensity (of an earthquake): A measurement of the effects of an earthquake on the environment expressed by the Mercalli scale in stages from I to XII.

Ion: An electrically charged molecule or atom that lost or gained electrons and therefore has a smaller or greater number of electrons than the originally neutral molecule or atom.

Ionization: The process of creating ions.

Iron Age: The period around 800 BC that followed the Bronze Age during which mankind began using iron to make implements and weapons. The earliest use of iron may go back to 2500 BC.

Ironstone: A term sometimes used to describe a hardened plinthite layer in tropical soils, primarily composed of iron oxides bonded to kaolinitic clays.

Isopach: A line drawn on a map through points of equal thickness of a designated stratigraphic unit or group of stratigraphic units.

Isopach map: A map that shows the thickness of a bed, formation, or other tabular body throughout a geographic area; or, a map that shows the varying true thickness of a designated stratigraphic unit or group of stratigraphic units by means of isopachs plotted normal to the bedding or other bounding surface at regular intervals.

Isotopes: Atoms of a given element that have the same atomic number but different atomic weights because of variations in the number of neutrons.

Jet stream: A high-velocity, high-altitude (25,000–40,000 ft or 7700–12,200 m) wind that moves within a relatively narrow oscillating band within the upper westerly winds.

Joint (in geology): A natural fissure in a rock formation along which no movement takes place.

Karst: A type of topography characterized by closed depressions (sinkholes), caves, and subsurface streams.

Landslide: A general term that denotes a rapid downslope movement of soil or rock masses.

Land subsidence: A gradual or sudden lowering of the land surface caused by natural or human-induced factors such as solution (see **Karst**) or the extraction of water or oil.

Lapse rate: The rate of change in temperature or pressure of atmospheric values with a change in elevation.

Latent energy: Heat energy that changes the state of a substance without increasing its temperature. An example is the melting of ice into liquid water and the subsequent evaporation to vapor. Latent energy is released when the processes are reversed.

Leachate: The solution obtained by the leaching action of water as it percolates through soil or other materials, such as wastes containing soluble substances.

Lithification: The conversion of unconsolidated material into rock.

Lithology: The description of rocks on the basis of characteristics such as color, structures, mineralogic composition, and grain size; or, the physical character of a rock.

Lithosphere: The outer solid layer of the earth that rests on the nonsolid asthenosphere. The lithosphere averages about 60 mi (100 km) in thickness.

Loess: Fine silt-like soil particles that have been transported and deposited by wind action. Some loess deposits may be hundreds of feet thick.

Magma: Naturally occurring molten rock that may also contain variable amounts of volcanic gases. It issues at the earth's surface as lava.

Magma chambers: Underground reservoirs of molten rock (magma) usually found beneath volcanic areas.

Mantle (in geology): The intermediate zone of the earth found beneath the crust and resting on the core. The mantle is believed to be about 1800 mi (2900 km) thick.

Member: A division of a formation, generally of distinct lithologic character or of only local extent. A specially developed part of a varied formation is called a member if it has considerable geographic extent. Members are commonly, though not necessarily, named.

Mercalli scale: Used to describe the effects of an earthquake's intensity on a scale of I to XII, ranging from "imperceptible" to "major catastrophe." It is not a quantified scale.

Metamorphism: A process that induces physical or compositional changes in rocks caused by heat, pressure, or chemically active fluids.

Meteorology: The scientific study of weather and atmospheric physics.

Millidarcy (md): The customary unit of fluid permeability, equivalent to 0.001 darcy.

Moho discontinuity: A zone between the earth's crust and mantle that shows a marked change in the travel velocity of seismic waves caused by density changes between these layers. It is named after Mohorovicic, a seismologist who discovered this discontinuity in 1909.

Mudflow: A downslope movement of water-saturated earth materials such as soil, rock fragments, or volcanic ash.

Mud logs: A record of continuous analysis of a drilling mud or fluid for oil and gas content.

Neutralization: Reaction of an acid or alkali with the opposite reagent until the concentrations of hydrogen and hydroxyl ions in the solution are approximately equal.

Nonrenewable resources: Resources (coal, oil, ores, and so on) that cannot be renewed once they have been used up. In contrast, wood, air, and water are renewable resources.

Overburden (spoil): Barren bedrock or surficial material that must be removed before the underlying mineral deposit can be mined.

Oxidation: The addition of oxygen to a compound. More generally, any reaction that involves the loss of electrons from an atom.

pH: The negative logarithm of the hydrogen-ion concentration. The concentration is the weight of hydrogen ions in grams per liter or solution. Neutral water, for example, has a pH value of 7 and a hydrogen ion concentration of 10.

Packer: In well drilling, a device lowered in the lining tubes that swells automatically or can be expanded by manipulation from the surface at the correct time to produce a watertight joint against the sides of the borehole or the casing, thus entirely excluding water from different horizons.

Percentage map: A facies map that depicts the relative amount (thickness) of a single rock type in a given stratigraphic unit.

Perched aquifer: A water body that is not hydraulically connected to the main zone of saturation.

Permafrost: Permanently frozen ground.

Permeability: The ability of a porous rock, sediment, or soil to transmit a fluid without impairing the structure of the medium; it is a measure of the relative ease of fluid flow under unequal pressure. The customary unit of measurement is the millidarcy.

Pesticide: Any chemical used for killing noxious organisms.

Plugging: The act or process of stopping the flow of water, oil, or gas in strata penetrated by a borehole or well so that fluid from one stratum will not escape into another or to the surface; especially, the sealing up of a well that is tube abandoned. It is usually accomplished by inserting a plug into the hole, by sealing off cracks and openings in the sidewalls of the hole, or by cementing a block inside the casing. Capping the hole with a metal plate should never be considered an adequate method of plugging a well.

Porosity: A property of a rock, soil, or other material containing interstices. It is commonly expressed as a percentage of the bulk volume of material occupied by interstices, whether isolated or connected.

Potentiometric surface: An imaginary surface representing the static head of groundwater and defined by the level to which water will rise in a well. The water table is a particular potentiometric surface.

Pressure: The total load or force acting on a surface; in hydraulics, without qualifications, usually the pressure per unit area, or the intensity of pressure above local atmospheric pressure, expressed in pounds per square inch or kilograms per square centimeter.

Primary porosity: The porosity that develops during the final stages of sedimentation or that is present within sedimentary particles at the time of deposition. It includes all depositional porosity of the sediments or rocks.

Resistivity: Refers to the resistance of a material to electrical current; the reciprocal of conductivity.

Refusal: During drilling, the maximum depth beyond which augers (drill bits) cannot advance; usually the top of the bedrock.

Resource: A concentration of naturally occurring solid, liquid, or gaseous materials in or on the earth's crust in such a way that economic extraction of the commodity is currently or potentially feasible.

Rotary drilling: A common method of drilling consisting of a hydraulic process involving a rotating drill pipe, at the bottom of which a hard-toothed drill bit is attached. The rotary motion is transmitted through the pipe from a rotary table at the surface; that is, as the pipe turns, the bit loosens or grinds a hole in the bottom material. During drilling, a stream of drilling mud is in constant circulation down the pipe and out through the bit, from where the mud and the cuttings from the bit are forced back up the hole outside the pipe and into pits where the cuttings are removed and the mud is picked up by pumps and forced back down the pipe.

Runoff: The portion of precipitation that flows over the surface of the land as sheet wash and stream flow.

Salinization: Excessive buildup of soluble salts in soil or water. This is often a serious problem in crop irrigation systems.

Saltation: A form of wind erosion where small particles are picked up by the wind and fall back to the earth's surface in a "leap-and-bound" fashion. The impact of the particles loosens other soil particles, making them prone to further erosion.

Sanitary landfill: A land site where solid waste is dumped, compacted, and covered with soil to minimize environmental degradation.

Sea level: An imaginary average level of the ocean as it exists over a long period of time. It is also used to establish a common reference for standard atmospheric pressure at this level.

Secondary porosity: The porosity developed in a rock formation subsequent to its deposition or emplacement, either through natural processes of dissolution or stress distortion or through artificial processes such as acidization or the mechanical injection of coarse sand.

Secondary wave (S): A body earthquake wave that travels more slowly than a primary wave (P). The wave energy moves earth materials at a right angle to the direction of wave travel. This type of shear wave cannot pass through liquids.

Sedimentation: The removal of solids from water by gravitational settling.

Seismic activity: Earth vibrations or disturbances produced by earthquakes.

Seismic survey: The gathering of seismic data from an area; the initial phase of seismic prospecting.

Seismograph: A device that measures and records the magnitude of earthquakes and other shock waves such as underground nuclear explosions.

Seismology: A science concerned with earthquake phenomena.

Seismometer: An instrument, often portable, designed to detect earthquakes and other types of shock waves.

Semi-Arid regions: Transition zones with very unreliable precipitation, located between true deserts and subhumid climates. The vegetation usually consists of scattered short grasses and drought-resistant shrubs.

Septic-tank system: An on-site disposal system consisting of an underground tank and a soil absorption field. Untreated sewage enters the tank, where solids undergo decomposition. Liquid effluent moves from the tank to the absorption field via a perforated pipe.

Shear: The movement of one part of a mass relative to another, leading to lateral deformation without resulting in a change in volume.

Shear strength: The internal resistance of a mass to lateral deformation (see **Shear**). Shear strength is mostly determined by internal friction and the cohesive forces between the particles.

Sinkhole: A topographic depression developed by a solution of limestone, rock salt, or gypsum bedrock.

Sludge: Mud obtained from a drill hole in boring; mud from drill cuttings. The term has also been used for the cuttings produced by drilling; a semifluid, slushy, and murky mass or sediment of solid matter resulting from treatment of water, sewage, or industrial and mining wastes and often appearing as local bottom deposits in polluted water bodies.

Slurry: A very wet, highly mobile, semiviscous mixture or a suspension of finely divided, insoluble matter.

Soil failure: Slippage or shearing within a soil mass because of some stress force that exceeds the shear strength of the soil.

Soil liquefaction: The liquefying of clayey soils that lose their cohesion when they become saturated with water and are subjected to stress or vibrations.

Soil salinization: The process of accumulation of soluble salts (mostly chlorides and sulfates) in soils due to the rise of mineralized groundwater or a lack of adequate drainage when irrigation is used.

Soil structure: The arrangement of soil particles into aggregates that can be classified according to their shapes and sizes.

Soil texture: The relative proportions of various particle sizes (clay, silt, sand) in soils.

Solution: A process of chemical weathering by which rock material passes into calcium carbonate in limestone or chalk by carbonic acid derived from rainwater containing carbon dioxide acquired during its passage through the atmosphere.

Sorting: A dynamic gradational process that groups sedimentary particles by size or shape. Well-sorted material has a limited size range, whereas poorly sorted material has a large size range.

Specific conductance: The electrical conductivity of a water sample at 25°C (77°F), expressed in micro-ohms per centimeter.

Specific gravity: The ratio of the mass of a body to the mass of an equal volume of water.

Spontaneous combustion: A type of fire started by the accumulation of oxidation heat until the kindling temperature of the material is reached.

Stage: The height of a water surface above an established datum plane.

Standing wave: An oscillating wave on the surface of an enclosed water body. The wave acts similarly to water sloshing back and forth in an open dish.

Stock: An irregularly shaped discordant pluton with less than 100 km² surface exposure.

Storage coefficient: In an aquifer, the volume of water released from storage in a vertical column of 1 ft² when the water table or other potentiometric surface declines 1 ft. In an unconfined aquifer, it is approximately equal to the specific yield.

Stratification: A structure produced by a series of sedimentary layers or beds (strata).

Stratigraphy: The study of rock strata, including their age relations, geographic distribution, composition, and history.

Stratosphere: The part of the upper atmosphere that shows little change in temperature with altitude. Its base begins at about 7 mi (11 km) and its upper limits reach about 22 mi (35 km).

Stream terraces: Elevated remainders of previous flood plains; they generally parallel the stream channel.

Stress: Compressional, tensional, or torsional forces that act to change the geometry of a body.

Structure-contour map: A map that portrays subsurface configuration by means of structure contour lines; a contour map or tectonic map. Synonyms include structural map, structure map.

Summit aridity: Dry conditions that may develop on convex hills as a result of excessive drainage and thin soil layers.

Surface casing: The first string of a well casing to be installed in a well. The length will vary according to the surface conditions and type of well.

Surficial deposit: Unconsolidated transported or residual materials such as soil, alluvial, or glacial deposits.

Surge: A momentary increase in flow in an open conduit or in pressure in a closed conduit that passes longitudinally along the conduit, usually due to sudden changes in velocity.

Swab: A piston-like device equipped with an upward-opening check valve and provided with flexible rubber suction caps, lowered into a borehole or casing by means of a wire line for cleaning out the drilling mud or for lifting the oil.

Talus debris: Unconsolidated rock fragments that form a slope at the base of a steep surface.

Tectonic: Pertaining to the forces involved in, or the resulting structures or features of, tectonics. Also called "geotectonic."

Till: Unstratified and unsorted sediments deposited by glacial ice.

Topsoil: The surface layer of a soil that is rich in organic materials.

Tornado: A highly destructive and violently rotating vortex storm that frequently forms from cumulonimbus clouds. Also referred to as a twister.

Total porosity: The measure of all void space of a rock, soil, or other substance. Total porosity is usually expressed as a percentage of the bulk volume of material occupied by the void space.

Toxin: A colloidal, proteinaceous, poisonous substance that is a specific product of the metabolic activities of a living organism and is usually very unstable, notably toxic when introduced into the tissues, and typically capable of inducing antibody formation.

Transmissivity: In an aquifer, the rate at which water of the prevailing kinematic viscosity is transmitted through a unit width under a unit hydraulic gradient. Though spoken of as a property of the aquifer, it also includes the saturated thickness and the properties of the contained liquid.

Transpiration: The process by which water absorbed by plants evaporates into the atmosphere from the plant surface.

Triangulation: A survey technique used to determine the location of the third point of a triangle by measuring the angles from the known end points of the base line to the third point.

Tsunami: A Japanese term that refers to a seismic sea wave that can be generated by severe submarine fault slippages or volcanic eruptions. The tsunami reaches great heights when it enters shallow waters, but it is unnoticeable on the high seas.

Turbulence (in meteorology): Any irregular or disturbed wind motion in the air.

Twister: An American term for a tornado.

Unconfined aquifer: A groundwater body that is under water table conditions.

Unconsolidated material: A sediment that is loosely arranged or whose particles are not cemented together, occurring at either the earth's surface or depth.

Urbanization: The transformation of rural areas into urban areas, also referred to as urban sprawl.

Vapor pressure: A part of the total atmospheric pressure that is exerted by water vapor and is usually expressed in inches of mercury or in millibars.

Vesicular: A textural term indicating the presence of many small cavities in a rock.

Viscosity: A property of a substance to have internal resistance to flow; its internal friction. The ratio of the rate of shear stress to the rate of shear strain is known as the coefficient of viscosity.

Vorticity (in meteorology): Any rotary flow of air such as in tornadoes, midlatitude cyclones, and hurricanes.

Wastewater: Spent water. Depending on the source, it may be a combination of the liquid and water-carried wastes from residences, commercial buildings, industrial plants, and institutions, together with any groundwater, surface water, and storm water that may be present. In recent years, the term wastewater has taken precedence over the term "sewage."

Water quality: The chemical, physical, and biological characteristics of water with respect to its suitability for a particular purpose.

Water table: The surface marking the boundary between the saturation zone and the aeration zone. It approximates the surface topography.

Weather: The physical state of the atmosphere (wind, precipitation, temperature, pressure, cloudiness, and so on) at a given time and location.

Well log: A log obtained from a well, showing information such as resistivity, radioactivity, spontaneous potential, and acoustic velocity as a function of depth, especially a lithologic record of the rocks penetrated.

Well monitoring: Measurement, by on-site instruments or laboratory methods, of the water quality of a water well. Monitoring may be periodic or continuous.

Well plug: A watertight and gastight seal installed in a borehole or well to prevent movement of fluids. The plug can be a block cemented inside the casing.

Well record: A concise statement of the available data regarding a well, such as a scout ticket; a full history or day-by-day account of a well, from the day the well was surveyed to the day production ceased.

Well stimulation: A term used to describe several processes used to clean the well bore, enlarge channels, and increase pore space in the interval to be injected, thus making it possible for wastewater to move more readily into the formation. Well-stimulation techniques include surging, jetting, blasting, acidizing, and hydraulic fracturing.

Windbreak: Natural or planted groups or rows of trees that slow down the wind velocity and protect against soil erosion.

Zone of aeration: The zone in which the pore spaces in permeable materials are not filled (except temporarily) with water; also referred to as unsaturated zone or vadose zone.

Zone of saturation: The zone in which pore spaces are filled with water; also referred to as phreatic zone.

REFERENCES

American Geological Institute. 1996. *Glossary of Geology and Related Sciences*. Washington, DC: AGI.

Soliman M.M. et al. 1988. *Multilingual Technical Dictionary on Irrigation and Drainage*. New Delhi: ICID.

UNESCO. 1990. *Hydrology and Water Resources for Sustainable Development in a Changing Environment*. New York: UNESCO.

Appendix C: Statistics and Stochastic Analysis in Hydrology

C.1 STATISTICAL ANALYSIS

C.1.1 GENERAL REMARK

The probability of occurrence of a particular extreme rainfall of a certain duration will be of importance in many hydraulic-engineering applications, such as those concerned with floods. A brief description of the terminology and a simple method for predicting the frequency of an event are described in this section. The analysis of annual series, though described here with rainfall as a reference, is equally applicable to any other hydrologic process, such as flood flow.

First, the terminology used in frequency analysis must be understood. The probability of occurrence of an event (e.g., rainfall) whose magnitude is equal to or greater than a specified magnitude X is denoted by P. The return period (also known as recurrence interval) is defined as

$$T = \frac{1}{P} \tag{C.1}$$

This represents the average interval between the occurrence of a rainfall of magnitude equal to or greater than X. Thus, if the return period of rainfall of 20 millimeters in 2 hours is 10 years at a certain station A, this implies that on average, rainfall of a magnitude equal to or greater than 20 millimeters in 2 hours occurs once in 10 years. However, it does not mean that every 10 years one such event is likely to occur, that is, periodicity is not implied. The probability of a rainfall of 20 millimeters in 2 hours occurring in any one year at station A is,

$$P = \frac{1}{T} = \frac{1}{10} = 0.1 \tag{C.2}$$

If the probability of occurrence of an event is P, the probability of nonoccurrence of the event in a given year is $q = (1 - P)$.

The purpose of frequency analysis of an annual series is to obtain a relationship between the magnitude of the event and its probability of exceedance. Probability analysis may be made either by the empirical method, which is called plotting position formulas (as given in Chapter 3), or by the analytical and statistical methods, which are discussed in this section.

C.1.2 BINOMIAL DISTRIBUTION

If the probability of an event occurring in a given year is P, the probability of the event not occurring in a given year is $q = (1 - P)$. The binomial distribution can be used to find the probability of occurrence of the event r times in n successive years. Thus,

$$P_{r,n} = {}^{n}C_{r}P^{r}q^{n-r} = \frac{n!}{(n-r)!r!}P^{r}q^{n-r} \tag{C.3}$$

where $P_{r,n}$ is the probability of a random hydrologic event (rainfall) of given magnitude and exceedance probability P occurring r times in n successive years. For example, the probability of an event of exceedance probability P occurring two times in n successive years is

$$P_{2,n} = \frac{n!}{(n-2)!2!}P^{2}q^{n-2}$$

1. The probability of the event not occurring at all in n successive years is

$$P_{0,n} = q^{n} = (1 - P)^{n}$$

2. The probability of the event occurring at least once in n successive years is

$$P_{1} = 1 - q^{n} = 1 - (1 - P)^{n} \tag{C.4}$$

EXAMPLE C.1

Analysis of the data on the maximum flood of a stream indicated that a discharge of 280 cubic meters per second had a return period of 50 years. Determine the probability of the flood equal to or greater than 280 cubic meters per second in the stream occurring (a) once in 20 successive years, (b) two times in 15 successive years, and (c) at least one in 20 successive years.

SOLUTION

Here $P = \dfrac{1}{50} = 0.02$ (using Equation C.2)

$$\text{(a) } n = 20, \ r = 1$$
$$P_{1,20} = \frac{20!}{19!1!} \times 0.02 \times (0.98)^{19}$$
$$= 20 \times 0.02 \times 0.68123 = 0.272$$

(b) $n = 15$, $r = 1$

$$P_{2,15} = \frac{15!}{13!2!} \times (0.02)^2 \times (0.98)^{18}$$

$$= 15 \times \frac{14}{2} \times 0.0004 \times 0.695 = 0.0292$$

(c) By Equation (C.4)

$$P_1 = 1 - (1 - 0.02)^{20} = 0.332$$

Chow (1964) has shown that most frequency distribution functions applicable in hydrologic studies can be expressed by the following equation, known as the general equation of hydrologic frequency analysis:

$$\bar{x}_T = \bar{x} + K\sigma \tag{C.5}$$

where $\bar{x}_T$ is the value of the variate X of a random hydrologic series with a return period T, $\bar{x}$ is the mean of the variate, σ is the standard deviation of the variate, and K is the frequency factor, which depends upon T and the assumed frequency distribution. Some of the commonly used frequency distribution functions for the prediction of extreme flood values are as follows:

- Gumbel's extreme value distribution
- Log-Pearson type-III distribution
- Log-normal distribution

Only the first two distributions are dealt with in this section, with an emphasis on application.

C.1.3 GUMBEL'S METHOD

Gumbel (1941) introduced this extreme distribution, which is commonly known as Gumbel's distribution. It is one of the most widely used probability distribution functions for extreme values in hydrologic and meteorologic studies for prediction of flood peaks, maximum rainfalls, maximum wind speed, and so on.

Gumbel defined a flood as the largest of the 365 daily flows. The annual series of flood flows constitutes a series of largest values of flows. According to his theory of extreme events, the probability of occurrence of an event equal to or larger than a value x_0 is

$$P(X \geq x_0) = 1 - e^{-e^{-y}} \tag{C.6}$$

where y is a dimensionless variable which is given by

$$y = \alpha(x - a)$$
$$a = x - 0.45005\sigma$$
$$\alpha = 1.2825/\sigma$$

Thus

$$y = \frac{1.2825(x - \bar{x})}{\sigma} + 0.577 \tag{C.7}$$

where $\bar{x}$ is the mean and σ is the standard deviation of the variate X. In practice, it is the value of X for a given P that is required and as such Equation C.6 is transported as

$$y_P = -\ln\left[-\ln(1 - P)\right] \tag{C.8}$$

Note that the return period $T = 1/P$ and designate

$$y_T = -\left[\ln \cdot \ln \frac{T}{T - 1}\right] \tag{C.9}$$

or

$$y_T = -\left[0.834 + 2.303 \log \frac{T}{T - 1}\right] \tag{C.9a}$$

Now rearrange Equation (C.7), the value of the variate X with a return period T is

$$x_T = \bar{x} + K\sigma \tag{C.10}$$

where

$$K = \frac{y_T - 0.577}{1.2825} \tag{C.11}$$

Note that Equation C.11 is the same form as the general equation of hydrologic frequency analysis (Equation C.5). Further, Equations C.10 and C.11 are the basic Gumbel's equations and are applicable to an infinite sample size (i.e., $N \to \infty$).

Since practical annual data series of extreme events such as floods and maximum rainfall depth have finite lengths of record, Equation C.11 is modified to account for finite N as given below for practical use.

C.1.3.1 Application of Gumbel's Equation

Equation C.10 gives the values of the variate X with a recurrence interval T and is used as

$$x_T = \bar{x} + K\sigma_{n-1} \tag{C.12}$$

where σ_{n-1} is the standard deviation of the sample

$$= \sqrt{\frac{\Sigma(x - \bar{x})^2}{N - 1}}$$

The frequency factor K is expressed as

$$K = \frac{y_T - \bar{y}_n}{S_n} \qquad \text{(C.13)}$$

where y_T is the reduced variate, a function of T and is given by

$$y_T = -\left[\ln \cdot \ln \frac{T}{T-1}\right] \qquad \text{(C.14)}$$

or

$$y_T = -\left[0.834 + 0.303 \log\left[\frac{T}{T-1}\right]\right]$$

$\bar{y}_n$ is the reduced mean, a function of sample size N, and is given in Table C.1 for $N \to \infty$, $\bar{y}_n \to 0.577$; S_n is the reduced standard deviation, a function of sample size N, and is given in Table C.2 for $N \to \infty$, $S_n \to 1.2825$.

These equations are used to estimate the flood magnitude corresponding to a given return based on an annual flood series by the following procedure:

1. Assemble the discharge data and note the sample size N. Here the annual flood value is the variate X. Find $\bar{x}$ and σ_{n-1} for the given data.
2. Using Tables C.1 and C.2, determine $\bar{y}_n$ and S_n appropriate to given N.
3. Find y_T for a given T by Equation C.14.
4. Find K by Equation C.13.
5. Determine the required x_T by Equation C.12.

The method is illustrated in Example C.2.

To verify whether the given data follow the assumed Gumbel's distribution, the following procedure may be adopted. The values of x_T for some return periods $T < N$ are calculated by using Gumbel's formula and plotted as x_T versus T on a convenient paper such as a semi-log, log–log, or Gumbel probability paper. The use of Gumbel probability paper results in a straight line for x_T versus T plot. Gumbel's distribution has a property which gives $T = 2.33$ years for the average of the annual series when N is very large. Thus, the value of a flood with $T = 2.33$ years is called mean annual flood. In graphical plots, this gives a mandatory point through which the line showing variation of x_T with T must pass. For the given data, values of return periods (plotting positions) for various recorded values, x values of the variate are obtained by the relation $T = (N+1)/m$ and plotted on the graph. A good fit of observed data with the theoretical variation line indicates the applicability of Gumbel's distribution to the given data series. By extrapolating the straight line x_T versus T, x_T values for $T < N$ can be easily determined (Example C.2).

TABLE C.1

Reduced Mean Y_n in Gumbel's Value Distribution

N	0	1	2	3	4	5	6	7	8	9
10	0.4952	0.4996	0.5035	0.5070	0.5100	0.5157	0.5157	0.5181	0.5202	0.5220
20	0.5236	0.5252	0.5268	0.5283	0.5296	0.5320	0.5320	0.5332	0.5343	0.5353
30	0.5362	0.5371	0.5380	0.5388	0.5396	0.5410	0.5410	0.5418	0.5424	0.5430
40	0.5436	0.5442	0.5448	0.5453	0.5458	0.5468	0.5468	0.5473	0.5477	0.5481
50	0.5485	0.5489	0.5493	0.5497	0.5501	0.5508	0.5508	0.5511	0.5515	0.5518
60	0.5521	0.5524	0.5527	0.5530	0.5533	0.5538	0.5538	0.5540	0.5543	0.5545
70	0.5548	0.5550	0.5552	0.5555	0.5557	0.5561	0.5561	0.5563	0.5565	0.5567
80	0.5569	0.5570	0.5572	0.5574	0.5576	0.5580	0.5580	0.5581	0.5583	0.5585
90	0.5586	0.5587	0.5589	0.5591	0.5592	0.5595	0.5595	0.5596	0.5598	0.5599
100	0.5600									

N = sample size.

TABLE C.2
Reduced Standard Deviation S_n in Gumbel's Value Distribution

N	0	1	2	3	4	5	6	7	8	9
10	0.9496	0.9676	0.9833	0.9971	1.0095	1.0206	1.0316	1.0411	1.0493	1.0565
20	1.0628	1.0696	1.0754	1.0811	1.0864	1.0915	1.0961	1.1004	1.1047	1.1086
30	1.1124	1.1159	1.1193	1.1226	1.1255	1.1285	1.1313	1.1339	1.1363	1.1388
40	1.1413	1.1436	1.1458	1.1480	1.1499	1.1519	1.1538	1.1557	1.1574	1.1590
50	1.1607	1.1623	1.1638	1.1658	1.1667	1.1681	1.1696	1.1708	1.1721	1.1734
60	1.1747	1.1759	1.1770	1.1782	1.1793	1.1803	1.1814	1.1824	1.1834	1.1844
70	1.1854	1.1863	1.1873	1.1881	1.1890	1.1898	1.1906	1.1915	1.1923	1.1930
80	1.1938	1.1945	1.1953	1.1959	1.1967	1.1973	1.1980	1.1987	1.1994	1.2001
90	1.2007	1.2013	1.2020	1.2026	1.2032	1.2028	1.2044	1.2049	1.2055	1.2060
100	1.2065									

C.1.3.1.1 Gumbel Probability Paper

Gumbel probability paper is an aid for convenient graphical representation of Gumbel's distribution. It consists of an abscissa specially marked for various convenient values of the return period T. To construct the T-scale on the abscissa, first construct an arithmetic scale of y_T values, say from −20 to +3, as in Figure C.1. For selected values of T, say 2, 10, 50, 100, 500, and 1000, find the values of y_T by Equation C.14 and mark off those positions on the abscissa. The T-scale is now ready for use (Figure C.1).

The ordinate of a Gumbel paper on which the variate x_T (flood discharge, maximum rainfall depth and so on) is plotted may have either an arithmetic scale or a logarithmic scale. Since, by Equations C.10 and C.11, x_T varies linearly with y_T, a Gumbel distribution will plot as a straight line on Gumbel probability paper. This property can be used advantageously wherever necessary for graphical extrapolation.

EXAMPLE C.2

The maximum recorded floods in a river for the period 1973–1999 are given in Table C.3. Verify whether the Gumbel extreme value distribution fits the recorded values. Estimate the flood discharge with recurrence interval of (a) 100 years and (b) 150 years by graphical extrapolation.

SOLUTION

The flood discharge values are arranged in descending order, and the plotting position recurrence interval T_p for each discharge is obtained as

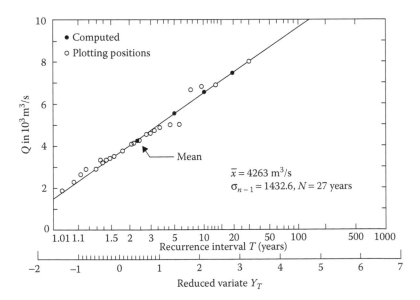

FIGURE C.1   Gumbel extreme probability paper.

$$T_p = \frac{(N+1)}{m} = \frac{28}{m}$$

The discharge magnitude Q is plotted against the corresponding T_p on a Gumbel extreme probability paper (Figure C.1). The statistics $\bar{x}$ and σ_{n-1} for the series, which are shown in Table C.3, are next calculated. Using these discharge magnitudes, x_T is calculated for a chosen recurrence interval by using Gumbel's formulas (Equations C.12 through C.14).

From Tables C.1 and C.2, for $N = 27$, $\bar{y}_n = 0.5332$ and $S_n = 1.1004$.

Substituting $T = 10$ years into Equation C.14

TABLE C.3
Calculation of T_p for Observed Data—Example C.2

Order Number	Flood discharge x (m³/s)	T_p (years)
1	7826	28.00
2	6900	14.00
3	6761	9.33
4	6599	7.00
5	5060	5.60
6	5050	4.67
7	4903	4.00
8	4798	3.50
9	4652	3.11
10	4593	2.80
11	4366	2.55
12	4290	2.33
13	4175	2.15
14	4124	2.00
15	3873	1.87
16	3757	1.75
17	3700	1.65
18	3521	1.56
19	3496	1.47
20	3380	1.40
21	3320	1.33
22	2988	1.27
23	2947	—
24	2947	1.17
25	2709	1.12
26	2399	1.08
27	1971	1.04

$N = 27$ years, $\bar{x} = 4263$ m³/s, $\sigma_{n-1} = 1432.6$ m³/s (can be obtained by using Excel).

$$y_T = -\left[\ln\cdot\ln\left(\frac{10}{9}\right)\right] = 2.25037$$

$$K = \frac{2.25037 - 0.5332}{1.1004} = 1.56$$

$$\bar{x}_T = 4263 + (1.56 \times 1432.6)$$

$$= 6499 \text{ m}^3/\text{s}$$

Values of x_T are similarly calculated for two more T values as shown below.

T (years)	$\bar{x}_T$ (obtained by Equation C.12; m³/s)
5.0	5522
10.0	6499
20.0	7436

These values are shown in Figure C.1. Because of the properties of the Gumbel's extreme probability paper, these points lie on a straight line. A straight line is then drawn through these points. Observed data fit well with the theoretical Gumbel's extreme value distributions.

Note that in view of the linear relationship of the theoretical x_T and T on a Gumbel probability paper, only two values of T and the corresponding x_T must be calculated. However, if Gumbel's probability paper is not available, a semi-log plot with log scale for T should be used, and a large set of (x_T, T) values is needed to identify the theoretical curve.

By extrapolation of the theoretical x_T versus T relationship, from Figure C.1 we get

For (a): At $T = 100$ years, $x_T = 9600$ m³/s
For (b): At $T = 150$ years, $x_T = 10,700$ m³/s

By using Equations C.12 through C.14, $x_{100} = 9558$ m³/s and $x_{150} = 10,088$ m³/s.

C.1.3.1.2 Confidence Limits

Since the value of the variate for a given return period x_T as determined by Gumbel's method can have errors due to the limited sample data used, the confidence limits of the estimate can be estimated. The confidence interval indicates the limits of the calculated value, between which the true value can be said to lie with a specific probability based on sampling errors only.

For a confidence probability C, the confidence interval of the variate x_T is bounded by values x_1 and x_2 given by

$$x_{1/2} = x_T \pm f(c)S_e \tag{C.15}$$

where $f(c)$ is the function of the confidence probability c as determined by using the table of normal variates:

c (in percent)	50	68	80	90	95	99
$f(c)$	0.674	1.00	1.282	1.645	1.96	2.58

$$S_e = \text{probable error} = b\frac{\sigma_{n-1}}{\sqrt{N}}$$

$$b = \sqrt{1+1.3K+1.1K^2}$$

where K is the frequency factor given by Equation C.13, σ_{n-1} is the standard deviation of the sample, and N is the sample size.

For a given sample and T, 80% confidence limits are twice as large as the 50% limits and 95% limits are three times as large as 50% limits.

EXAMPLE C.3

Data covering a period of 92 years for a stream yielded the mean and standard derivation of the annual flood series as 6437 and 2951 cubic meters per second respectively. Using Gumbel's method, estimate the flood discharge with a return period of 500 years. What are the (a) 95% and (b) 80% confidence limits for this estimate?

SOLUTION

From Table C.2 for $N = 92$ years, $y_n = 0.5589$ and

$$y_{500} = -\left[\ln\cdot\ln\left(\frac{500}{499}\right)\right]$$
$$= 6.21361$$
$$K_{500} = \frac{6.21361 - 0.5589}{1.2020} = 4.7044$$
$$x_{500} = 6437 + 4.7044 \times 2951 = 20320 \text{ m}^3/\text{s}$$
$$b = \sqrt{1+1.3(4.7044)+1.1(4.7044)^2}$$
$$= 5.61$$
$$S_e = \text{probable error} = 5.61 \times \frac{2951}{\sqrt{92}} = 1726$$

where $S_n = 1.2020$ from Table C.3

(a) For 95% confidence probability, $f(c) = 1.96$ and by Equation C.15

$$x_{1/2} = 20{,}320 \pm (1.96 \times 1726)$$
$$x_1 = 23{,}703 \text{ m}^3/\text{s} \quad \text{and} \quad x_2 = 16937 \text{ m}^3/\text{s}$$

Thus, the estimated discharge of 20,320 cubic meters per second has a 95% probability of lying between 23,700 and 16,940 cubic meters per second.

(b) For 80% confidence probability, $f(c) = 1.282$, and by Equation 7.23

$$x_{1/2} = 20,320 \pm (1.282 \times 1726)$$
$$x_1 = 22,533 \text{ m}^3/\text{s} \quad \text{and} \quad x_2 = 18,107 \text{ m}^3/\text{s}$$

The estimated discharge of 20,320 cubic meters per second has an 80% probability of lying between 22,530 and 18,110 cubic meters per second.

C.1.4 Log-Pearson Type-III Distribution

This distribution is extensively used in United States for projects sponsored by the U.S. government. In this distribution, the variate is first transformed into logarithmic form (base 10), and the transformed data is then analyzed. If x is the variate of a random hydrologic series, then the series of the z variate are first obtained, where

$$z = \log x \tag{C.16}$$

For this z series, for any recurrence interval T, Equation C.5 gives

$$z_T = \bar{z} + K_z \sigma_z \tag{C.17a}$$

where K_s is a frequency factor, which is a function of recurrence interval T and the coefficient of skew C_s,

σ_z is the standard deviation of the Z variate sample
$$= \sqrt{\Sigma(z-z)^2/N - 1}$$

and

C_s is the coefficient of skew of variate Z
$$= \frac{N\Sigma(z-\bar{z})^3}{(N-1)(N-2)(\sigma_z)^3} \tag{C.17b}$$

where $\bar{z}$ is the mean of the z values, N is the sample size equal to the number of years of record. The variation of $K_z = f(C_s, T)$ is given in Table C.4.

After finding z_T by Equation C.17, the corresponding value of x_T is obtained by Equation C.16 as

$$x_T = \text{antilog} (z_T) \tag{C.18}$$

Sometimes, the coefficient of skew C_s is adjusted to account for the size of the sample by using the following relation, proposed by Hazen (1930):

$$C_s = C_s\left(1 + \frac{8.5}{N}\right)$$ (C.19)

TABLE C.4
$K_z = f(C_s, T)$ for Use in Log-Pearson Type-III Distribution

Coefficient of Skew(C_s)	Recurrence Interval T (in years)						
	2	10	25	50	100	200	1000
3.0	−0.396	1.180	2.278	3.152	4.051	4.970	7.250
2.5	−0.360	1.125	2.262	3.048	3.845	4.652	6.600
2.2	−0.330	1.284	2.240	2.970	3.705	4.444	6.200
2.0	−0.307	1.302	2.219	2.912	3.605	4.298	5.910
1.8	−0.282	1.318	2.193	2.848	3.499	4.147	5.660
1.6	−0.254	1.329	2.163	2.780	3.388	3.990	5.390
1.4	−0.225	1.337	2.128	2.706	3.271	3.828	5.110
1.2	−0.195	1.340	2.087	2.626	3.149	3.661	4.820
1.0	−0.164	1.340	2.043	2.542	3.022	3.489	4.540
0.9	−0.148	1.339	2.018	2.498	2.957	3.401	4.395
0.8	−0.132	1.336	1.998	2.453	2.891	3.312	4.250
0.7	−0.116	1.333	1.967	2.407	2.824	3.223	4.105
0.6	−0.099	1.328	1.939	2.359	2.755	3.132	3.960
0.5	−0.083	1.323	1.910	2.311	2.686	3.041	3.815
0.4	−0.066	1.317	1.880	2.261	2.615	2.949	3.670
0.3	−0.050	1.309	1.849	2.211	2.544	2.856	3.525
0.2	−0.033	1.301	1.818	2.159	2.472	2.763	3.380
0.1	−0.017	1.292	1.785	2.107	2.400	2.670	3.235
0.0	0.000	1.282	1.751	2.054	2.326	2.576	3.090
−0.1	0.017	1.270	1.716	2.000	2.252	2.482	2.950
−0.2	0.033	1.258	1.680	1.945	2.178	2.388	2.810
−0.3	0.050	1.245	1.643	1.890	2.104	2.294	2.675
−0.4	0.066	1.231	1.606	1.834	2.029	2.201	2.540
−0.5	0.083	1.216	1.567	1.777	1.955	2.108	2.400
−0.6	0.099	1.200	1.528	1.720	1.880	2.016	2.275
−0.7	0.116	1.183	1.488	1.663	1.806	1.926	2.150
−0.8	0.132	1.166	1.448	1.606	1.733	1.837	2.035
−0.9	0.148	1.147	1.407	1.549	1.660	1.749	1.910
−1.0	0.164	1.128	1.366	1.492	1.588	1.664	1.880
−1.4	0.225	1.041	1.198	1.270	1.318	1.351	1.465
−1.8	0.282	0.945	1.035	1.069	1.087	1.097	1.130
−2.2	0.330	0.844	0.888	0.900	0.905	0.907	0.910
−3.0	0.396	0.660	0.666	0.666	0.667	0.667	0.668

where C_s is the adjusted coefficient of skew. However, the standard procedure for using log-Pearson Type-III distribution adopted by the U.S. Water Resources Council does not include this adjustment for skew.

When $C_s = 0$, the log-Pearson Type-III distribution reduces to log-normal distribution. The log-normal distribution plots as a straight line on logarithmic probability paper.

EXAMPLE C.4

For the annual flood series data given in Example C.2, estimate the flood discharge for a return period of (a) 100 years and (b) 200 years by using log-Pearson Type-III distribution.

SOLUTION

The variate $z = \log x$ is first calculated for all the discharges. Then, the statistics $\bar{z}$, σ_z, and C_s are calculated as shown in Table C.5.

TABLE C.5
Variate z—Example C.4

Years	Flood x (m³/s)	z = log x
1973	2947	3.4694
4	3521	3.5467
5	2399	3.3800
6	4124	3.6153
7	3496	3.5436
8	2947	3.4694
9	5060	3.7042
1980	4903	3.6905
1	3751	3.5748
2	4798	3.6811
3	4290	3.6325
4	4652	3.6676
5	5050	3.7033
6	6900	3.8388
7	4366	3.6401
8	3380	3.5289
9	7826	3.8935
1990	3320	3.5211
1	6599	3.8195
2	3700	3.5682
3	4175	3.6207
4	2988	3.4754
5	2709	3.4328
6	3873	3.5880
7	4593	3.6621
8	6761	3.8300
1999	1971	3.2947

The flood discharge for a given T is calculated as below. K values for given T and $C_s = 0.043$ are read from Table C.4.

T (years)	K_z	$K_z \sigma_z$	z_T (Equation C.18)	$x_T = $ antilog z_T (m³/s)
100	2.358	0.3365	3.9436	8782
200	2.616	0.3733	3.9804	9559

C.1.5 PARTIAL-DURATION SERIES

In the annual hydrologic data series of floods, only one maximum value of flood per year is selected as the data point. It is likely that in some catchments there is more than one independent flood in a year and many of them are an appreciably high magnitude. To enable all of the large-fold peaks to be considered for analysis, floods of a magnitude larger than an arbitrarily selected base value are included in the analysis. Such a data series is called a partial-duration series.

In the partial-duration series, all events considered must be established as independent. Here, the partial-duration series is adopted mostly for rainfall analysis where the conditions of independence of events are easy to establish. Its use in flood studies is rather rare. The recurrence interval of an event obtained by annual series (T_A) and by the partial-duration series (T_P) arc related by

$$T_P = \frac{1}{\ln T_A - \ln(T_A - 1)} \qquad (C.20)$$

This equation shows that the difference between T_A and T_P is significant for $T_A < 10$ years and is negligibly small for $T_A > 20$.

C.1.6 REGIONAL FLOOD FREQUENCY ANALYSIS

When the available data at catchments is too short to conduct a frequency analysis, a regional analysis is adopted. In this analysis, a hydrologically homogeneous region is considered from a statistical point of view. Available log-time data from neighboring catchments are tested for homogeneity, and a group of stations satisfying the test are identified. This group of stations constitutes a region, and all the station data of this region are pooled and analyzed as a group to find the frequency characteristics of the region. The mean annual flood, Q_{ma}, which corresponds to a recurrence interval of 2.33 years, is used for nondimensionalizing the results. The variation of Q_{ma} with drainage area and the variation of Q_T/Q_{ma} with T, where Q_T is the discharge for any T, are the basic plots prepared in this analysis. Details of this method are available in Chapter 3.

C.1.7 LIMITATION OF FREQUENCY STUDIES

The flood-frequency analysis described in Sections C.1.4 through C.1.6 is a direct means of estimating the desired flood based upon the available fold-flow data of a catchment. The results of the frequency analysis depend upon the length of data. The minimum number of years of record required to obtain satisfactory estimate depends upon the variability of the data and hence on the physical and climatological characteristics of the basin. Generally, a minimum of 30 years of data is considered essential. Shorter records are also used when this is unavoidable. However, frequency analysis should not be adopted if the record is less than 10 years.

Flood-frequency studies are most reliable in climates that are uniform from year to year. In such cases, a relatively short record gives a reliable picture of the frequency distribution. Increasing lengths of flood records affords a viable alternative method for flood-flow estimation in most cases.

C.1.8 DESIGN FLOOD

In the design of hydraulic structures, it is not practical from an economic standpoint to provide for the safety of the structure and the system at the maximum-possible flood in the catchment. Small structures such as culverts and storm drainages can be designed for less severe floods as the consequences of a higher-than-design flood may not be very serious. A high flood can cause temporary inconveniences like a disruption of traffic and only very rarely causes severe property damage and loss of life. On the other hand, storage structures such as dams demand greater attention to the magnitude of floods used in the design. The failure of these structures causes large loss of life and property damage downstream from the structure. The type and importance of the structure and the economic development of the surrounding area therefore determine the design criteria when choosing the flood magnitude. This section highlights the procedures adopted for selecting the flood magnitude for the design of some hydraulic structures.

C.2 STOCHASTIC HYDROLOGY

C.2.1 GENERAL REMARKS

In statistics, the word stochastic is synonymous with random, but in hydrology it is used to refer to a partially random time series. Stochastic hydrology fills the gap between deterministic models and probabilistic hydrology. Deterministic hydrology assumes the time variability to be totally explained by other variables as processed through an appropriate model. Probabilistic hydrology is not concerned with the time sequence but only with the probability, or chance, that an event will be equaled or exceeded. In stochastic hydrology, the time sequence is all-important.

A simple example of a stochastic process is that of drawing colored balls from an urn. The important feature is the actual order in which the balls are drawn from the urn. Valuable information is contained in the sequence of withdrawals: red, red, black, green, black, white, green, green, and so on. Average probability statements, by contrast, are concerned only with the relative numbers of the different colored balls withdrawn. Stochastic representation preserves sequencing of events.

Typical hydrologic time series are the quantitative descriptions of stream flow or precipitation history at a given point in time. There is a finite amount of information contained in any hydrologic time series. The most complete description of this information is in the continuous observed time record. However, the same record can be described in terms of mechanisms (mathematical relationships) with various degrees of precision. Time series that differ from the observed time series but retain many properties of the original series can be generated by mathematical functions. Each generated sequence is constructed so that the events have the same probability of occurrence as the observed sequence. Such time series are obtained by stochastic generation techniques.

Stochastic hydrology is meaningful only in a design or operational decision-making sense. In hydrologic design, the designer usually wishes to see how the particular facility will perform to represent future hydrologic inputs. The designer is not in a position to know what future flows or future precipitation events will be, but he or she can assume that future events will have the same stochastic properties as the observed historical record. This assumption forms the principal basis of stochastic hydrology, namely, the generation of equiprobable input traces, each trace having similar statistical properties. Each input sequence yields a sequence of outputs from the system under investigation. Thus, a stochastic analysis using many input traces yields a probability distribution of system response. The probability distributions of system response are used for design and operation decision making.

Stochastic methods were first introduced into hydrology to solve the problem of reservoir design. The required capacity of a reservoir depends on the sequence of flows, especially a sequence of low flows. If a reservoir operates on an annual cycle, that is, it is filled and is partially or wholly emptied each year, it may be possible to assess its reliability (the probability that it will deliver its intended yield each year) on the basis of an analysis of the historic flow record if this record is sufficiently long. However, if the reservoir operates with carryover storage, that is, it provides storage to meet required withdrawals over a period of several dry years, the historic record is unlikely to provide adequate information on reliability, since it is generally too short to define the probability of a series of subnormal years. Stochastic methods provide a means of estimating the probability of sequences of dry years during any specified period of future time. Even if the historic record suggests that a reservoir will operate on an annual cycle, the possibility of two or more dry years in sequence exists and thus stochastic analysis should be part of the hydrologic study for all reservoirs dependent on natural stream inflows. Stochastic methods in combination with deterministic methods also offer the prospect of improving estimates of flood frequency, since this task also depends on long records for reliable estimates.

Attempts to solve the problem of short record length by various statistical means were probably first initiated by Hazen (1914), who suggested combining records from several stations into a single long record. Sudler (1927) wrote historic flows on cards and dealt the cards randomly from the deck to construct a synthetic record of 1000 years. This procedure produces a variety of sequences of historic flows for storage analysis. The advent of the computer has made more sophisticated techniques

(Thomas and Fiering 1962) possible, collectively known as stochastic hydrology, or the generation of synthetic hydrologic time series.

C.2.2 THE STOCHASTIC PROCESS

In stochastic analysis, the process is assumed to be stationary, that is, statistical properties of the process do not vary with time. Thus, the properties of the historic record can be used to derive long synthetic sequences, which can be more effectively used in planning than the short historic sequence. The synthetic sequences must resemble the historical sequences, that is, they must display similar statistical properties.

Some properties of hydrologic time series can be investigated in a time domain through correlogram analysis. In some situations, it is convenient to work in a frequency domain and use spectral analysis tools to identify the principal harmonics contained in the series. However, the short length of hydrologic time series limits the usefulness of spectral analysis. Correlogram and spectral analysis enable some deterministic trends to be located (Box and Jenkins 1970). When trends have been detected and subtracted from the original series, the residual series is examined. The probability distributions of the elements of the residual series are usually of interest. For example, if a monthly time scale is the basic unit used in analysis, the probability distributions of the flow volumes (or residual flow volumes) for each month are of interest.

A time series can be modeled mathematically as a combination of deterministic and residual random components. One purpose of analyzing a time series is to determine the particular forms of the deterministic and random residual terms. The form of a stochastic generation equation can be very simple (preserving mean, variance, and lag-1 serial correlation) or more sophisticated. More sophisticated generators attempt to preserve low-frequency (as well as high-frequency) fluctuations in the time series; the more simple generators (Fiering 1967) are restricted to preserving high-frequency fluctuations.

In stochastic design applications, the designer is interested in the total system response. Although many properties are needed to describe a historic sequence fully, the stochastic analysis needs to consider only important properties for the system under study. This is paramount in any type of mathematical system simulation, and reflects the important coupling of inputs, system demands, and system operation. It is therefore important to identify the most suitable generation scheme necessary for the problem at hand.

In most stream flow–volume generation schemes, a first-order Markov structure can be used, that is, the idea that any event is dependent only on the preceding event. A simple Markov generating function for annual flow volume Q_i is

$$Q_i = \overline{Q} + \rho(Q_{i-1} - \overline{Q}) + t_i \sigma \sqrt{1 - \rho^2} \qquad \text{(C.21)}$$

Deterministic   Random

where t is a random variate from an appropriate distribution with a mean of zero and unit variance, σ is the standard deviation of Q, ρ is the lag-1 serial correlation coefficient, and $\overline{Q}$ is the mean of Q. The subscript i designates flows in series from year 1 to year n. If the parameters $\overline{Q}$, σ, and ρ can be determined from the historic record and a starting value of Q_{i-1} is assumed, a simple computer algorithm can be used to generate a long series of Q values using values of the random variate t derived sequentially by the computer. The Q_i values are obtained by Monte–Carlo sampling from the probability distribution t. The computation, of course, could be made manually by using a table of random numbers as a source of t, but this would be time-consuming.

If seasonal or monthly values of Q are desired, the procedure must allow for the characteristic seasonal variation as follows:

$$Q_{i,j} = \overline{Q}_j + \rho_j \frac{\sigma_j}{\sigma_{j-1}} (Q_{i-1,j-1} - \overline{Q}_{j-1}) + t_i \sigma_j \sqrt{1 - \rho_j^2} \qquad \text{(C.22)}$$

where the subscript j defines the seasons or months. For monthly synthesis j varies from 1 to 12 through the year. The subscript i is a serial designation from month 1 to month n, as in Equation C.21. ρ_j is the serial correlation coefficient between Q_j and Q_{j-1}. Other symbols are as they were for Equation C.21. Equation C.22 is used to determine $\overline{Q}$, σ, and ρ for each month or season. An initial value of $Q_{i-1,j-1}$ is assumed. It is advisable to start at the beginning of the local water year when the flows are low, although this is not necessary. To avoid the influence of start-up bias, sufficient flow values should be generated to remove the influence of the assumed initial flow. Usually, 12–20 time increments are sufficient: the sequence beyond the warm-up period is the input to the system under investigation. The coding of Equation C.22 for a computation method is relatively simple with values of t_i produced by a random-number generator having the appropriate probability distribution.

The simple single-season generator (Equation C.22) is appropriate for large reservoirs where seasonal variation of flow cannot materially affect the storage requirement or where flow is highly seasonal with the high-flow season preceding the high-use season. In all other cases, the multiseason generator (Equation C.22) should be used.

C.2.3 DISTRIBUTION OF T

Unless the random variate t is drawn from an appropriate distribution, Equation C.22 cannot reproduce the historical distribution, although it may satisfactorily duplicate the mean and variance. If the historic distribution is normal, the problem is simple; one draws from a normal random-number generator. If stream flow volumes are distributed log-normally, generation is effected by using transformed variables that are normally distributed. Exponentiation yields log-normal sequences.

There is no *a priori* basis to choose a distribution, and no single distribution is applicable to all streams. In addition, the relatively short historic flow records that are usually available do not clearly define the properties (parameters) of the distribution. Therefore, select a distribution that fits the observed data within the bounds of acceptance criteria. This selection is influenced by the fact that certain distributions are amenable to generation techniques, while others are extremely difficult to work with. For example, a three-parameter, log-normal distribution might fit the observed data as satisfactorily as a more complicated general gamma distribution, but the former is more ideal because it can be easily and economically programmed. Probability distributions can be fitted using Bayesian or classical methods. For example, if there is good physical reason to expect that flow volumes are of an additive nature, a normal distribution could be anticipated.

Figure C.2 shows the distributions of annual flows for three streams representing widely varied climatic regimes. The data suggest distributions for some rivers in United States ranging from nearly normal (Delaware and Flathead Rivers) to

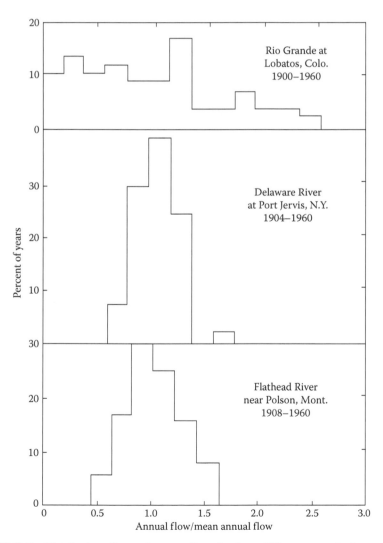

FIGURE C.2 Distribution of annual stream flows for three different watersheds.

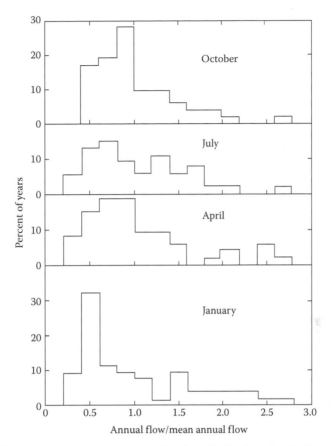

FIGURE C.3 Comparison of selected monthly flow distributions for the Flathead River at Poison, Montana (7096 mi² [18,379 km²]; data from 1908–1960).

approximately exponential (Rio Grande). Figure C.3 shows a similar comparison of distributions for selected months at a single station. Distribution selection is most difficult for streams where zero flows are a frequent occurrence. There is little alternative to selecting the distribution to which the historic data seem to conform. If there is serious doubt as to the best distribution, it is wise to generate two (or more) synthetic sequences from different distributions to determine the effect on the project plan resulting from differences in the sequences.

C.2.4 DEFINITION OF *F* PARAMETERS

A probability distribution can be defined by a few parameters. For example, a normal distribution can be completely defined by two parameters: the mean and variance. Most important probability distributions in hydrology require two or more parameters to define them. These parameters can be directly related to quantities such as the arithmetic mean, variance, and skew of the observed data. However, at best, it is only possible to estimate the true value of the parameters. The observed time series is only a short sample of the total time series. If the true arithmetic mean of the population is μ, μ can only be approximated by the sample mean $\overline{X}$. Thus, for a normal distribution $N(\mu, \sigma_1^2)$ having population mean μ and population variance σ_1^2, the distribution is approximated by $N(\overline{X}, \sigma^2)$.

There are two principal methods in use for computation of parameters. The most commonly used is the method of moments (Benjamin and Cornell 1970). A second method, the method of maximum likelihood, has received increasing attention with the wide availability of computers. Because most generation schemes in use employ parameters calculated by the method of moments, only moment-derived parameters are discussed here.

For any probability distribution, the arithmetic mean is approximated by

$$\overline{X} = \sum_{i=1}^{n} \frac{X_i}{n} \tag{C.23}$$

where n is the number of items in the sample. The variance must be corrected for serial correlation, if present, as follows:

$$\sigma^2 = \psi \frac{\sum_{i=1}^{n} X_i^2 - n\overline{X}^2}{n-1} \tag{C.24}$$

where ψ is a function of serial correlation ρ and length of record and can be obtained from Equations C25 and C26.

$$\psi = \frac{\dfrac{n-1}{n}}{1 - \dfrac{1-\rho^2}{n(1-\rho)^2} + \dfrac{2\rho(1-\rho^n)}{n^2(1-\rho)^2}} \tag{C.25}$$

If the sequences $\{Y_i\} = \{X_i\}$, $i = 1, 2, \ldots, n-1$, and $\{Z_{i-1}\} = \{X_i\}$, $i = 2, 3, \ldots, n$ are formed and have means $\overline{Y}$ and $\overline{Z}$ and variances σ_y^2 and σ_z^2, respectively, then the serial correlation coefficient of the series is

$$\rho = \frac{1}{n-2} \frac{\sum_{j=1}^{n=1} (Y_j Z_j) - (n-1)\overline{Y}\overline{Z}}{\sigma_y \sigma_z} \tag{C.26}$$

The calculated value of ρ should not be accepted without some thought. Anderson (1962) suggested a test for the significance of the correlation coefficient, which is displayed in Figure C.4. For short records, values of $\rho \leq 0.3$ are not statistically different from zero. However, even a small correlation reduces the effect of the random term and increases the persistence term of the generating function. At the same time, the variance is increased (Equation C.25). If the value of ρ is too large, these effects lead to an overestimation of the storage required.

There is no obvious physical explanation for the negative values of ρ. If such negative values are statistically not distinguishable from zero, they should be set

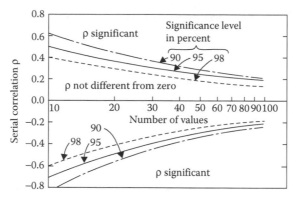

FIGURE C.4 Significance of serial correlation coefficient as a function of sample size.

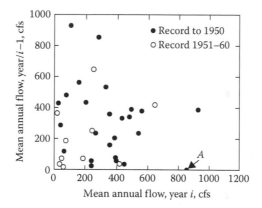

FIGURE C.5 Serial plot of annual flows for Yegua Creek. Data from USGS.

to zero. However, if negative correlations are thought to be significant, the analyst should immediately check the flow record for deterministic causes. If such a cause is found, it must be incorporated into the generation model. A scattergram is helpful for examining the correlation coefficients. A plot of Q_i versus $Q(i-1)$ may indicate (Figure C.5) that only a few chance flow sequences materially influence the correlation. For Yegua Creek, 26 years of data yield $\rho = 0.29$, which is not significantly different from zero by Andersen's test. If point A is omitted, $\rho = -0.10$, while with 35 years of data (omitting point A) $\rho = -0.007$. Similarly, a chance occurrence of two high years in sequence may lead to a large positive value of ρ. Test calculations can be made, omitting outliers from the main body of data to define the probable values of ρ more accurately. Fairly high values of intermonthly correlation may often occur, but interannual correlations greater than 0.4 are most unlikely, and if the last months of the year are poorly correlated, the interannual correlation must be near zero under the assumption of a Markov relationship.

If the log-normal distribution is selected, the parameters should be calculated from

$$\mu_x^2 = \exp\left(\sigma_y^2 + 2\mu_y\right) \tag{C.27}$$

$$\sigma_x^2 = \exp\left(2\sigma_y^2 + 2\mu_y\right) - \exp\left(\sigma_y^2 + 2\mu_y\right) \tag{C.28}$$

$$\rho_x = \frac{\exp\left(\sigma_y^2 \rho_y - 1\right)}{\exp\left(\sigma_y^2\right) - 1} \tag{C.29}$$

where μ_x, σ_x, and ρ_x are calculated with Equations C.23, C.24, and C.26 using the observed data and μ_y, σ_y, and ρ_y are the parameters to be used. The common procedure of transforming the data to logarithms and calculating the parameters of the transformed data can lead to substantial bias in the generated flows because the statistical parameters of the stream flow are not preserved. The transformed parameters σ_y^2 and μ_y are calculated by solving Equations C.27 and C.28 simultaneously. Then, ρ_y can be found from Equation C.29. Detailed analysis is unwise when parameters are estimated from a very short record, because gross errors could be made in selecting the underlying distribution. Furthermore, there is considerable uncertainty in the parameter estimates. If an analysis is necessary, the record should be extended as far as possible by a deterministic approach such as simulation. Table C.6 compares parameters for Arroyo Seco near Paso Robles, California, as derived from a 65-year record and two 32-year portions of the record. The data suggest that record length should exceed 65 years for adequate definition of parameters. However, note that Arroyo Seco exhibits high variability in runoff volumes. Variability should be examined when deciding what database length is needed to yield sufficient accuracy to perform stochastic analysis. Selection of data sample size must be viewed in terms of the total system under consideration.

It has been suggested that multilag models should be used in lieu of the first-order Markov model, which assumes that any flow is dependent only on the immediate previous flow. A correlogram (plot of correlation coefficient for various lags; Figure C.6) for a Markov process is a smooth exponential declining function. A correlogram calculated from a record is usually a jagged function, showing abrupt changes. Although it is assumed that the irregularities of the correlogram are real features of the regime, there is little physical basis for such an assumption. For example, if with annual data, lag 5 has a high correlation, then lag 4 should be at least as high. Burges (1970) showed that irregular correlograms can be obtained from short samples drawn from a randomly generated flow record. The irregular correlogram results from either chance variations in a short sample, from some underlying trends, or from both.

There are sometimes instances of significant multiperiod carryover effects, primarily as the result of groundwater storage. In the short term storage, soil-moisture carryover may also have a significant effect. Thus, use of multiple lags may sometimes be appropriate in monthly models and occasionally in annual models. However, considerable thought should be given to use of multiple lags, and a reasonable physical understanding of their significance should be available before their adoption.

TABLE C.6
Parameters of the Historic Stream Flow Record for Arroyo Seco, near Soledad, California

Month	Mean[a]			Variance			Serial Correlation		
	1903–1967	1903–1934	1936–1967	1903–1967	1903–1934	1936–1967	1903–1967	1903–1934	1936–1967
October	0.040	0.050	0.032	0.004	0.005	0.003	0.539	0.736	0.172
November	0.224	0.233	0.216	0.259	0.300	0.233	0.003	0.012	0.003
December	0.824	0.717	0.952	1.269	0.799	1.776	0.028	0.034	0.078
January	1.693	1.925	1.460	5.014	7.577	2.661	0.260	0.187	0.437
February	2.360[b]	2.412	2.370	5.802	5.389	6.456	0.209	0.280	0.129
March	1.979	2.123	1.864	5.007	5.959	4.316	0.442	0.140	0.770
April	1.191	1.035	1.296	2.246	0.798	3.715	0.470	0.382	0.626
May	0.427	0.423	0.429	0.127[b]	0.133	0.130	0.6250	0.541	0.732
June	0.172	0.173	0.171	0.025	0.030	0.021	0.947	0.926	0.981
July	0.058	0.058	0.058	0.005	0.005	0.004	0.966	0.960	0.978
August	0.020	0.022	0.019	0.001	0.001	0.001	0.963[b]	0.9640	0.965
September	0.018	0.022	0.014	0.001	0.001	0.001	0.684	0.660	0.732

[a] Data are expressed as inches of runoff over the catchment (244 mi^2 or 632 km^2).
[b] These values fall outside the range of the shorter periods because the water year 1934–1935 is included in the full period and not in the short periods.

Source: Burges, S. J. 1970. Use of stochastic hydrology to determine storage requirements for reservoirs: A critical analysis. Program in Engineering Economic Planning, Report EEP-34, Stanford University, Stanford.

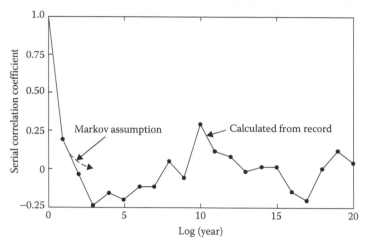

FIGURE C.6 Correlogram of annual flows for the Susquehanna River at Danville, Pennsylvania ($n = 51$, $\rho_1 = .19$) and the correlogram for a first-order Markov process with $\rho = 0.19$.

C.2.5 THE HURST PHENOMENON

While studying long-term storage requirements on the Nile, Hurst (1951) found that the range R_n can be expressed as

$$R_n = \sigma_n \left(\frac{n}{2}\right)^h \tag{C.30}$$

where σ_n is the standard deviation and n is the length of the series. The range R_n is defined as the difference between the greatest cumulative excess inflow above the mean flow and the greatest cumulative deficiency below the mean inflow. The exponent h is called the Hurst coefficient. For a Markov process, $h = 0.5$. Hurst tested 800 time series (stream flow, varves, tree rings, precipitation, and temperature) and found $0.5 < h < 1$ with a mean value of $h = 0.73$ and a standard deviation of 0.09. One implication of $h > 0.5$ is that there is long-term persistence in natural time series and the Markov process is not a valid model.

Mandelbrot and Wallis (1969) suggested the use of fractional Gaussian noise to generate sequences with $h > 0.5$. Autoregressive integrated moving averages (ARIMA; Box and Jenkins 1970) and a broken-line process (Mejia et al. 1972) have also been suggested. It is not practical to give an adequate description of these processes here, and interested readers should refer to the bibliography.

Long-term trends in climate are clearly indicated by geologic evidence, but no physical explanation for the kind of persistence suggested by Hurst's results exists. The cause must be something other than the carryover processes of hydrology and must be sought in terms of long-term climatic change. Currently available evidence does not suggest significant climatic change in the time scale of 100 years or less, within which project-planning horizons normally fall. Since the purpose of stochastic analysis is usually to generate many short sequences representing possible alternative short-term futures, the matter of long-term persistence may be of no importance. For most purposes, it would be incorrect to generate a series of several thousand years with persistence effects on the same scale. If there were identifiable trends expected to continue into the planning period of a project, the only correct treatment would be to add a trend term to the generating function. Attempting to treat a known trend by a random function would surely not be proper. Since a Markov process will generate sequences that are either more critical or less critical than the historic sample, and since the difficulty of defining the correct value of h from a short historic record is recognized, there is a considerable argument for the continued use of the Markov assumption until clear proof of a better method is available. For more details on stochastic analysis, refer to the references.

REFERENCES

Anderson, R. L. 1962. Distribution of the serial correlation coefficient. *Ann Math Statist* 13:1–13.

Benjamin, J. R., and C. A. Cornell. 1970. *Probability, Statistics, and Decision for Civil Engineers.* New York: McGraw-Hill.

Box, G. E. P., and G. M. Jenkins. 1970. *Time Series Analysis: Forecasting and Control.* San Francisco, CA: Holden-Day.

Burges, S. J. 1970. Use of stochastic hydrology to determine storage requirements for reservoirs: A critical analysis. Program in Engineering Economic Planning, Report EEP-34, Stanford University.

Chow, V. T. 1964. *Handbook of Applied Hydrology.* New York: McGraw-Hill.

Fiering, M. 1967. *Streamflow Synthesis.* Cambridge, MA: Harvard University Press.

Gumbel, E. J. 1941. *The Limiting Form of Poisson's Distribution.* New York: New School for Social Research.

Hazen, A. 1914. Storage to be provided in impounding reservoirs for municipal water supply. *Trans ASCE* 77:1539–1669.

Hurst, H. E. 1951. Long-term storage capacity of reservoirs. *Trans ASCE* 116:770–808.

Mandelbrot, B. B., and J. P. Wallis. 1969. Computer experiments with fractional gaussian noises. *Water Resour Res* 5:228–67.

Mejia, J. M., I. Rodriguez-Iturbe, and D. R. Dawdy. 1972. Streamflow simulation: 2. The broken line process as a potential model for hydrologic simulation. *Water Resour Res* 8:931–41.

Sudler, C. E. 1927. Storage required for the regulation of streamflow. *Trans ASCE* 91:622–704.

Thomas, H. A., Jr., and M. B. Fiering. 1962. Mathematical synthesis of streamflow sequences for the analysis of river basins by simulation. In *Design of Water Resources Systems,* ed. A. Maass, et al. Cambridge, MA: Harvard University Press.

BIBLIOGRAPHY

Franz, D. D. 1969. Hourly rainfall synthesis for a network of stations. Stanford University, Department of Civil Engineering, Technical Report 126.

Hufschmidt, M., and M. Fiering. 1962. *Simulation Techniques for the Design of Water-Resource Systems.* Cambridge, MA: Harvard University Press.

Kottegoda, N. T., and R. Rosso. 1997. *Statistics, Probability, and Reliability for Civil and Environmental Engineers.* New York: McGraw-Hill.

Matalas, N. C. 1963. Autocorrelation of rainfall and streamflow minimums. US Geological Survey, Paper 434-B, 16.

Ott, R. F. 1971. Streamflow frequency using stochastically generated hourly rainfall. Stanford University, Department of Civil Engineering, Technical Report 151.

Pattison, A. 1964. Synthesis of rainfall data. Stanford University, Department of Civil Engineering, Technical Report 40.

Thomas, H. A., Jr., and M. B. Fiering. 1963. The nature of the storage yield function. In *Operations Research in Water Quality Management*, Cambridge, MA: Harvard University Water Program.

U.S. Army Corps of Engineers. 1993. Hydrologic frequency analysis. Report EM 1110-2-1415. Washington, DC: U.S. Army Corps of Engineers.

Appendix D: Software Manual for Hydrograph Development Using a Simplified Model in Arid and Semi-Arid Regions

D.1 INTRODUCTION

Most attempts to derive the unit hydrograph (UH) have been aimed at determining the time to peak, peak flow, and time base. These factors and other empirical relations, in addition to the requirement that runoff volume must equal 1.00 centimeter, permit the sketching of the complete UH. The key item in most studies has been the basin lag, most frequently defined as the time interval from the centroid of rainfall to the hydrograph peak. Another method is the use of a dimensionless UH either developed from a well-defined basin with enough hydrologic data or by using a well-known dimensionless UH, such as that of the Soil Conservation Service (SCS). In the current model, a set of equations based on the SCS dimensionless UH is used to derive both the UH and the total hydrograph (TH). However, certain conditions described in this section are required to be met to obtain the most dependable results.

D.2 MODEL DESCRIPTION

The program Visual Basic was used to formulate the execution file, which opens the main window frame that includes all the aforementioned input data. The main frame, which is called the Project Form—shown in Figure D.1, is composed of five columns on the upper area of the frame. These columns include the following items:

The first column (T_{rat}) lists the time ratio t/t_p (where t is the time and t_p is the time to peak) of the dimensionless UH.
The second column (Q_{rat}) lists the discharge ratio q/q_p (q is the discharge and q_p is the peak discharge) of the dimensionless UH.
The third column (t) lists the time values, in hours.
The fourth column (UH) lists the ordinates of the UH, in cubic meter per second.
The fifth column lists the ordinates of the TH, in cubic meter per second.

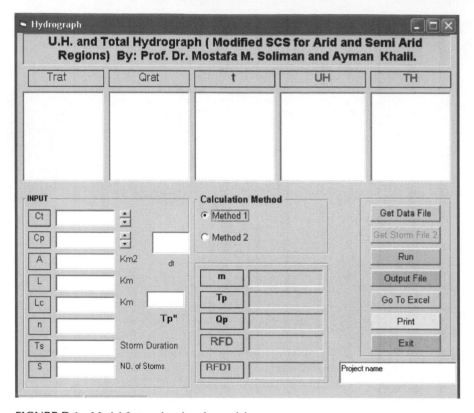

FIGURE D.1   Model frame showing the model components.

TABLE D.1

Coefficients of C_t and C_p

Catchment Coefficient	For Steep Slope	For Moderate Slope	For a Valley Slope
C_t	0.2–0.6	0.7–1.4	1.5–2.2
C_p	3.3–3	2.9–1.7	1.6–0.8

The input block (lower left) contains 10 labels, placed one below the other, each denoting the following parameters:

C_t is the lag coefficient depending on the slope, as given in Table D.1.
C_p is the peak flow coefficient (Table D.1).
A is the area of the catchment, in square kilometers.
L is the maximum length, in kilometers, of the main stream from the outlet to the catchment boundary.
L_c is the length, in kilometers, along the main stream from the outlet to a point nearest to the centroid of the catchment.

n is the number of values selected from the dimensionless UH (optional).

T_s is the storm duration in hours (it is advisable that T_s should neither be less than 0.25 nor more than 0.5 hours for more dependable results).

S is the number of storms.

$dt = (T_s/2)$ is the time increment.

T_p'' is the time to peak, calculated from the average time lag T_1 obtained by both the Natural Resources Conservation Service (NRCS, formerly SCS) and kinematic wave methods, as explained in the following example.

All the blank boxes near the labels should be filled by typing the required values of the catchment characteristics (as will be described in Section D.2), except the value of dt, which will be calculated by the model. The block next to the input block shows two circles at the very top for the calculation methods. The upper one is clicked first for the calculation of the UH. The lower circle should be clicked for the calculation of the TH. The lower labels—marked m, T_p, Q_p, RFD, and RFD1—denote the following values (obtained by the model):

m is the number of ordinates of the developed TH.

T_p is the time to peak.

Q_p is the peak flow of the UH.

RFD is the runoff depth of the UH, in centimeters.

RFD1 is the runoff depth of the TH, in centimeters.

The calculations of T_p and Q_p are based on formulas similar to those of the Snyder method, except that C_t and C_p are selected according to the slope and catchment characteristics. The values of C_t and C_p are preliminarily selected from Table D.1; subsequently, they will be adjusted after obtaining the value of T_p''. The empty boxes near the last labels, T_p, Q_p, and RFD, will be filled on pressing the command button (Run).

The seven buttons to the right are named as follows: "Get Data File," "Get storm File 2," "Run," "Output file," "To Excel," "Print," and "Exit." The lower box to the right is used to write the name of the project. The two rectangles at the bottom left corner are applied for the date and time of the model operations.

The following steps should be followed to derive the UH and TH using the given software.

D.2.1 Step 1 (Optional)

Prepare a text file having two lines, where the first line is used to include the values of the t/t_p ratio and the second line is used to enter the selected values of the q/q_p ratios from the dimensionless UH. Each value should be separated by a space. These values can be selected from the SCS dimensionless UH used in this model, as given in the following example. From the author's experience, the number n of ordinates should not be less than 14. The number of ordinates before $t/t_p = 1$ should not be less than four, and the remaining ordinates should be provided for larger time ratios. Save this file as (input1.txt). The ordinates of the dimensionless UH are used only for comparison and not for calculation of the UH. This means that this step is optional. As mentioned before, the UH is obtained from equations representing the SCS dimensionless UH. These equations have been developed by the author of this book.

D.2.2 STEP 2

As a first choice, one can select the values of C_t and C_p according to the Table D.1, which is considered as a preliminary selection.

The successful selection of values will only come about by experience, after a series of trials relating to the problem and after the calculation of T_p'', as explained in the following example.

D.2.3 STEP 3

Fill the appropriate boxes with the values of the catchment area A (in square kilometers), the catchment length L (in kilometers), and the length from the catchment centroid to the point nearest to the outlet L_c (in kilometers). It is possible to get the centroid of the catchment by simply using the moment method around an axis, which passes through the catchment outlet and is perpendicular to a straight line connecting both ends of the centerline of the stream. Then, divide the catchment into a number (k) of elemental areas. Each elemental area, a_i, can be approximated into a rectangle or triangle, for which the area can be easily evaluated. In addition, one can calculate the catchment area A and L_c' as follows:

$$L_c' = \frac{\sum_1^k a_i l_i}{A} \quad \text{and} \quad A = \sum_i^k a_i$$

where L_c' is the distance the centroid of the catchment to the outlet. In case the stream does not have an approximately straight line, L_c can be obtained from a point at the stream near the centroid. These data can be obtained easily from geographic information systems, watershed-modeling systems (WMS), or similar programs for a sophisticated watershed.

Fill the number n, the UH storm duration T_s in hours, and then the number of storms S. Note that the storm duration of the UH should neither be less than 0.25 nor more than 0.5 hours for more dependable results, as mentioned before. However, the THs obtained from both durations are very close, as shown in Figure D.2, for one of the subcatchments of the Wadi Watir, Sinai, Egypt, for the same storm.

D.2.4 STEP 4

Once the final data are properly selected, the calculation of the UH can begin. This is first done by pressing the circle near "method 1" and then pressing the upper box 'Get Data File'. A window will appear that includes all the text files you prepared in the model's folder. Select 'input1.txt' and then click "open." The lists of

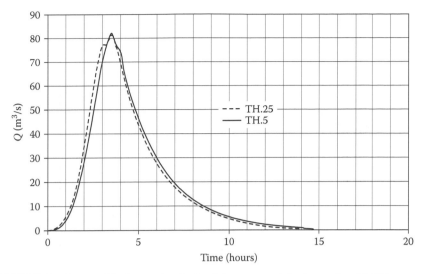

FIGURE D.2 Two total hydrographs obtained from two unit hydrographs of 0.25 hours and 0.5 hours durations.

dimensionless values appear in the two lists T_{rat} and Q_{rat}. Then, press "Run"; the third and the fourth lists will appear in the t and UH columns, respectively. In addition, the values for T_p, Q_p, and RFD will appear, and RFD1 shows "0" at this time. As a check; if your selection of C_t and C_p are appropriate, the RFD should approximately equal the value 1. If the RFD value is either more or less than one, you must go back and select a new value for C_p alone, in case the difference is small. This step can be repeated until the RFD value approaches unity. Checking the value of T_p is also required to be done by comparing its value with that of T_p'' given in the attached excel file (T_p''.xls). The author recommends using the values of T_1 and T_c from the NRCS (SCS) and kinematic wave equations, respectively (refer to Equation 5.12, Chapter 5, in this book). In case the values are not very close, you can get the average T_p'' from both the equations. This is clearly demonstrated in the following example. You can adjust the T_p value by changing C_t until T_p and T_p'' are close. Once this step is completed, you can go to step 5.

D.2.5 STEP 5

Prepare another text file and save it as (input2.txt) file. This file contains a single reading of the storm depth, in centimeters, if you are calculating a TH from one storm. If you are using more than one storm, write the excess rainfall depths one below the other, one for each storm. It is also better to divide any storm lasting greater than one hour into equal divisions. Each division may stand for one storm with a 0.5-hour duration, which is adopted in the UH calculations. Thus, you will have a number of consequent storms. Then, press the circle near "method 2."

D.2.6 STEP 6

On pressing the "method 2" circle, the "Get Storm File 2" box appears. Click this button. A window will appear again; then, select (input2.txt) file from the list and press "open." Press the box "Run" to run the module. The ordinates of the TH will fill the fifth list, TH. Notice that a value for m denoting the number of TH ordinates appears. The new value for RFD1 also appears. This value indicates the total runoff depth of the TH, in centimeters. This value should be the same as the total excess rainfall depth in the input file, thus acting as a further check to your calculations.

D.2.7 STEP 7

You can get the output values of both the UH and the TH in an output.txt file. This can be done by saving an output file that includes it in a folder of the model. Press the box "Output File" and click Save. Finally, you have saved an output file having three columns in the model's folder. The first column stands for the time, whereas the second and the third columns indicate the discharges of the UH and the TH, respectively.

D.2.8 STEP 8

It is also possible to get the time and the ordinates of the UH and TH in a worksheet directly as the last step without going through an output file; this is achieved by pressing the "To Excel" button. If you press "Insert" on the excel toolbar and select the chart type, you will have both the UH and the TH curves plotted on the worksheet.

You can also have a printout of the UH and TH ordinates if you press the Print button.

Once you attain your objective, you can press the Exit button; you do not have to save the information on the Model Form so long as you have already saved it either in the output file or in the excel worksheet.

Note: Before running the model, it is recommended to reload the computer with the most recent version of Microsoft, MS Office 2003, or above. The recent versions that include excel are capable of easily plotting the UH and TH hydrographs.

On downloading the software, ensure that all the forms are maintained in the model's folders. The model and its manual will be downloaded from the publisher's site in a Zip file format, which includes all the forms. In case any form is lost, it can be extracted again.

For better results from the model output, the catchment size should be less than 100 km². However, the model is capable of yielding results for catchment sizes as large as 300 km². In case the catchment size is larger than this value, it can be divided into several small subcatchments to get better results for the flood hydrographs after undertaking the necessary routing, as given in Chapter 8. The model is also considered suitable for storms in all arid and semi-arid regions. The following example applies the model to one of the subcatchments of the Wadi Watir in Sinai, Egypt.

EXAMPLE 1

The subcatchment B of the Wadi Watir in Sinai, Egypt, is shown in Figure D.3. The rainfall over a period of 15 years for 2-hour storms recorded by an intermediate gage station is given in Table D.2. The following data were obtained from a WMS:

$$A = 64.6 \text{ km}^2; L = 12.4 \text{ km}; L_{ca} = 6.64 \text{ km}$$

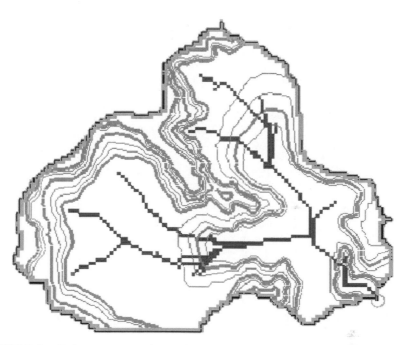

FIGURE D.3   4B Subcatchment of Wadi Watir Sinai Egypt.

TABLE D.2
Rainfall During a 2-hour Storm in the Subcatchment 4B

Year	Rainfall (mm)	Year	Rainfall (mm)
1981	18.3	1989	12.7
1982	11.0	1990	27.0
1983	26.8	1991	24.5
1984	26.6	1992	19.3
1985	25.3	1993	28.2
1986	15.6	1994	17.5
1987	14.8	1995	16.9
1988	13.7		

Catchment's maximum slope = 0.1 m/m; centroid slope = 0.015 m/m
You are required to find the following:

1. Analyze the rainfall during the 2-hour storms given in Table D.2 and find the rainfall that will give the maximum probable flood in the subcatchment for a return period of 50 years.
2. Assuming uniform rainfall distribution for the 2-hour storm, determine the excess rainfall and duration using the initial and constant-loss method given in Table D.3.
3. Determine the UH and the maximum probable hydrograph for the catchment 4 B using both the Soliman and the WMS–Hydrologic Engineering Center (HEC-1) softwares and then compare the results.

SOLUTION

1 Rainfall Analysis for Subcatchment B

The return period of 50 years was first developed by the Weibull method, as given in Table D.4, and from Figure D.4

TABLE D.3
General Characteristics of the Surface Geology of the Subcatchment 4B

Area (%)	Soil	Average CNSTL (mm/h)	Average STRTL (mm)
10	Alluvium cover		
90	Limestone and shale cover	1.8	18

TABLE D.4
Return Period T (in Years) and Rainfall Based on Table D.2

m	$T = n + 1/m$	Rainfall (Descending Order)(mm)
1	16.000	28.2
2	8.000	27
3	5.333	26.8
4	4.000	26.6
5	3.200	25.3
6	2.667	24.5
7	2.286	19.3
8	2.000	18.3
9	1.778	17.5
10	1.600	16.9
11	1.455	15.6
12	1.333	14.8
13	1.231	13.7
14	1.143	12.7
15	1.067	11

P_{av} = 19.88 mm, the average rainfall in the whole period

σ = 5.94, the standard deviation

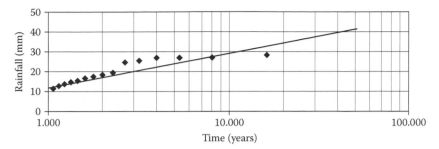

FIGURE D.4   Return period against rainfall.

Using the Gumble method to get the probable maximum precipitation (PMP)

$$PMP = P_{av} + K6_{(n-1)}$$

where P_{av} is the average precipitation = 19.88 mm and $\sigma_{(n-1)}$ = 5.94
   K = 3.3 for a 50-year period (from Table 3.7, Chapter 3)

$$PMP = 19.88 + 3.3 \times 5.94 = 39.482 \text{ mm}$$

It was found that the return period of 50 years from Figure D.4 gave a rainfall value
of 40.00 millimeters, which is almost the same as the PMP; therefore, 40 millimeters will be the value taken for our analysis.

2 Rainfall Losses

From Table D.3,
   Starting loss or STRTL = 18 millimeters
   Constant loss or CNSTL = 1.8 mm/h = 0.9 mm/0.5 h
   The excess rainfall is distributed in the following sketch for each 0.5 hour.
   Rainfall distribution during the 2-hour storm

TABLE D.4a
Showing Total Initial Loss, Contant Loss and Rainfall Excess for the 2 hr Storm

Accumulated Rainfall mm	0	10	20	30	40
			RFE* = 2 mm	RFE = 9.1 mm	RFE = 9.1 mm
		Total Initial loss =	18 mm	Constant loss = 0.9 mm/0.5 hr	
Time hours	0	0.5	1.0	1.5	2.0
*RFE = Rainfall Excess each 0.5 hr					

It is obvious from the rainfall distribution that there are three divisions where
the depth of the rainfall excess (in centimeters) in each division becomes 0.2, 0.91,
and 0.91 centimeters, respectively. In the model, we shall consider that each division of the rainfall excess represents a 0.5-hour storm. This means that the number
of storms S = 3.

To derive T_p'', we shall consider two equations, namely, the NRCS and Dynamic-wave equations. The NRCS equation will yield the time lag T_l as

$$T_l = \frac{L^{0.8}(S'+1)^{0.7}}{1900 \times Y^{0.5}} \; h$$

where L is the catchment length to be divided, in feet.

$$S' = \frac{1000}{CNII} - 10$$

Y is the average catchment slope (%), and CNII is the SCS curve-number condition II.

The T_l value is for rural natural catchments. In case there is an impervious area, a reduction factor should be considered. For example, if a catchment contains 40% impervious area, T_l should be multiplied by 0.7. For more details about this factor, refer to Chin, 2006 (see Chapter 5 in this book). Therefore, $T_p'' = T_s/2 + T_l$.

The second equation is the kinematic wave equation (Equation 5.12 in Chapter 5). This equation gives T_c, from which you can get T_p'' as follows:

$$T_p'' = \frac{T_s}{2} + 0.6 \; T_c$$

The two values of T_p'' from the previous two equations for the catchment under consideration are given in Table D.5.

TABLE D.5

Values of T_p'' from the NRCS Model and Equation 5.12, Chapter 5, Demonstrated in the Worksheet File T_p''.xls in the Model's Folder

1: NRCS		Formula $T_l = L^{0.8}\,[(S+1)^{0.7}/$ $(1900 \times Slope\%^{0.5})]$
CN	=	88
S	=	1.36363636
L (m)	=	12400
Slope (%)	=	5
L (ft)$^{0.8}$	=	4868.96243
T_l (h)	=	2.09268704
T_s (h)	=	0.5
T_p'' (h)	=	$T_s/2 + T_l = 2.342687$
2: Kinematic Wave Equation		
n	=	0.05
I (mm/h)	=	20
S_o	=	0.05
T_c (h)	=	4.08938443
T_p'' (h)[a]	=	2.70363066

[a] $T_p'' = T_s/2 + 0.6\,T_c$.

The two T_p'' values are 2.34 and 2.70 hours. It is recommended, therefore, to take the average value as,

$T_p'' = (2.34 + 2.70)/2 = 2.52$ hours (write this value in the text box T_p'' in the Model Form to guide you for adjustment of T_p).

The file "input1.txt" can be obtained from the SCS dimensionless UH as mentioned before. This step is considered optional and was meant here for comparison.

The input file "input2.txt" can be obtained from the rainfall excess data. The three storms are written in the text file as follows:

> 0.2
>
> 0.91
>
> 0.91

where the values are in centimeters.

Both the files "input1.txt" and "input2.txt" are attached with the model software for guidance.

3 Model Preparation

Once all the data needed are prepared, press the model folder and then press on Hydrograph3.exe file. The model frame will appear as shown in Figure D.1. As a first choice, one may select values of C_t and C_p that suit the catchment characteristics, as given in column 2 of Table D.1. Therefore, we can select $C_t = 1$ and $C_p = 2.5$ as a first choice. Fill the rest of the blank text boxes from the available data, as given in Figure D.5. Follow the steps described below to get the UH and TH.

1. Be sure that the "method 1" circle is filled, then press the "Get Data File" button and open "input1.txt" file from the drive where you stored all the model files. The first and the second lists will be filled. Then, press the "Run" button. The third and the fourth lists will be filled with both the time and UH ordinates, as given in Figure D.6. Notice that RFD and T_p need to be adjusted.

2. Change the value of C_p first, until RFD approaches 1 centimeter. Then, change C_t until T_p approaches T_p''. You must click the "Run" button each time you make a change. Figure D.7 shows all the values after the necessary adjustment. This will end the method 1 stage.

3. Press the "method 2" circle. A point will appear in the method 2 circle and disappear from the method 1 circle. In addition, the "Get Storm File 2" button becomes clearer and the "Get Data File" fades away. Press "Get Storm File 2" and open input2.txt file from the drive where you stored all your files. Then, press the "Run" button.

4. The fifth list will be filled with the TH ordinates, as shown in Figure D.8. Notice that RFD1 = 2.04 centimeters, which is slightly greater than the total excess rainfall of 2.02 centimeters. It is possible to do another adjustment. However, the difference is negligible. Therefore, we can go ahead and plot the output.

5. Save a text file in a model folder and name it as "output.txt." Click the "Output File" button, and the window that contains the output file appears. Click the output file and then press the "Save" button in the window. Thus,

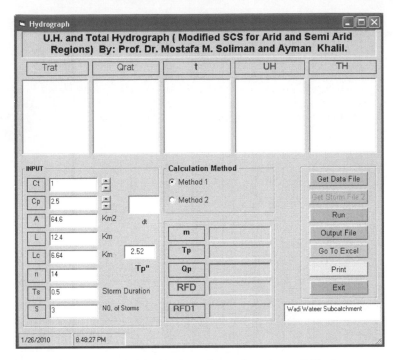

FIGURE D.5   Filling the rest of the blank text boxes from the available data.

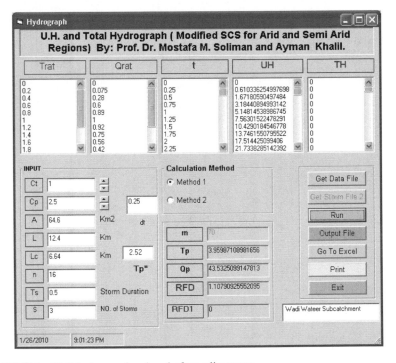

FIGURE D.6   Unit hydrograph values before adjustment.

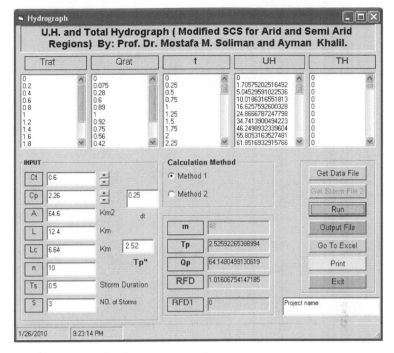

FIGURE D.7   Final adjustment for UH development.

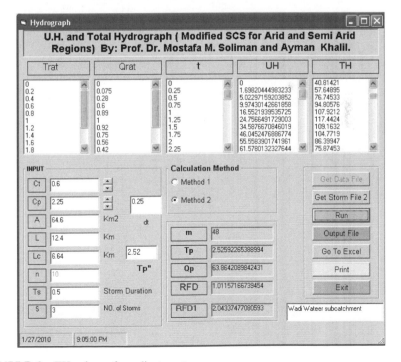

FIGURE D.8   TH values after adjustment.

you have saved the UH and TH data in the output file. You can use this file with the excel platform to plot the UH and TH.

6. You can do the same operation of step 5 by clicking on the "Go To Excel" button. An excel worksheet appears, containing the following three columns—the time, and the UH and TH ordinates. By pressing the "Insert" button on the excel tool bar and selecting the chart type, the UH and TH will be plotted. You can add a name and all the legends needed to the graph, as given in Figure D.9.

7. For comparison, the WMS-HEC-1 software is used to get the TH for the same data in the problem. Figure D.10 shows the two THs, one obtained

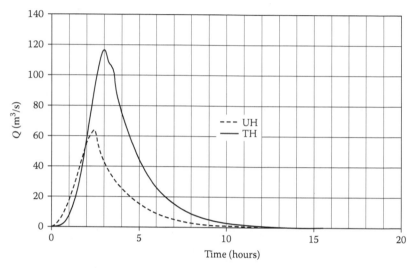

FIGURE D.9   Unit hydrograph and total hydrograph by Soliman model.

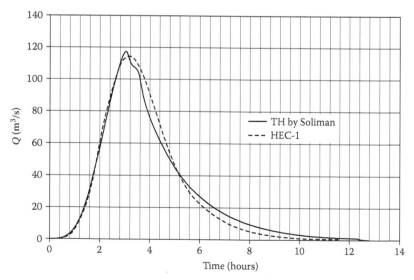

FIGURE D.10   Comparison between Soliman model and HEC-1.

by the Soliman model and the other by the HEC-1 model. It is obvious that the difference is negligible. The values of both the hydrographs are also tabulated in Table D.6.

TABLE D.6

Unit Hydrograph and Total Hydrograph by the Soliman Model and the HEC-1 Hydrograph List

	Soliman Model Values		HEC-1 Values	
T (h)	UH (m³/s)	TH (m³/s)	T (h)	TH (m³/s)
0	0	0	0	0
0.25	1.69820445	0.33964089	0.25	0
0.5	5.02297159	1.00459433	0.5	0.395
0.75	9.97430143	3.54022622	0.75	2.247
1	16.552194	7.88134289	1	6.551
1.25	24.7566492	15.5733099	1.25	14.164
1.5	34.5876671	26.5509338	1.5	26.185
1.75	46.0452477	40.8142128	1.75	42.327
2	55.5583902	57.6489525	2	60.896
2.25	61.5780132	76.7453308	2.25	79.738
2.5	63.864209	94.8057556	2.5	96.186
2.75	49.9203002	107.921227	2.75	107.8
3	43.8391535	117.442398	3	113.458
3.25	38.4987944	109.163223	3.25	113.431
3.5	33.8089825	104.771858	3.5	108.948
3.75	29.69047	86.3994675	3.75	101.122
4	26.0736627	75.8745346	4	90.327
4.25	22.8974444	66.6317215	4.25	78.308
4.5	20.1081439	58.5148354	4.5	66.141
4.75	17.6586279	51.3867264	4.75	54.713
5	15.5075049	45.1269455	5	45.016
5.25	13.6184254	39.6297112	5.25	37.475
5.5	11.9594682	34.8021355	5.5	31.43
5.75	10.5026003	30.5626392	5.75	26.396
6	9.22320391	26.8395863	6	22.183
6.25	8.09965985	23.5700645	6.25	18.564
6.5	7.11298269	20.6988277	6.5	15.517
6.75	6.24649969	18.1773567	6.75	12.943
7	5.48556915	15.9630432	7	10.855
7.25	4.81733296	14.0184717	7.25	9.059
7.5	4.23049938	12.3107824	7.5	7.564
7.75	3.71515217	10.8111181	7.75	6.31
8	3.26258306	9.49413872	8	5.274
8.25	2.86514462	8.33759022	8.25	4.417
8.5	2.51612098	7.32192945	8.5	3.701
8.75	2.20961439	6.42999315	8.75	3.103

(Continued)

TABLE D.6 (*Continued*)

	Soliman Model Values		HEC-1 Values	
T (h)	*UH* (m³/s)	*TH* (m³/s)	*T* (h)	*TH* (m³/s)
9	1.94044555	5.64670992	9	2.592
9.25	1.70406608	4.95884371	9.25	2.173
9.5	1.49648167	4.35477209	9.5	1.818
9.75	1.3141846	3.82428622	9.75	1.533
10	1.15409443	3.35842276	10	1.301
10.25	1.01350598	2.94930935	10.25	1.112
10.5	0.89004361	2.59003305	10.5	0.945
10.75	0.78162107	2.27452254	10.75	0.792
11	0.68640625	1.99744689	11	0.645
11.25	0.60279022	1.75412369	11.25	0.504
11.5	0.52936006	1.54044139	11.5	0.368
11.75	0.46487495	1.35278928	11.75	0.244
12	—	1.10634732	—	—
12.25	—	0.97157532	—	—
12.5	—	0.48171765	—	—
12.75	—	0.42303619	—	—

CONCLUSION

In conclusion, the Soliman model is simpler and more flexible than many other models. Most of the other models do not give any flexibility in the equations used to derive the time to peak or even the peak values. The hydrologist should use the equations included in the model without any change. Some models give only one T_p for different rainfall events and durations for the same catchment, which is not appropriate.

There is another benefit in using the Soliman model: you can follow any operation or calculation step-by-step and you can discover any error very easily in any operation. This benefit is not clearly demonstrated in many models, where many operations are dealt as black boxes. This is very important for both students and professionals who want to know what they are doing. Another benefit with the author's model is that one can get both the UH and TH curves simultaneously on the spread sheet,

Above all, this model is also designed to specifically serve arid and semi-arid regions.

Index

A

Acoustic Doppler arrangements, 127
Active satellites. *See* Polar-orbiting satellite
Actual evapotranspiration, 77
Analytical methods, evaporation
      estimation by, 72–75
Antecedent runoff conditions (ARC), 88, 90
Anticyclones, 38
Aqueducts, hydraulic design, 239–240
Aquifer
    migration of pollutants in, 218–219
    slug test in, 204
ARC. *See* Antecedent runoff conditions
Area-elevation curve, 104–105
Arid regions, 1, 5
    flood-control methods in, 6–7
Arithmetic mean, 50
Artificial groundwater recharge, 206–208
Atmospheric pressure, 328
Automatic stage recorders, 122–123
Average depth of precipitation, 50–52
Average unit hydrograph, 142

B

Bank storage, 99, 100
Base-flow separation, methods of, 138–139
Basic modified equation, 189–191
Basin lag, 147, 152–153
Binomial distribution, 345–346
Blaney–Criddle formula, 78
Bligh's coefficients for soils, 243
Bligh's formula, 242–243, 245
Box culverts, 237
Bridge, hydraulic design of, 233–235
Buried culverts, 318

C

Calibrated parameter values, 296
Calibration for simulation model, 293–296, 299
Capillary water, 183–184
Catchment characteristics, 99–101
Cetyl alcohol, 76
Channel, definition of, 133
Channel routing, 165–166
Channel slope, 103–104
Check dams, 247–248
Chemical films, 76

Chemical gauging, 125–126
Chen and Wong formula, 109, 113
Chezy application, 240
Cistern system, 248–251
Class A evaporation pans, 69–70
Clear over fall weir, 129–130
Climate
    human activity and, 28–30
    solar variability and, 30–32
Cloud seeding, 39–40
Condensation, 22
Confined aquifer, 183, 185–187, 199, 210–212
    partially penetrating well in, 200
    transient flow in, 188
    unsteady-state flow in, 188–189
Constant loss method, 287, 288
Constant rate injection method, 125
Contaminants, 9, 217–219
Continuity equation, 163–165, 167, 174,
      184, 186
Control works, 240–245
Convective precipitation, 38–39
Conversion table, units, 325–326
Critical hydraulic gradient (HG), 242
Crop-management factor, 224
Crossing works, 232
Culverts, 314–318
    hydraulic design of, 235–238
    under Wadis, 319–320
Current meters, 124
Curve-number model, 86–91
Cut-throat flumes, 131–132
Cyclone, 37–38
Cyclone Gonu analysis, 280–283

D

Dalton-type equation, 70
Dam, detention, 6
Dam reservoirs, 241
Darcy's law, 184, 186, 241
Darcy–Weisbach equation, 108
DDF relationships. *See* Depth-duration-
      frequency (DDF) relationships
Depletion curve. *See* Recession curve
Depression storage, 81, 99
Depth–duration component, 60
Depth–duration–frequency (DDF)
      relationships, 58–60
Depth–frequency component, 60

de Saint–Venant equations, 157
Design flood, 360
Design storm, 52, 249
Dew point, 23
Dew point hygrometers, 25
Digital elevation model (DEM), 258
Digital terrain models (DTMs), 255
Dilution method, 125–126
Dimensionless hydrograph
    of channel flow, 160
    of overland flow, 156
Dimensionless unit hydrograph, 150–154
Direct surface flow, 99
Discharge calculation using rational
            method, 117
Discharge hydrograph routing effects, 164
Discharge measurement, 124–128
Dispersion, 218–219
Double-mass analysis, 49–50
Double-ring infiltrometer, 95
Downstream
    boundary conditions, 304
    effect on, 323
Drainage basin, 99–100
    characteristics of, 101–105
    shape, 102
Drainage density, 101
Drainage divide, 99, 101
Drawdown equation, 193–197
Drizzle, 36
Drought, 4
Dynamic wave equations, 164, 165

E

Earth radiation, 19–20
Effective radius, 201
Effluent streams, 106
Egyptian Water Resources Institute
            (WRRI), 42
Electromagnetic method, 126–127
Empirical evaporation equations, 70–72
Energy-budget method, 73–74
Engineering hydrology, 231
Environmental impacts, 9
Ephemeral streams, 106
Equation of motion, 184
Equation of state, 184, 186
Erosion control, 7–8
Erosion-control practices, parameters
            influencing on, 224–225
Estimation
    of evaporation, 68–69
    by analytical methods, 72–75
    reservoir, 75–76
    of missing precipitation data, 48–49
Evaporated water, 10

Evaporation, 67
    empirical equations, 70–72
    estimation of, 68–69
Evaporation losses, reduction of, 75–76
Evaporation station network, 70
Evaporimeters, 68–69
Evapotranspiration, 77–80
Extratropical cyclones, 38

F

Faraday's principle, 126
Fayum-type weir. See Clear over fall weir
Field capacity, 77
Field plots, 77
Fill methodology, 321
Flash floods, 3, 6, 46, 248
Float-operated stage recorder, 122
Float-type gauge, 42
Floods
    events, characteristics of, 291–292
    hydrographs, 311
    protection scheme, 320
    protection works, 313–318
Flood control, 6–7
Flood frequency analysis, 359–360
Flood hydrograph
    using rational method, 118
    values, 136
    and unit hydrograph, 143
Flooding-type infiltrometers, 95
Flood plain, 268, 283
Flood protection works, design of, 314
Flood routing, 175–181, 256. See also
            Routing
Flow equation, 166
Flow length, calculating, 109
Flow measuring structures, 128–132
Flumes, 128
Force, 328
Forms of precipitation, 35–36
Fossil fuels, 28
Free-aquifer conditions, modified
            equations for, 191–192
Free water, 184
Frequency analysis of point rainfall, 53–56
Front, 37

G

Geographic information system (GIS),
            15, 256
Geographic variation of wind, 27–28
Geostationary satellites, 45–46
GIS. See Geographic information system
Glaze, 36

Gravity (vadose) water, 183
Green–Ampt
    infiltration equation, 82–86
    infiltration parameters, 84
    loss-rate parameters, 93
    method, 288, 290–291
Groundwater
    flow equation, 186
    flow theories, 184–186
    pollution, 217–219
    recharge, 206–210
        application, 214–217
    steady-state flow in aquifers, 186–187
    unsteady-state flow in confined aquifers,
        188–189
Groundwater divide, 101
Groundwater flow, 134
Groundwater table (GWT), 99, 106, 107
Gumbel's equation
    application of, 348–349
    confidence limits, 354–356
Gumbel's method, 347–348
Gutter design, 160

H

HADR. See High Aswan Dam Reservoir
Hail, 36
Hair hygrometer, 25
Heading up, calculating, 233–235
Head loss, 238–239
HEC-1. See Hydrologic Engineering
        Center model
High Aswan Dam Reservoir (HADR)
    constraints of, 177–178
    development of, 175–177
    results, 178–181
    simulation model, 178
    water levels of, 179–181
High-intensity storm, 114
Holton infiltration equation, 81–82
Humidity, 21–23
    measurement of, 25
    units, 23
Hurst phenomenon, 370–371
Hydraulic characteristics,
        basic equations of, 104
Hydraulic jump, 245
Hydraulic routing, 163–165
Hydraulics of wells, 201–203
Hydraulic structures, 232
Hydrographs
    of overland flow, 154–160
    separation, 138–139
Hydrograph analysis, 136

Hydrograph curve on semilog paper, 137
Hydrologic cycle, 9–11
Hydrologic Engineering Center (HEC-1) model,
        91, 249–250, 257, 260
Hydrologic Engineering Center model-
        Hydrologic Modeling System
        (HEC-HMS), 292, 307
Hydrologic model, 13–15, 257
    construction of, 307–308
Hydrologic parameter, 308, 311
Hydrologic routing, 163, 165
Hydrologic soil group, 89, 93
Hydrologic systems, 11–12
Hydrometeorological data, 300
Hygrometer, 25

I

IDF. See Intensity-duration-frequency
Infiltration
    Green–Ampt equation, 82–86
    groundwater recharge by, 206, 209
    Holton equation, 81–82
    indexes, 96
    measurement of, 95–96
    method, 116
Infiltrometers, 95–96
Influent streams, 106
Infrared (IR) image, 45
Initial loss method, 287, 288
Initial loss (STRTL), 92
    plus uniform loss rate (CNSTL), 91–95
Intensity–duration–frequency (IDF), 57–58
    curve fitting of data, 303
    for Wadi El-Arish, 272
Intercept coefficients, 110
Interception loss, 80–81
Interflow, 99, 134
Intermittent streams, 107
International Hydrological Program (IHP-V) of
        UNESCO, 12
Irish crossing, 232, 233
Irrigation, 5–6, 77–79, 106, 121
Isochrones, 114
Isohyetal method, 50–52
Izzard equation, 111

K

Kerby equation, 111–114
Kerby–Hathaway equation. See Kerby
        equation
Kinematic-wave equation, 107–109
Kirchhoff's law, 20
Kirpich equation, 110–111

L

Land remote-sensing satellite system (LandSat), 175, 300
Land surfaces, precipitation on, 10
Lane formula, 243
Lane's coefficients for soils, 244
Lapse rate, 21
Leaky aquifer, 183, 198
Linear unit hydrograph, 141
Log-Pearson type-II distribution, 356–359
Long-base weirs, 128
Long wave radiation, 19
Low flow, 99
Low-intensity storms, 114
Lysimeters, 77

M

Manning equation, 107–109, 160
Mass-transfer method, 75
Mean elevation, 104
Measurement
    of evapotranspiration, 77–78
    of infiltration, 95–96
    of precipitation, 40–46
Median elevation, 104
Meteorological radars, 43
Meteorological stations, 260–262
Meyer's formula, 71
Missing precipitation data, estimation, 48–49
Modeling for hydrology, 13–15
Modified Puls reservoir routing, 172
Modified universal soil-loss
    equation (MUSLE), 223
Momentum equation, 163, 164
Multibasin recharge method, 208
Multipurpose dams, 241
Municipal water supplies, 4
Muskingum–Cunge channel routing, 168–170.
    See also Channel routing
Muskingum–Cunge method, 169–170
Muskingum method, 165, 166
Muskingum routing equation, 166–168
MUSLE. *See* Modified universal soil-loss
    equation

N

Nagamish subbasin, 249
National Oceanic and Atmospheric
    Administration (NOAA), 31
Natural Resources Conservation
    Services (NRCS)
    curve-number model, 86–91

dimensionless unit hydrograph, 150–152
    method, 109–110
Neder–Mead method, 293
Next Radar Generation (NEXRAD) systems, 45
Nonrecording gauges, 41

O

Optimization of simulation model, 293
Orographic precipitation, 39
Overland flow, 99, 109, 133
    hydrograph of, 154–160
    over parking lot, 159
    time of concentration, 99, 109, 110

P

Pan coefficient, 69–70
Parshall flume, 131
Partial-duration series, 58, 359
Peak-over-threshold (POT)
    frequency analysis on, 58, 59
Peak runoff, 114
Pellicular water, 183, 184
Perched water, 183, 184
Percolation, 241–243
Perennial streams, 107
Permanent wilting point, 77
Pipe aqueduct, 240
Pipe culverts, 237
PMP. *See* Probable maximum precipitation
Point rainfall, 53–56
Polar-orbiting satellites, 44–45
Pollution control, 9
Power-law profile, 27–28
Precipitation
    average depth over area, 50–52
    data interpretation, 48–50
    forms of, 35–36
    gauge network, 46–47
    on land surfaces, 10
    measurement of, 40–46
    types of, 37–40
Precipitation enhancement. *See* Cloud seeding
Pressure, 328
Pressure conversions, 328
Prism storage, 165, 166
Probable maximum precipitation
    (PMP), 60–63

R

Radioactive traces, 126
Rain, 36
    gauges, 40, 41

Rainfall, 35
   depth, 57
   duration, 57, 140
   estimation methods, 288–291
   meteorological stations, 260–262
   radar measurement of, 43–44
   temporal distributions of, 63
   types of, 36
Rainfall losses, 67
   components of, 68
   methods for estimating, 81
      Green–Ampt infiltration equation, 82–86
      Holton infiltration equation, 81–82
      initial- and uniform method, 91–92
Rainfall runoff, 115, 247
Rainfall-runoff regime, 262
Rainwater, disposal of, 100
Rational method, 116–118
Recession curve, 135–136
Recording gauges, 41, 43
Recovery equation, 192–193
Recovery method, 211
Rectangular section aqueducts, 240
Regression curve, 137
Relative density of substances, 326–327
Relative humidity, 23–25
   variation of, 26
Reservoir
   evaporation, 75–76
   operation, 227, 228
   sedimentation, 225–229
Reservoir routing, 172
Reservoir suitability index (RSI), 258
Retardance coefficient, 156
Return period, 53–56, 63
River foothill method, 250
Rohwer's formula, 71–72
Roughness coefficients, 169
Routing, 163
   coefficients, 167
   of flood hydrograph at HADR, 177
   period, 166
   selection of, 181–182
Runoff, 99–100
   amount of, 141
   areal distribution, 140
   estimation of, 115–118
   factors affecting, 114–115
Runoff coefficient, 116–117
Runoff simulation model, 287
   development of, 292–293

S

Safe percolation length, 242
Safe velocity, 241
Saint Venant equations, 163

Saturation vapor pressure, 22, 23
   of water, 71
Scale effect, 132
Screen, 202
Sediment-yield theories, 221–225
Seepage. *See* Percolation
Semi-arid regions, 1, 5
   flood-control methods in, 6–7
Semiconfined aquifer, 212–213
Semipervious aquifer, 198
Shallow concentrated flow, 109, 110
   intercept coefficients for, 110
Sheet flow, 109, 155
Short wave radiation, 19
Silt deposition, 8
Simulation model
   calibration, 293
   development, 292
   optimization, 293
   validation, 296
Siphon, 238–239
Sleet, 36
Slope length-and-gradient factor, 224
Slug test, 203–205
Small Wadis, 318–320
Snow, 36
Snyder method, 147–149
Soil, 7, 77–78, 81–82, 242
Soil Conservation Service (SCS), 151, 262, 373
   loss method, 262, 289–290
Soil-erodibility factor
   determination of, 223
   for sand, silt, and clay, 228
Soil texture classification, 92
Solar radiation, 19–20
Solar variability, 30–32
Soliman model, 386–388
Space Environment Center (SEC), 31
Specific capacity, 201
Staff gauges, 121–122
Stage data, 123
Stage hydrograph, 123
Standing wave weir, 130–131
Statistical analysis in hydrology, 345
   binomial distribution, 346–347
Steady-state flow in aquifers, 186–187
Stilling well installation, 122–123
Stochastic hydrology, 360–362
   definition of f parameters, 365–368
   distribution of random variate t, 363–365
   stochastic process, 362–363
Storage
   equation, 165
   reservoir routing methods, 173
   system, rainfall harvesting, 246–247
   vs. water level in reservoir, 172

Storage-indication curve, 174
Storage loop, 168, 170, 171
    river-routing, 168
Storage works, 241, 245
Storm
    convection, 38
    cyclonic, 37
    design, 52
        duration, 57
        frequency analysis of point rainfall,
            53–56
    orographic, 39
Stream density, 101
Stream discharge, 126
Stream-flow hydrographs, 133–137
Stream order, 103
Streams
    classifications, 106–107
    hydraulic characteristics, 104–105
    longitudinal section of, 103
    pollution, 9
Subbasin cisterns, 250
Subbasin outlet, 250
Subbasins
    hydrographs, 264–265
    physiographic parameters of, 259
    Wadi El-Arish, 274
    of Wadi Nagamish, 258
Sublimation, 22
Subsoil water, 183
Subsurface water, 183–184
Sudr region, 59
    rainfall depth for, 60
Suez Canal, tunnels, 238
Surface area reduction, 75
Surface detention, 99, 154
Surface float, 125
Surface hydrologic modeling system, 14
Surface retention loss, 81
    for various land surfaces, 94
Surface runoff, 108, 133–134
Surface water, 10
Synthetic unit hydrograph, 147

T

Telemetry system, 122
Temperature
    measurement of, 20
    relative humidity, 23–25
    units, 23, 24
Temporal distributions of rainfall, 63
Theis curve, 189
Theis nonequilibrium equation, 189, 192
Thermometer, 25
Thiessen method, 50, 51

Thin plate structures, 128
Time–intensity pattern, 140
Time of concentration, 107
    Izzard equation, 111
    Kerby equation, 111–114
    kinematic-wave equation, 107–109
    Kirpich equation, 110–111
    NRCS methods, 109–110
Tipping-bucket type gauge, 41–42
Topsoil, loss of, 7–8
Tracers, 125
Transpiration, evaporation and, 76–80
Triangular unit hydrograph.
    SCS, 151
Turbulent flow, 134

U

Ultrasonic flow meters, 127
Ultrasonic method, 127–128
Unconfined aquifer, 183, 185–187
    slug test in, 204
Underflow, 99
Uniform loss rate (CNSTL), 91–93
Unit hydrograph, 138–141
    derivations from complex
        hydrograph, 142–144
    derivations from dimensionless
        unit hydrograph, 154
    derivations from simple
        hydrograph, 141–142
    transposing, 152–154
    for various durations, 144–147
Univariate gradient method, 293
Unsteady-state flow
    in confined aquifers, 188–189
    in semiconfined aquifers, 197–198
Unsteady-state well formulas, 210–213
Uplift, 243–245
Upstream boundary conditions, 304
Urbanization, Wadi Adai, 283–285

V

Vapor pressure, 22
Vegetal cover, 222
Velocity distribution, 202–203
Velocity measurement by float, 124–125
Vertical temperature gradient, 21

W

Wadi Adai
    cyclone Gonu analysis, 280–283
    rainfall analysis, 276–278
    runoff analysis, 278–279
    urbanization, 283–285

Wadi El-Arish
    catchment characteristics, 268–270
    rainfall analysis, 270–271
    runoff analysis, 272–276
Wadi El Meleha
    PMP for, 63
    rainfall storms at, 56
Wadi hydrology, 12
Wadi Nagamish, 255
    check dams in, 248, 249
    cistern system in, 251
    hydrologic studies of, 259–260
    rainfall analysis, 260–262
    subbasins of, 258
Wadi Salboukh, 299
    fill methodology, 321
    hydrograph of, 311
    hydrologic analysis, 307
    rainfall input, 308–311
Wadi Sudr, 291–292
Water
    contaminants, 9
    droplets, 35
    saturation vapor pressure of, 71
    vapor, 21–23
Water-budget method, 73

Water level hydrograph, 123
Watershed
    routing, 256
    slope length-and-gradient factor for, 224
    urbanization, impact of, 268
    Wadi Sudr, 291–292
Watershed modeling system (WMS), 255
    application of program, 262–266
Water table, 183
Water-table condition, 213
Weather satellites, 44–46
Wedge storage, 165, 166
Weighing-bucket gauge, 42
Weight, 328
Weirs, safeguarding, 128–129, 241, 245
Well
    effects of partial penetration, 198–200
    hydraulics of, 201–203
Well screen, 202
Wet-bulb depression, 25, 26
Wien's law, 20
Wind, 25
    geographic variation of, 27–28
    measurement of, 25–27
World Meteorological Organization
        (WMO), 47